GLOBE PHYSICAL SCIENCE

Bryan Bunch

Senior Author and Editor, Globe Science Series

•

Marie E. Marshall

GLOBE FEARON EDUCATIONAL PUBLISHER
A Division of Simon & Schuster
Upper Saddle River, New Jersey

Bryan Bunch, a former high school teacher and editor of school science materials, currently teaches mathematics at Pace University in Pleasantville, New York. Mr. Bunch is president of Scientific Publishing, Inc., and the author of handbooks on current science and medicine, and the co-author of *The Timetables of Science* and *The Timetables of Technology*. He is an editor/author of many other science reference books and teaching materials.

Marie E. Marshall, is currently the Principal of Hicksville Middle School in Hicksville, New York. Over the past 15 years, she has held the titles of assistant principal, supervisor, and science chairperson in this large suburban school district. Concerned about issues of science literacy and the development of process and investigative skills, Mrs. Marshall has presented workshops for both the New York State Reading Association and the International Reading Association, as well as for the New York State and National Science Teachers Association. She has served as president of the Nassau County Science Supervisors Association and has acted as a consultant for many local and regional school districts.

AUTHOR FOR: TO DO YOURSELF
Edward J. Metzendorf, Jr., Principal, Plainview-Old Bethpage School District.

Executive Editor: Barbara Levadi
Project Editor: Laura Baselice
Production Manager: Penny Gibson
Production Editor: Nicole Cypher
Marketing Manager: Sandra Hutchison
Photo Research: Jenifer Hixson
Cover Design: Nicole Cypher
Cover Photograph: Michael Abbey/Photo Researchers

Printed in the United States of America
 6 7 8 9 10   02 01
ISBN: 0-8359-1180-2

GLOBE FEARON EDUCATIONAL PUBLISHER
A Division of Simon & Schuster
Upper Saddle River, New Jersey

Table of Contents

UNIT 1 THE NATURE OF MATTER

UNIT 2 ELEMENTS

UNIT 3 COMPOUNDS

UNIT 4 MIXTURES

UNIT 5 CHEMICAL REACTIONS

UNIT 6 NUCLEAR CHANGE

UNIT 7 THE NATURE OF ENERGY

UNIT 8 WORK AND MACHINES

UNIT 9 MAGNETISM AND ELECTRICITY

UNIT 10 HEAT, LIGHT, AND SOUND

To the Student

Welcome to *Globe Physical Science*

Globe Physical Science is different from other science books you have used. Each lesson starts with a question that will be answered in the lesson. The question is followed with **Exploring Science,** a story of one of the many interesting or exciting parts of physical science.

After that, you start the lesson itself. Here is where you will learn about the ideas of physical science. We stick to the ideas that are most important. You will not need to learn a lot of dull details. Here also there are activities called **To Do Yourself.** They provide a way to see how physical science works first hand. There is also a **Review** that will help you remember what you have learned. Both **To Do Yourself** and **Review** provide places where you may write answers in the book.

There are also reviews at the end of each unit. **Summing Up** sections in some of the units review everything you have studied in the book so far. Other units feature **Careers in Physical Science,** short descriptions of how people work using the ideas of physical science. You can also find out what is needed to get into these careers.

Those are the parts of *Globe Physical Science*. The whole book was put together with you in mind. Try it. You'll like it

Marie E. Marshall
Bryan Bunch

1

THE
NATURE
OF
MATTER

What Is Matter?

Exploring Science

Thanksgiving Matter. A giant mouse is reflected in the glass of a skyscraper! A huge dinosaur casts its shadow on a concrete sidewalk! A spaceman hovers over thousands of startled children!

These are not scenes from a science fiction movie. This is the Macy's Thanksgiving Day Parade in New York City.

For a few hours, millions of viewers are entertained by these wonders. Few are aware of the years of work that were needed to develop these super balloons.

Let us look at one member of the Macy's Balloon Family, Garfield. This balloon is made of a coated fabric and has seams. When collapsed, the balloon has a mass of 130 kilograms and takes up 15 cubic meters of space. This is about the size of a very small room.

When blown up to full size, the balloon takes up 550 cubic meters of space. It takes about 1 kilogram of helium to blow the Garfield balloon up to full size. A slow leak from a seam would be a disaster. Thus, the seams are made very strong.

● What is the mass of Garfield as it floats over the city?

Balloons float because the mass of the inflated balloon is less than the mass of the air that would otherwise occupy the space taken up by the inflated balloon.

Matter

Glass, concrete, rubber, and helium are very different materials. Yet, they are all *matter*. But what could the helium in Garfield have in common with these other materials? What are the basic **properties** or characteristics, of matter? The clues are in the parade story.

MASS: Helium, although one of the "lightest" substances in the universe, has mass. **Mass** is a measure of *how much* matter an object contains. Whether the object is on the ground, floating above the buildings, or orbiting in space, its mass remains the same.

A steel tank like the ones used to hold helium has a mass of almost 50 kilograms. When filled with helium, its mass can increase by about .01 kilograms.

Have you have ever tried to lift a large pane of glass, a rubber tire, or a concrete block? If so, you know that these materials have mass. These examples show a common property of all samples of matter: *Matter has mass.*

VOLUME: Each of the objects mentioned above takes up space. The space that an object takes up is called its **volume.**

When collapsed, the shell of Garfield takes up 15 cubic meters of space. In other words, the volume of the rubber shell is 15 cubic meters. Now we have a second common property of matter: *Matter has volume*.

Measuring Matter: We can now say that **matter** is anything that has mass and takes up space (has volume). These properties of matter can be measured. Scientists all over the world use the same units to measure mass and volume. The units are part of the standard international language of measurement. The table shows some of these mass and volume units and their symbols.

	Unit Name	Unit Symbol
Mass:	gram	g
	kilogram	kg
Liquid Volume:	liter	L
	milliliter	mL
Solid Volume:	cubic meter	m^3
	cubic centimeter	cm^3

The three basic units of measurement in this table are the gram, the liter, and the meter. Three *prefixes* are also used. These prefixes are kilo-, milli-, and centi-. The prefix **kilo-** means

To Do Yourself Massing the Air in a Balloon

You will need:

Balance, tape, balloon

1. Place a ring of tape on the center of the pan of the balance, sticky side out. Place a balloon on the pan also.
2. Find the mass of the empty balloon and the tape. Record your answer to the nearest tenth of a gram in the table.
3. Inflate the balloon until it is nearly full. Tie a knot in the neck of the balloon.
4. Stick the inflated balloon to the tape on the balance. Find the mass and record the answer in the table.
5. Subtract the mass of the empty balloon and tape from that of the inflated balloon and tape. The difference will give you the mass of the air in the balloon.

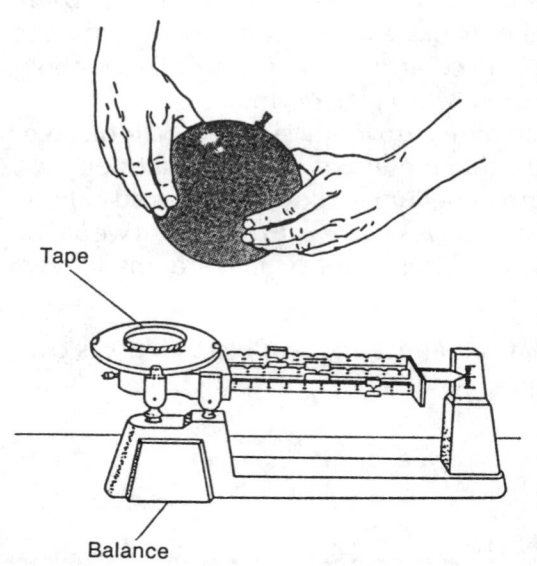

Tape

Balance

Data Table: Finding Mass

Empty Balloon and Tape	Inflated Balloon and Tape	Mass of Air
g	g	g

Questions.

1. Does the air in the balloon have mass?_____

2. Is air matter? Explain._____

1000. A kilogram is 1000 grams. The kilogram is used to measure large masses. The prefix **centi-** means 1/100. The cubic centimeter is used to measure the volume of small solid objects. The prefix **milli-** means 1/1000. The **liter** is used to measure the volume of a liquid. The milliliter is used to measure small liquid volumes.

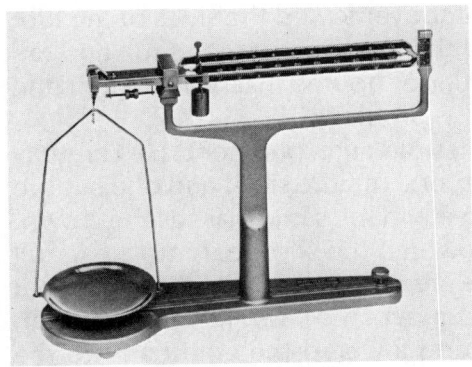

A balance is used to find mass.

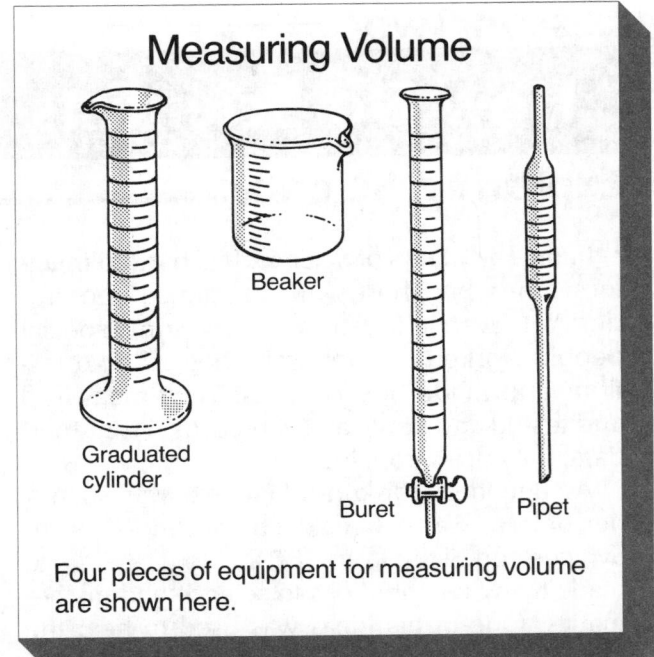

Measuring Volume

Beaker

Graduated cylinder

Buret   Pipet

Four pieces of equipment for measuring volume are shown here.

Review

I. Fill in each blank with the word that fits best. Choose from the words below.

gram   matter   volume   mass   centimeter
cubic centimeter

A moon rock was brought to earth. It is called _____

because it has mass and _____ . A unit that could be used

to measure its mass is the _____ . A unit that could be

used to measure its volume is the _____ .

II. Which statement seems more likely to be true?

     a. _____ When a balloon is filled with helium its mass decreases.

     b. _____ When a balloon is filled with helium its volume increases.

III. Write the most likely mass next to the name of each object.

     A. _____ pencil   1 g

     B. _____ notebook   2 kg

     C. _____ paper clip   10 g

     D. _____ chair   500 g

IV. Answer in sentences.

In what ways are gold, water, and oxygen alike?

How Do Scientists Learn About Matter?

Exploring Science

In the Early Days of Science. It is hard to imagine a time when there were no science laboratories. Yet, even in those early days, some special people worked as scientists. They asked questions and made observations. They measured and tested and kept careful records. From their data, they drew conclusions.

Archimedes (ahr-kuh-MEE-deez) was such a person. He was a Greek scholar who lived in Syracuse on Sicily from 287 B.C. to 212 B.C. He made many contributions to science and mathematics. One of his ideas was used to help the king of Syracuse catch a crooked craftsman.

The king had hired the craftsman to make a crown from a bar of pure gold. When the crown was finished, its mass was the same as that of the gold bar. But the king suspected that the craftsman had cheated him by using silver in place of some of the gold. The king asked Archimedes to find out if the crown was pure gold without harming the crown.

One day Archimedes went to the public baths to relax and think about the crown. As he lowered himself into a full tub, he saw that some of the bath water overflowed the sides of the tub. It is said that he jumped out of the tub and ran home naked. He shouted that he had found the solution to his problem.

Archimedes did an experiment. He carefully cut two bars, one of pure silver and one of pure gold. Each bar had the same mass as the crown. He gently lowered the silver bar into a pan full of water. He measured the volume of water that ran over the sides of the pan. He refilled the pan with water and lowered the gold bar into the pan. This time, a smaller amount of water ran over the sides of the pan. Finally, he lowered the crown into the pan full of water. Less water overflowed than with the silver bar, but more than with the gold bar. From this evidence, Archimedes concluded that the crown was not made of pure gold.

● When would the overflow produced by the crown be equal to that produced by the bar of pure gold?

The Ways of the Scientist

In his experiment, Archimedes worked as a scientist. Was he a chemist? A **chemist** studies matter and its properties. Was he a physicist? A **physicist** (FIZ-ih-sist) studies the forces of nature and the forms of energy.

Archimedes studied a property of gold and silver. That was chemistry. But the method he used depended on the force that made the water overflow. That was physics. Thus, Archimedes should be called a **physical scientist.** That is, a scientist who investigates problems involving both matter and energy.

What a scientist studies is not as important as how he or she studies it. How can we work as scientists? How can we try to answer questions so that our results can be trusted?

STATING A PROBLEM. A scientist must know what question he or she is trying to answer. For Archimedes, the question was: "Is the crown made of pure gold?"

Finding the answer to one question often leads to new questions. Thus, one experiment may lead to other experiments. For example, while working to answer the question about the crown, Archimedes wondered about related questions. He did many other experiments using different objects and liquids.

CHOOSING A HYPOTHESIS. Based on what we know, we make a hypothesis. A **hypothesis** (hy-POTH-ih-sis) is an "educated guess" as to a solution to a problem. Archimedes may have thought of other ways to solve the problem of the crown. He may have known that gold and silver show differences if they are melted or hammered. But he could not use these tests because they would have damaged the crown.

GATHERING INFORMATION. Sometimes we don't know enough to suggest even one hypothesis. Then we must gather new information. We do this by reading the work of other scientists, and by observing the world around us. His observation of the overflowing bath water led Archimedes to this hypothesis: "Equal masses of pure gold and pure silver will take the place of, or **displace,** different amounts of water."

DESIGNING AN EXPERIMENT. An experiment is a test of a hypothesis. It must be planned so that the results can be repeated. Archimedes used samples of the same mass and purity. He used the same pan and the same water supply. We say that he controlled certain variables. A **variable** (VAR-ee-uh-bul) is something that could change. A **controlled variable** is one the scientist works to keep from changing during an experiment.

OBSERVING AND RECORDING. Scientific observations are more than just "seeing." We must use all of our senses. We must also use tools that help us observe. Many of these tools measure variables. Measurements make our observations more accurate. Archimedes made measurements. These measurements told him how much water overflowed.

It is important to record what we observe. This collection of facts is called **data.** Data help up to organize our observations. It allows us to see relationships. And it helps others to compare their results to ours when they repeat our experiments.

DRAWING CONCLUSIONS. When a scientist completes an experiment, a solution to the

How a Scientist Works

The most important steps in the way a scientist works are shown here.

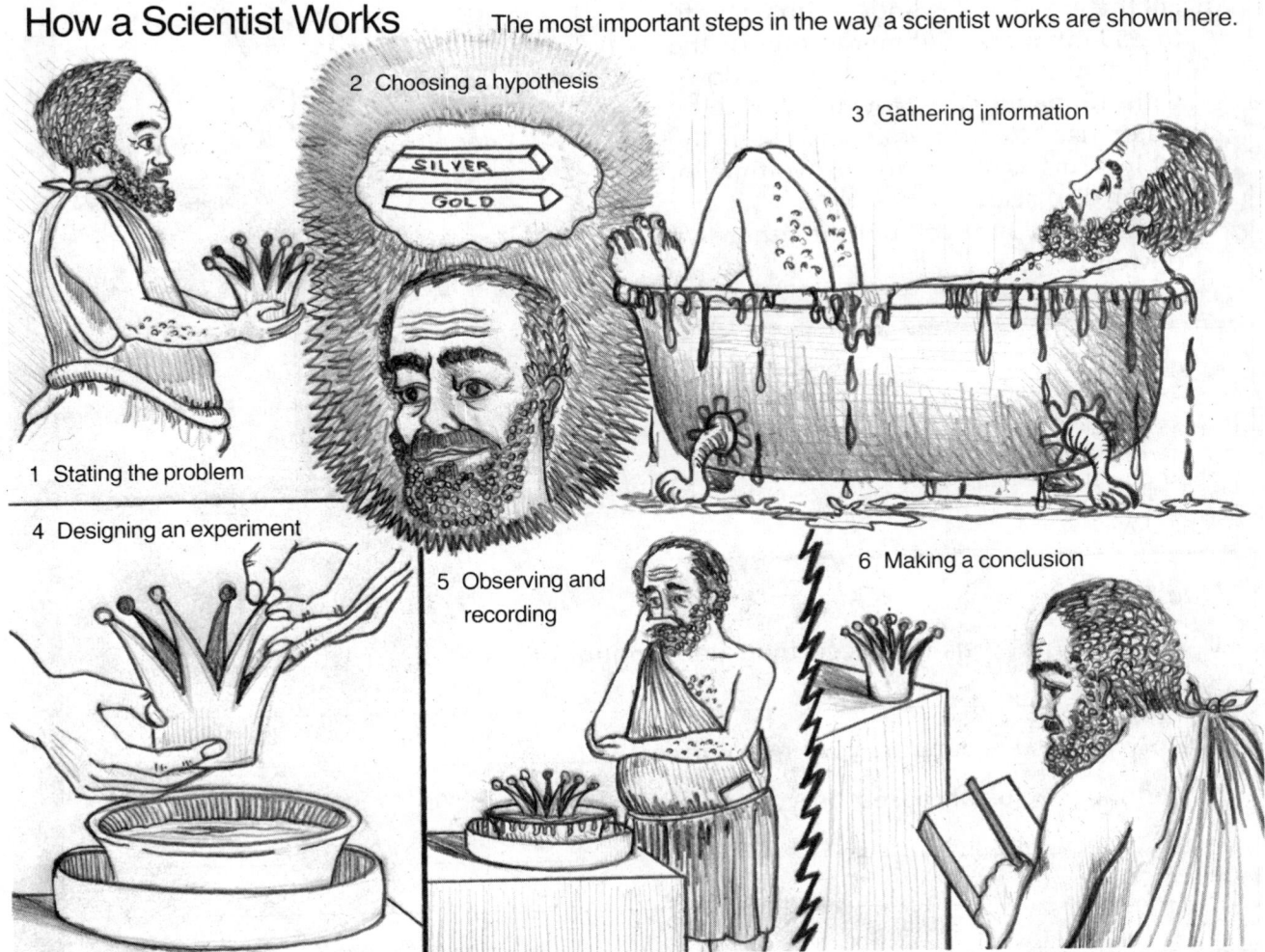

1 Stating the problem

2 Choosing a hypothesis

SILVER

GOLD

3 Gathering information

4 Designing an experiment

5 Observing and recording

6 Making a conclusion

original problem can be suggested. Scientists draw **conclusions** based on data. Archimedes' conclusion was that the crown was not made of pure gold.

Sometimes two scientists can make the same observations. But they may draw different conclusions. For example, Archimedes said the crown was not made of pure gold. Another scientist may have concluded that "The crown is made of some other metal."

To Do Yourself   What Determines How Much Water an Object Displaces?

You will need

Wooden dowels of different sizes (with and without nails), balance, 100-mL graduated cylinder, paper clip, water

1. Write your hypothesis by answering this question: What determines how much water an object displaces when it is submerged, its mass or its volume?
2. Find the mass of each dowel and record it in the Data Table.
3. Fill the graduated cylinder with water to the 50-mL mark. Submerge one of the wooden dowels by pushing it down with the paper clip. How much did the water rise? Record your answer in the table. This is equal to the volume of water displaced.
4. Repeat step 3 for each of the other two dowels.

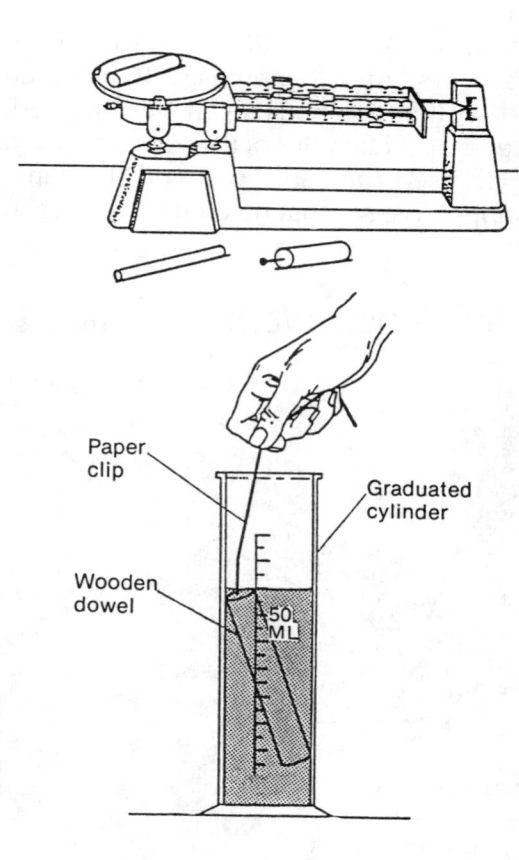

Paper clip

Graduated cylinder

Wooden dowel

50 ML

Data Table

	mass	water displaced
short dowel		
short dowel with nails		
long dowel		

Questions

1. Which two dowels displaced the same amount of water? _____

_____

2. Do your results support your hypothesis? _____

3. Write a conclusion based on your experimental results. _____

_____

_____

Review

I. Fill in each blank with the word that fits best. Choose from the words below.

**energy   hypothesis   experiment   matter   conclusions   measurements
observations**

A chemist studies _____ . To solve a problem, first a

_____ is stated. Careful _____ are made

and _____ are recorded. Finally, a _____
may be drawn.

II. Which statement seems more likely to be true?

a. _____ Observations are facts.

b. _____ Conclusions are facts.

III. Answer in sentences.

A. To control a variable means that you keep it from changing. Suppose you
had to find out which displaced more water, a 10-gram cube of lead or a
10-gram cube of wood. Which variables would you control? Why?

mass _____

_____

water supply _____

_____

container _____

_____

substance _____

_____

B. Why is accurate measurement so important in an experiment?

_____

_____

_____

Lesson Three

What Is Density?

Exploring Science

Science Fantasy. A scientist father accidentally exposes his little son to a new technology. The boy begins to get larger...and larger...and larger...

This was the idea behind the 1992 movie *Honey, I Blew Up the Kid.* The audience laughed as the giant child ran away, damaging anything in his path. He was shown to have the strength of a fairy-tale giant, like the one in "Jack and the Beanstalk."

A real scientist would know that this idea is pure entertainment, not science. If the boy weighed 44 pounds, his mass in the metric system is about 20 kilograms. Remember the Garfield balloon. Its mass was 130 kilograms. If the blown-up boy were the size shown in the

film, he would have floated away. He could have crushed nothing. The ground would not rumble when he walked. He would have been a harmless human balloon.

● What variables are considered in this discussion of the movie?

Density

The blown-up child has a greater size, or volume, than a regular baby has. Even though his weight has not changed, the blown-up boy is in some way "lighter." He has the same mass in more volume.

Scientists would say that the density of the blown-up boy is less than that of the regular boy. **Density** is the mass contained in one unit volume of a substance. All matter has density. Density is the relationship between two properties common to all matter: mass and volume.

$$\text{Density} = \frac{\text{mass}}{\text{volume}} \quad \text{or} \quad D = \frac{M}{V}$$

Five cubic centimeters of pure silver has a mass of 52.5 grams. We can use the formula to find the density of pure silver:

$$\text{Density} = \frac{52.5 \text{ g}}{5 \text{ cm}^3} = \frac{10.5 \text{ g}}{1 \text{ cm}^3}$$

The density of pure silver is 10.5 g/cm^3.

The density of pure gold is 19.3 g/cm^3. This means that one cubic centimeter of pure gold has a mass of 19.3 grams. Ten cm^3 of pure gold

will have a mass of 10 x 19.3 g, or 193 grams. Suppose you were given a gold bar with a volume of 5 cm^3 and a mass of 90 g. How would you find out if it was pure gold?

Think back to Archimedes' experiment. We know that the density of gold is greater than that of silver. We also know that the two bars had the same mass. From this information, we can conclude that the gold bar had a smaller volume than the silver bar. Since its volume was less, the gold bar displaced less water than either the silver bar or the crown. If the crown had been pure gold, it would have displaced the same amount of water that the gold bar displaced.

Understanding density can also help us to understand why some objects float in water while others sink. For example, a cork will float in water, but a piece of gold will sink. Let's look at the densities of these materials: water = 1.0 g/cm^3; gold = 19.3 g/cm^3; cork = 0.25 g/cm^3. Notice that the density of cork, which floats, is less than that of water. The density of gold, which sinks, is greater than that of water. From this data, we can state a hypothesis: Matter with a density less than that of water (1.0 g/cm^3) will float in water. Matter with a density

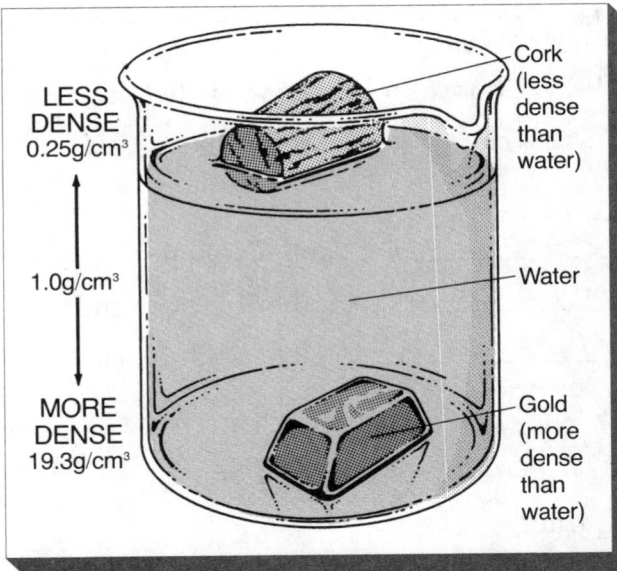

LESS DENSE 0.25g/cm³

1.0g/cm³

MORE DENSE 19.3g/cm³

Cork (less dense than water)

Water

Gold (more dense than water)

greater than 1.0 g/cm³ will sink in water.

Look at the table of densities. Make two lists showing those substances that float in water and those that sink.

Table of Densities

gold	19.3 g/cm³	butter	0.86 g/cm³
cork	0.26 g/cm³	gasoline	0.7 g/cm³
silver	10.5 g/cm³	oak (wood)	0.75 g/cm³
mercury	13.6 g/cm³	aluminum	2.7 g/cm³

Mercury is a liquid at room temperature. Use the items in the table to identify which one will sink in a bowl of mercury. Hint: it is the only one that has a density greater than that of liquid mercury.

To Do Yourself   Finding the Density of Soap

You will need:

A balance, soap, beaker, water

1. Your teacher will give you the volume of your soap sample to record in the table.
2. Find the mass of the soap. Record your answer to the nearest tenth of a gram in the table.
3. Find the density of the soap to the nearest tenth of a gram by dividing the mass by the volume. Record your answer in the table.
4. Place the soap sample in a beaker of water.

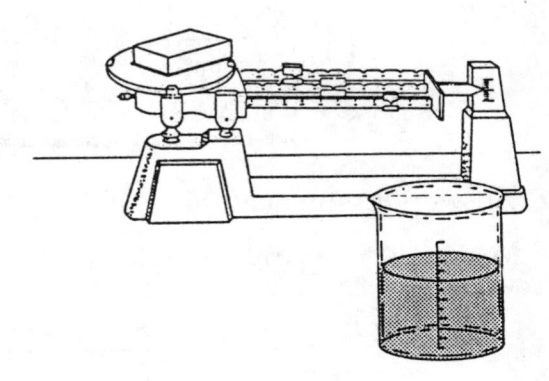

Data Table

Mass	Volume	Density
g	cm³	g/cm³

Questions

1. Does the soap sample float or sink? _____

2. Why does the soap float (or sink)? _____

Review

I. Fill in each blank with the word that fits best. Choose from the words below.

cm   6 g/cm³   density   sink   cm³   12 g/cm³   g   mass   float

A metal block with a mass of 48 _____ and a volume of

4 cm³ would have a density of _____ . Another block of the

same material would have the identical _____ . Both blocks

would _____ in water.

II. Which statement seems more likely to be true?

    A. _____ Solid rubber has a density greater than 1 g/cm³.

    B. _____ Solid rubber has a density less than 1 g/cm³.

III. Complete this table from information in the lesson.

SUBSTANCE	DENSITY
Aluminum	
Gold	
Mercury	
Gasoline	
Pure Water	

IV. Answer in sentences.

What information do you need to predict if a sample of matter will float in water?

_____

_____

_____

Lesson Four

What Are the Phases of Matter?

Exploring Science

Drama Aboard Apollo 13. Liquid oxygen, or LOX, is familiar to most space fans. It is a form of oxygen that reacts with liquid fuels to propel modern rockets. LOX is **cryogenic** (kreye-uh-JEN-ik) liquid. *Cryo-* means frost. LOX forms at very low temperatures and is stored at -183°C.

This pale-blue liquid is very dangerous when it is disturbed. In April 1970, a serious incident occurred on the *Apollo 13* spacecraft as it was on its way to the moon. The three astronauts aboard the spacecraft were talking with Mission Control when the drama began. It is believed that the temperature control broke for a supply of LOX that was being used as part of the energy system for the spacecraft. The LOX changed to oxygen gas so rapidly that the tank it was in exploded. Not only did the astronauts lose their power supply, they also lost part of the oxygen used for breathing. The crew was told to move at once to the part of the craft that was to have taken two of them from lunar orbit to the moon. They could use this lunar module as a "lifeboat" because it had its own energy and breathing system. It also had enough power to direct the craft back to earth. But now there were three astronauts in a module designed for two. Each astronaut needed a kilogram of oxygen a day. The emergency trip back to earth took more than three days.

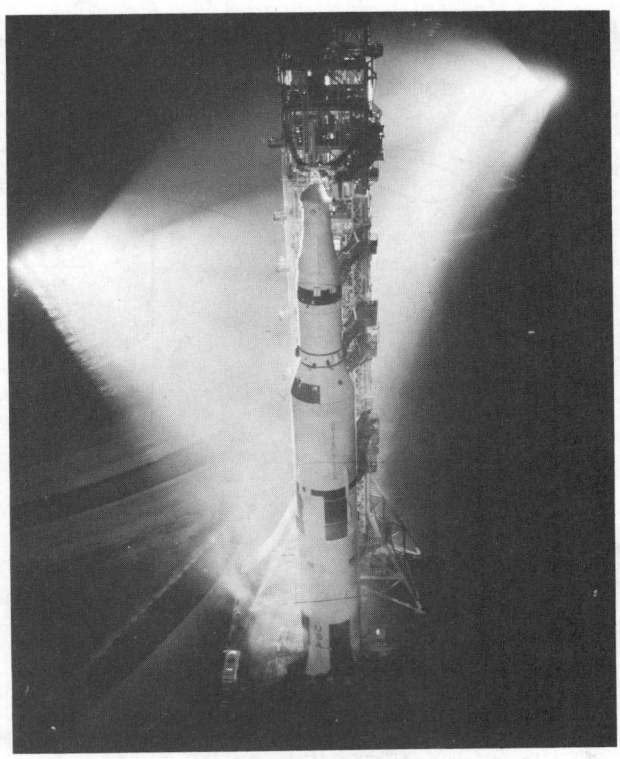

The *Apollo 13* rocket, ready for lift-off.

● How much oxygen was needed by the crew for breathing during the emergency trip home?

Phases of Matter

The *Apollo 13* accident involved a rapid change from liquid oxygen to oxygen gas. Under ordinary conditions matter can exist in three forms, or **phases**. These phases are solid, liquid, and gas.

Matter in the **solid** phase has definite volume and definite shape. This is why you can't fit a round peg into a square hole.

Matter in the **liquid** phase has definite volume, but no definite shape. It takes the shape of its container. For example a liter of milk can be shaped like a milk carton. When poured into a glass the milk will take the shape of the glass.

Matter in the **gas** phase has no definite shape and no definite volume. A gas will fill any container into which it is placed. It will also take on the shape of that container.

Now let us take a brief look at a fourth phase of matter. **Plasma** is a phase of matter in which the particles that make up the matter are no longer atoms. Electrically charged parts of atoms are separated by very high temperatures. It is rare on earth. About the only common examples of plasma that we might see are neon and fluorescent lights. The glow of these lights is given off by a plasma. Elsewhere in the universe, where super-high temperatures are common, plasma is an ordinary phase.

As conditions change, the phases of matter can change. Temperature change is an important cause of changes in phase. The following table lists some of the terms that are used to describe phase changes. We will discuss change of phase more fully in a later unit.

TERM	PHASE CHANGE
melting	solid to a liquid
freezing	liquid to a solid
evaporation	liquid to a gas
condensation	gas to a liquid

The Three Phases of Water

SOLID

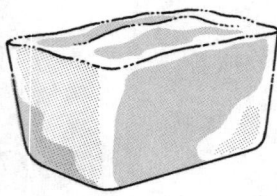

Solids have definite volume and definite shape.

LIQUID

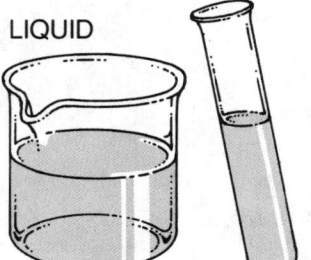

Liquids have definite volume, but no definite shape.

GAS

Gases have no definite volume and no definite shape.

To Do Yourself   Changing the Phase of Ice

You will need:

Balance, crushed ice, small beaker

1. Place about 25 milliliters of crushed ice in a small beaker. Make sure the outside of the beaker is dry.
2. Find the mass of the beaker and ice to the nearest tenth of a gram. Record your answer in the data table.
3. Without moving the balance, remove the beaker. Hold the beaker in your hand for a few minutes.
4. If the ice is undergoing a change, wait until the change is complete.
5. Dry the outside of the beaker again. Place it back on the balance.
6. Find the mass of the beaker and its contents. Record the mass in the data table.

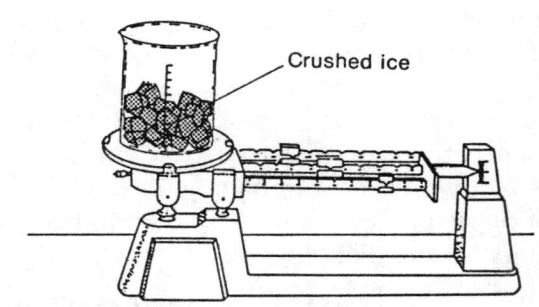

Crushed ice

Data Table

Mass of Beaker and Ice	Mass After Phase Change	Change in Mass
g	g	g

Questions

1. What caused the ice to change? _____

2. Did the mass of the beaker and its contents change? _____

3. What can you conclude? _____

Review

I. Fill in each blank with the word that fits best. Choose from the words below.

liquid   melts   condensation   plasma   solid   freezing

evaporation   gas   temperature   phase

When LOX is prepared, a gas changes to a _____ .

When ice changes to a liquid, it _____ . A "special"

phase of matter found at super-high temperatures is _____ .
Ordinary phase changes are usually caused by a change in

_____ . A change from a liquid to a solid is called

_____ .

II. Which statement seems more likely to be true?

A. _____ A gas cannot be contained.

B. _____ A gas assumes the shape of its container.

III. Match the phase with its description.

1. _____ solid   **a.** no definite volume

2. _____ liquid   **b.** definite shape

3. _____ gas   **c.** definite volume,
                                    no definite shape

IV. Answer in sentences.

How has technology changed our knowledge of phases of matter?

_____

_____

_____

How Does Technology Make Use of Matter's Properties?

Exploring Science

Space Spin-Off. What do the "Bionic Bat," a tired marathon runner, and an astronaut have in common? They all use the special properties of **metallized** (MET-uh-lized) **plastics**—plastics coated with metal.

The Bionic Bat uses the material's low density and strength. The runner benefits because it is lightweight and keeps in 80% of the body's heat. The astronaut benefits from all of these properties.

Metallization is the process in which super-fine metal mist is used to coat a thin plastic. Surprisingly, this process was discovered in the 19th century. It was used then for decoration.

Today, the metal is heated in a special chamber. At temperatures greater than 1100°C, the metal becomes a gas, or **vapor**. The vapor rises and coats a fast-moving plastic film. The result is a material that can be thinner than cellophane and has many useful properties. The technology for this exciting material came from NASA research.

● What do you predict would happen if the temperature in the vapor chamber dropped to 500°C.

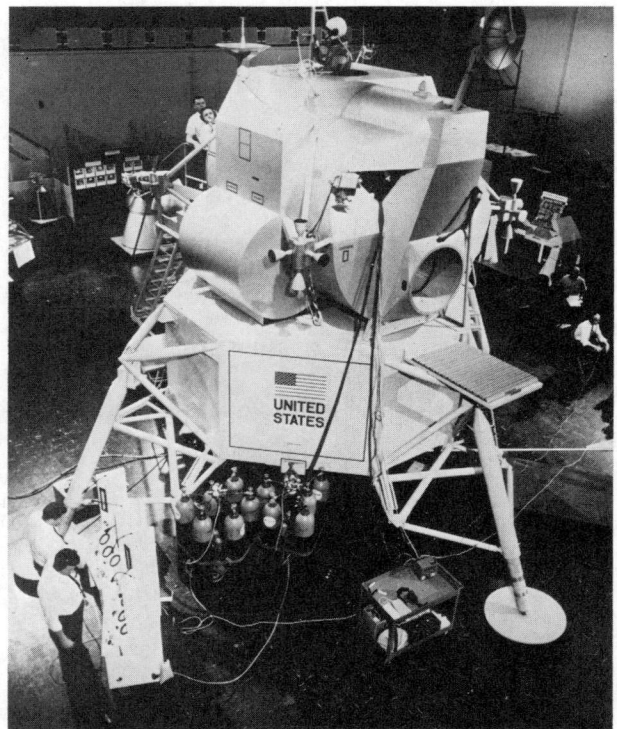

The spacecraft that the United States landed on the moon was made from metallized plastics. What properties of metallized plastics made them useful for this type of vehicle?

The *Gossamer Albatross* was built by Paul MacCready. It has a wingspan of 30 meters and a mass of only 30 kilograms. Metallized plastics helped to make this vehicle light in mass.

Technology and Properties of Matter

The example of metallized plastics shows several ideas about technology.

● **Technology** (TEK-nawl-uh-gee) is applied science, or basic science knowledge being used to meet everyday needs.

● An advance in technology is usually made to meet some special need.

● Once a product or a process is developed, new uses may be found for it.

● Knowledge of the special *properties,* or characteristics, of matter sometimes suggests ways to use it.

Recall that *all* matter has the properties of mass and volume. However, certain materials have one or more special properties that make them especially valuable. Copper is such a material.

Copper is **malleable** (MAL-ee-uh-bul). This means it can be hammered into thin sheets or bent into different shapes without breaking. Thus, it can be used for pipes and building material, among other things.

Another valuable property of copper is that it is **ductile** (DUK-tyle). This means that copper can be drawn out into a thin wire. This property is very important in the technology of electronics. Copper has one more property that makes it especially valuable. It can carry, or **conduct,** electricity.

Often there is a need for substances that have opposing properties. In electronics, for example, there is a need for materials that are good electrical conductors. There is also a need for substances that are poor electrical conductors. These poor conductors are used to keep electricity from escaping. Such materials are called electrical **insulators.**

Another property that can be valuable is the ability to conduct heat. **Heat conductivity** (kon-DUK-tiv-ih-tee) is a measure of how fast heat travels in a substance. Consider how important this property is in pots and pans. Copper and iron are good conductors of heat. So is aluminum. This ability explains why these substances are used to make pots and pans.

Poor conductors of heat are also needed. They act as heat insulators. Air is a good heat insulator. The thin fibers used in home insulation trap air. This air keeps heat from passing through. In the past, asbestos (as-BES-tus) was widely used as a heat insulator. Today we know that asbestos particles are very harmful to humans. Therefore, insulators are being manufactured from new materials.

What properties does copper have that makes it a useful material for each of these objects?

To Do Yourself   Conduction as a Special Property of Matter

You will need:

4 Styrofoam cups with lids, 4 Celsius thermometers, bent strip of copper and one other metal, water, hot plate, heat-resistant gloves, heat-resistant pad.

1. Set up the materials as shown in Figure

2. Boil enough water to fill two of the cups. **Caution: Handle hot materials with great care. Wear heat-resistant gloves. Do not set hot materials directly on a desk or table top. Place them on a heat-resistant pad.** Fill two cups with boiling water and two with cold water.

3. Connect one hot cup with one cold cup by means of the assembled lids with the copper strip and thermometers.

4. Repeat step 2 for the remaining two cups. Use the lids with the other metal strip.

5. Wait 15 seconds and record the starting temperature of each cup in the data table.

6. Record the temperature in each cup every 3 minutes for a total of 15 minutes.

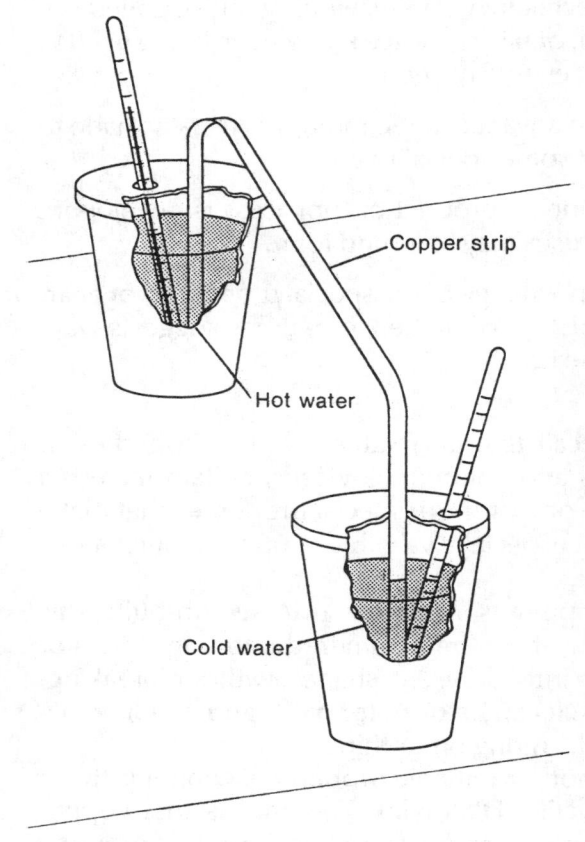

Copper strip

Hot water

Cold water

Data Table

	Temperature 3 Minutes	Temperature 6 Minutes	Temperature 9 Minutes	Temperature 12 Minutes	Temperature 15 Minutes
Cup 1	° C	° C	° C	° C	° C
Cup 2	° C	° C	° C	° C	° C

Questions

1. Which metal strip connected the cups with the greatest temperature change? _____

2. Which metal strip transferred the most heat from the hot cup to the cold cup? _____

3. Which metal strip is the better heat conductor? _____

_____

Review

I. Fill in each blank with the word that fits best. Choose from the words below.

**malleable   insulators   ductile   luster   properties   conductors
products**

Although most metals are _____ of heat, some are better than others. Some can be pulled into wires because they are

_____ . Others can be hammered because they are

_____ . Some _____ have special uses.

They must be made from materials that have special _____ .

II. Which statement seems more likely to be true?

A. _____ New technology builds on old technology.

B. _____ Technology is independent from basic science.

III. From your knowledge, list one special property of each substance: gold, rubber, styrofoam.

_____

_____

_____

IV. Answer in sentences.

What inventions or products have been developed because of America's concern for saving energy?

_____

_____

_____

Review What You Know

A. Use the clues below to complete the crossword.

Across

1. Unit for very large masses
6. Basic unit of length
7. Experimenting
9. A property of all matter
10. An electrically charged high-temperature phase
12. Anything with mass and volume
13. Prefix meaning 1000
14. Science used for everyday needs

Down

2. The phase of helium at room temperature
3. Smaller than a cm
4. Basic units of mass
5. An object with a density of 1.5 g/cm^3 _____ in water
8. Change from liquid to gas
9. Changing from a solid to a liquid
11. Unit for very small liquid volumes

B. Write the word (or words) that best completes each statement.

1. _____ The characteristics of a substance are its
   a. density **b.** properties **c.** mass

2. _____ All matter has mass and **a.** color **b.** malleability
   c. volume

3. _____ Before we can design an experiment we must
   state a(n) **a.** hypothesis **b.** conclusion **c.** trial

4. _____ An experiment tests for only one **a.** variable
   b. constant **c.** observation

5. _____ A scientist who studies matter is a(n)
   a. physicist **b.** chemist **c.** meteorologist

6. _____ Mass is measured with a **a.** ruler **b.** graduated
   cylinder **c.** balance

7. _____ The basic unit of mass is the **a.** gram **b.** liter
   c. meter

8. _____ Technology usually develops from **a.** a need
   b. an accident **c.** a conclusion

9. _____ The prefix for 1/1000 is **a.** kilo- **b.** centi-
   c. milli-

10. _____ The mass of a rock is _____ on the moon than
    it is on Earth. **a.** greater **b.** less **c.** the same

11. _____ The density of pure water is **a.** 1 g/cm^3
    b. 10 g/cm^3 **c.** 1 g/cm

12. _____ The density of cork is **a.** equal to water
    b. greater than water **c.** less than water

13. _____ Density is equal to **a.** mv **b.** $\dfrac{v}{m}$ **c.** $\dfrac{M}{V}$

14. _____ The density of pure gold **a.** is always the same
    b. changes with volume **c.** is less than 1g/cm^3

15. _____ At room temperature, gold is a **a.** gas **b.** solid
    c. liquid

16. _____ Condensation is a change from a **a.** solid to
    liquid **b.** liquid to gas **c.** gas to liquid

17. _____ An example of a phase change is **a.** ice melting
    b. paper ripping **c.** a cork floating

18. _____ Phases can be changed using a(n) **a.** balance
    b. thermometer **c.** heat source

19. _____ One characteristic of the plasma phase is
    a. electrical charges **b.** low temperatures **c.** bright colors

20. _____ A ductile substance can be **a.** drawn into wire
    b. hammered **c.** without volume

21. _____ A heat conductor **a.** creates heat **b.** carries
    heat **c.** stops heat

22. _____ Technology uses **a.** insulators only
    b. conductors only **c.** all special properties of matter

23. _____ Technology applies **a.** chemistry only **b.** physics
    only **c.** knowledge from all sciences

24. _____ Technology produces **a.** new mathematical
    formulas **b.** new scientific theories **c.** new products

C. Apply What You Know

Study the cartoon. Then answer the questions.

1. Fill in each blank with the word that fits best. Choose from the words below.

   observation   conclusion   density   problem   mass   volume
   hypothesis

   a. A(n) _____ is stated in box 1.

   b. The boy in box 3 is measuring _____.

   c. The girl in box 4 is measuring _____.

   d. Each of the people in boxes 3 and 4 are making a(n) _____.

   e. In box 5, the people are finding the _____.

   f. A(n) _____ is made in box 6.

2. Fill in each blank using information from the cartoon.

   a. The formula for density is _____.

   b. The mass of the ring is _____.

   c. The volume of the ring is _____.

   d. The unit for density is _____.

   e. The density of the metal in the ring is _____.

3. Circle the letter of the phrase that should appear in the last box of the cartoon. Explain your choice on the lines below. Answer in complete sentences.

   a. The ring is made of pure gold.

   b. The ring is not made of pure gold.

   c. There is no gold in the ring.

_____

_____

_____

_____

D. Find Out More

   1. Look through a recent newspaper or magazine. Make a list of all of the advertised products. Then ask a librarian to get you a newspaper or magazine from 20–30 years ago. Make a list of all of the advertised products in the old newspaper or magazine. Compare the two lists. How are the products similar? How are they different? How has science and technology changed our lives over the years? Make a chart of your findings. Report your findings to the class.

   2. Using a graduated cylinder (or beaker) and a balance, find the densities of some common materials. Put your observations in a chart. Compare your chart with a chart prepared by another student.

UNIT

2

ELEMENTS

Lesson One

What Is an Element?

Exploring Science

Buckyballs. If there are any doubts that scientists have a sense of humor, consider the case of the buckyball.

Carbon is one of the most common forms of matter. Until recently, we knew that it occurred in nature in two pure types: diamond and graphite. Then, in 1985, there was a discovery at Rice University. Chemists created a new pure carbon substance.

The carbon formed in clusters of 60 atoms. Rick Smalley and Harold Kroto, who found the odd carbon, observed that the atoms in each cluster were arranged like the domes created by architect R. Buckminster Fuller. So the new material was called buckminsterfullerene. Models of a cluster looked like a soccer ball with a dozen white faces outlined in black. So the chemists started to refer to the amazing figure as a "buckyball."

Since that discovery, buckyballs have been studied all over the world. We read of buckytubes, buckyfibers, and buckycharms. They have been tested as electrical wires, as tiny magnets, and as artificial diamonds. Also, very

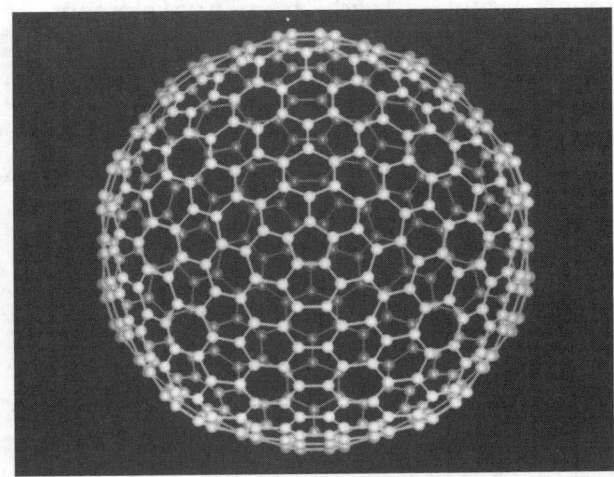

recently, buckyballs have been considered as blockers for the HIV virus.

In 1992, buckminsterfullerene was found in nature. An American scientist was studying a black Russian rock sample. He found buckyballs that may be 500 million years old.

● What is the difference between the 1985 and 1992 discoveries?

Elements

In Unit 1, you learned that matter is anything that has mass and volume. In other words, just about everything you can think of is matter. How many different kinds of matter can you think of? Hundreds? Thousands? Maybe even more. You may be surprised to learn that there are just over one hundred *basic* kinds of matter!

Each of the basic kinds of matter is called an element. An **element** (EL-uh-ment) is the simplest kind of matter. Elements cannot be broken down into other kinds of matter by ordinary chemical means.

Many of the common elements have been known for centuries. About 30 elements were listed as early as the 1700's. Among those elements were gold, silver, lead, oxygen, hydrogen, and iron. Many attempts were made to break these elements down into other kinds of

Gold (top) and sulfur (bottom) are elements.

matter. For example, iron was pounded, heated, and treated in many experiments. It could not be changed to a simpler kind of matter. Thus, iron was properly listed as an element.

Many very common kinds of matter are not elements. Water, for example, is not an element. Water is made from two elements—hydrogen and oxygen. Table salt is another common substance that is not an element. Table salt is made from the elements sodium and chlorine.

Water and table salt are not elements. Water is made from the elements hydrogen and oxygen. Table salt is made from the elements sodium and chlorine.

There are more than 100 known elements. Many of them are not found in nature. Many of those named after scientists (einsteinium) or countries (polonium) have been "made" in the laboratory. The first of these lab-made elements was called technetium (TEK-nee-she-um). The name technetium comes from the root word "teknetos," which means artificial.

From the earliest days, scientists from all parts of the world studied the elements. These scientists found that they needed a simple way to write the names of the elements. They also needed to write in a common "language." After much discussion, scientists developed a set of chemical symbols.

A **chemical symbol** is a shorthand way of writing the name of an element. Chemical symbols are always one or two letters long. A one-letter symbol is always written as a capital letter. For example, O is the symbol for the element oxygen. H is the symbol for hydrogen.

In a two-letter symbol, the first letter is a capital, the second is a small letter. For example, the symbol for calcium is Ca. The symbol for helium is He.

The symbols for some elements are sometimes surprising. Many of the earlier-known elements were known by their Latin names. For example, iron was known as *ferrum*. The chemical symbol for iron is Fe. The symbol for sodium, Na, comes from its Latin name, *natrium*. Ag, the symbol for silver, comes from *argentum*.

No two elements have the same symbol. Symbols are assigned by an international group of chemists. So, scientists all over the world use the same symbols for the elements.

In the future, more elements will probably be made in the laboratory. However, scientists believe that all of the natural elements have been identified. Even careful study of rocks from the moon and Mars has not turned up any new natural elements.

Symbols of Common Elements

Element	Symbol	Element	Symbol	Element	Symbol
Hydrogen	H	Silicon	Si	Silver	Ag
Helium	He	Phosphorous	P	Tin	Sn
Carbon	C	Sulfur	S	Iodine	I
Nitrogen	N	Chlorine	Cl	Platinum	Pt
Oxygen	O	Potassium	K	Gold	Au
Fluorine	F	Calcium	Ca	Mercury	Hg
Neon	Ne	Iron	Fe	Lead	Pb
Sodium	Na	Nickel	Ni	Uranium	U
Magnesium	Mg	Copper	Cu		
Aluminum	Al	Zinc	Zn		

I. Fill in each blank with the word that fits best. Choose from the words below.

simple   element   hydrogen   complex   substance   oxygen   symbol

The _____ for oxygen is O. Oxygen is a(n)

_____ . Water contains _____ and

oxygen. Sugar contains carbon, hydrogen, and oxygen, so it is not an

element. Sugar is more _____ than sodium.

II. Which statement seems more likely to be true?

A. _____ ES is the symbol for einsteinium.

B. _____ Es is the symbol for einsteinium.

III. Look at the table of symbols for common elements. List the elements whose symbols you think come from their Latin names.

IV. Answer in sentences.

If you "discovered" three new elements, what names and symbols would you choose? Explain.

_____

_____

_____

What Is an Atom?

Exploring Science

A Very Special Microscope. For centuries, scientists have talked about how small the pieces of an element could be. These smallest pieces of matter were called atoms. Atoms were so small, they had never been seen. Many scientists did not believe they existed. Those who did believe, argued about what atoms looked like.

The light microscope did not help scientists study atoms. Atoms were much too small to be seen with a light microscope. Light waves, 5000 times larger than a single atom, just passed around the atom.

We now have a new tool. A new type of microscope was developed in the 1980's. It is called the ARM. The ARM allows us to see some of the details of the atoms found in solids. For example, the atoms of zirconia (zur-CONE-ee-uh) clearly appear as black and white dots on a gray background. Scientists are looking forward to the data which the ARM will provide.

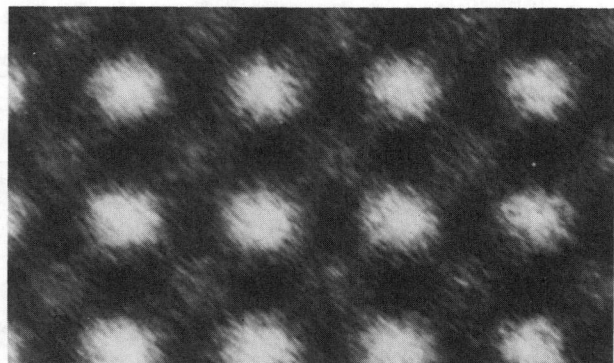

The atoms of zirconia clearly appear as black and white spots on a gray background when viewed through the ARM.

Will high schools have these microscopes by the year 2000? It is unlikely, because each ARM is about three stories tall and costs about three million dollars.

● How many different kinds of atoms are in zirconia?

Atoms

All matter is made up of atoms. An **atom** is the smallest part of an element. Therefore, an atom of helium is the smallest piece of helium. An atom of silver is the smallest piece of silver. An atom of lead is the smallest piece of lead.

Suppose you were able to see the atoms in a pure sample of an element. You would find that all of the atoms are almost identical. A sample of pure gold, for example, would contain one kind of atom. But, 14 karat gold, which is not pure gold, would have gold atoms and the atoms of other metals also.

Zirconia is not an element. The ARM showed that there is more than one kind of atom in zirconia. If the sample had been of the element zirconium, however, the picture would have shown all white figures. All of the atoms would have been alike.

The ARM picture shows that the atoms of solids are very close together. Although scientists cannot see the atoms of liquids and gases at this time, they think that the atoms of liquids are farther apart than those of solids. They also believe that the atoms of gases are farther apart than those of liquids.

Liquid oxygen, remember, is the element oxygen in its liquid phase. It contains only one kind of atom. The atoms are the same as those of oxygen gas. However, the atoms in liquid oxygen, or *LOX*, are closer together than those in oxygen gas. The atoms of solid oxygen are even more tightly packed.

What is in the space between atoms? Nothing. This absence of matter is what scientists call a **vacuum** (VAK-yoo-um). One of the experiments planned for the ARM is to see what will happen if one atom is removed from a solid like zirconia. Perhaps very little will happen. Then again, maybe a great deal will happen. Only the future will tell us the answer.

SOLID	LIQUID	GAS
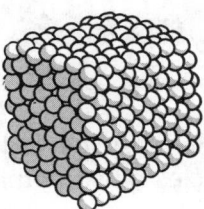		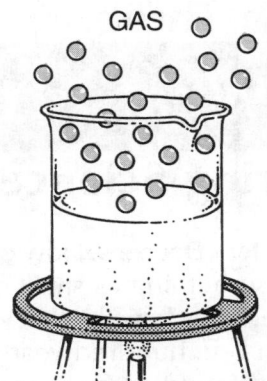
The atoms of a solid are packed close together.	The atoms of a liquid are farther apart than the atoms of a solid.	The atoms of a gas are farther apart than the atoms of a liquid.

Review

I. Fill in each blank with the word that fits best. Choose from the words below.

**solid   atoms   liquid   zirconia   gas   matter   element
zirconium**

All _____ is made up of small particles. These

particles are called _____ . The atoms of a

_____ are farther apart than the atoms of a

liquid. The atoms of a _____ are very close

together. _____ is made up of more than one
element.

II. Which statement seems more likely to be true?

A. _____ Atoms in a gas can move more easily than atoms in a solid.

B. _____ Atoms in a solid can move more easily than atoms in a gas.

III. The solid metal brass is a mixture of the elements copper and zinc. Place a **T**
(true) or an **F** (false) next to each of the following statements.

_____ Brass contains one kind of atom.

_____ Copper contains only one kind of atom.

_____ Copper atoms are the same as zinc atoms.

_____ The atoms of brass are tightly packed.

IV. Answer in sentences.

How do you think that microscopes of the future might help chemists? Explain
your answer.

What Is the Structure of an Atom?

Exploring Science

The Mystery Deepens. How can scientists find out what something as small as an atom looks like inside? They have been trying to do this for more than one hundred years. In 1911, Ernest Rutherford found some very interesting clues to the mystery.

Rutherford and his assistants fired tiny particles at a very thin sheet of gold foil. These particles were positively charged. Most of them passed straight through the foil. Some passed through and came out at a slight angle. But the big surprise was that about one in 20,000 particles bounced straight back. What did they hit?

Rutherford knew that two bodies with the same electrical charge **repel** each other. They push each other apart. So he suspected that his positively charged particles hit something with a positive charge.

What about the particles that went straight through the foil? They showed that matter is mostly empty space. The particles that did not pass straight through the foil, must have been close to something that pushed or pulled them from their paths. Only a few of the particles hit something solid and bounced back. Were atoms made of even smaller particles?

● What might have happened if the positively charged particles had hit something with a negative charge?

Rutherford's Experiment on Scattering Particles

Particle source

Beam of particles

Thin gold foil

Scattered particles

Some particles bounce straight back

Most particles pass straight through

Subatomic Particles

Rutherford fired particles with a positive electrical charge at the gold foil. There are also particles with negative electrical charges. Particles with the same electrical charges **repel,** or push away from, each other. Particles with opposite electrical charges **attract,** or move toward, each other.

The work of Rutherford and other scientists has provided a great deal of information about atoms. Because of this work, scientists have found that:

● The atom itself is not a solid piece of matter.

● Individual atoms have no electrical charge. They are electrically **neutral** (NOO-trul).

● Atoms have a dense center. This dense center is called a **nucleus** (noo-KLEE-us). The nucleus of an atom has a positive charge.

● An atom is made up of different kinds of particles. These particles are called **subatomic particles.**

● All atoms except for simple hydrogen are made from different numbers of the subatomic particles called *protons, neutrons,* and *electrons.* Hydrogen usually does not have a neutron.

PROTONS. A **proton** is a subatomic particle found in the nucleus of an atom. Protons have a positive charge. The charge of one proton is said to be +1. Its mass is very small. The mass of a proton is given in **atomic mass units** instead of grams. The mass of one proton is one atomic mass unit (abbreviated a.m.u.).

If an atom has five protons in its nucleus, the nucleus has a total charge of +5. Its mass is 5 a.m.u. If an atom has seven protons, its charge is +7, and its mass is 7 a.m.u.

NEUTRONS. **Neutrons** (NOO-tronz) are also found in the nucleus of an atom. Neutrons have no electrical charge. They are neutral. The mass of a neutron is about the same as that of a proton, 1 a.m.u.

If the nucleus of an atom contains both 5 protons and 5 neutrons, its atomic mass is 10 a.m.u. (because 5 + 5 = 10). Its electrical charge, however, would still be +5, because neutrons do not carry an electrical charge.

ELECTRONS. **Electrons** (ee-LEK-tronz) move in a path, or **orbit,** around the nucleus. Electrons have a negative charge. Each electron has a charge of −1. Electrons are much smaller than protons and neutrons. Because the mass of an electron is so small, it is not counted in the total mass of the atom.

Let's add 5 electrons to our atom of 5 protons and 5 neutrons. Because electrons are so small, they do not add to the mass of the atom. Therefore, the mass is still 10 a.m.u. But, the 5 electrons have a total charge of −5. This charge of −5 cancels out the charge of +5 from the protons. Thus, our atom is electrically neutral.

Helium with two protons has an atomic number of 2. Helium also has two neutrons, so it has an atomic mass of 4 a.m.u. The two electrons of helium orbit the nucleus.

To Do Yourself   Electrifying a Drinking Straw

You will need:

Two plastic drinking straws, thread, plastic food wrap, small piece of wool

1. Tie a piece of thread around the middle of one drinking straw. Hang the straw in a horizontal position as shown.
2. Steady the straw with one hand. Bring one end of the other straw close to one end of the hanging straw. Record your observations.
3. Rub one end of the hanging straw 12 times with a piece of food wrap. Quickly do the same to the other straw. Bring the 2 ends close together. Record your observations.
4. Rub one end of the hanging straw with a piece of wool. Rub one end of the other straw with food wrap. Bring the 2 ends close together. Record your observations.

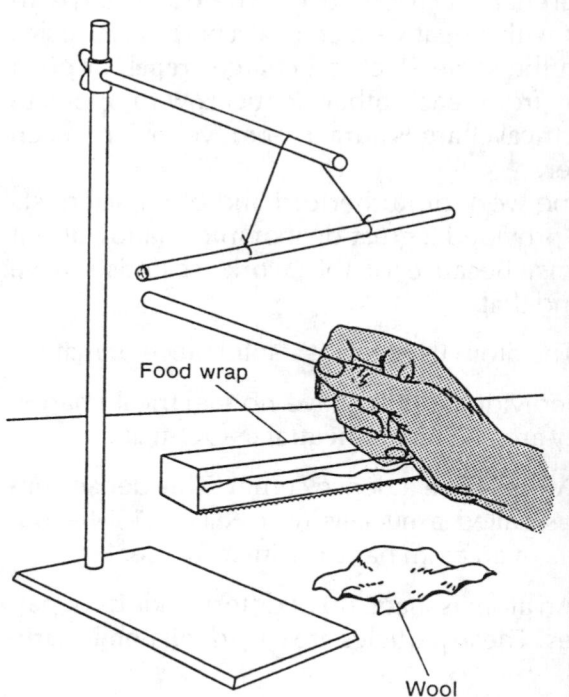

Food wrap

Wool

Questions

1. What happened when 2 straws rubbed with the same material were brought together? _____
   _____

2. What happened when 2 straws rubbed with different materials (food wrap and wool) were brought together? _____
   _____

3. What conclusion can you make about the charges on the 2 straws rubbed with the same material? Explain your answer. _____
   _____

4. What conclusion can you make about the charges on the 2 straws rubbed with different materials? Explain your answer. _____
   _____
   _____
   _____
   _____

Review

I. Fill in each blank with the word that fits best. Choose from the words below.

nucleus   protons   neutrons   atoms   electrons   a.m.u.

A helium atom has two positive charges, so it has two _____ .

The helium atom is neutral so it must also have two _____ .

The protons and neutrons are found in the _____ of the

atom. A hydrogen atom has one proton and no _____ , so

its atomic mass is one _____ .

II. Which statement seems more likely to be true?

A. _____ The mass of the helium atom is 4 a.m.u.

B. _____ The mass of the helium atom is 6 a.m.u.

III. Show the charge and mass of each nucleus.

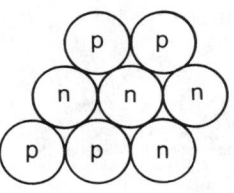   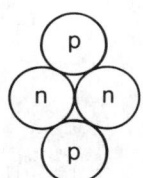   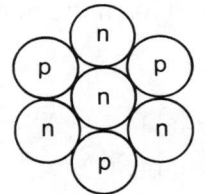

a. _____   b. _____   c. _____

_____   _____   _____

_____ Which of the above is the nucleus for helium?

IV. Answer in sentences.

The name atom originally meant indivisible. Do you think that this is still a good name? Explain your answer.

_____

_____

_____

How Can Atoms Differ?

Exploring Science

Martian Atoms. In 1979, a team of scientists in Antarctica found a grayish-brown chunk of rock. They would have passed it by, but something about the rock was strange. This 8-kilogram rock sample did not come from the Earth. It came from space as a meteorite (MEE-tee-ore-yte). But where exactly did it come from? The moon? Asteroids? Mars?

The rock sample was studied for several years. The most important information about the rock came from a machine called a **mass spectrometer** (spek-TRAWM-uh-ter). The mass spectrometer tells if a sample contains atoms of different masses. It also counts the number of atoms of each mass.

The information from the mass spectrometer agreed with the scientists' original belief that the rock was not from Earth. The rock had come from outer space. It contained some elements found on Earth, such as nitrogen and neon. However, these elements were not present in the same amounts that are usually found in Earth rocks. The amounts of these elements were different from those found in moon rocks also. Yet, the rock looked like the rocks brought back to Earth by our astronauts.

Scientists still don't know where the meteorite came from. Much of the data indicates that it may have come from Mars. But how did it get from the surface of that planet to Antarctica?

● Was it proven that the rock did not originally come from Earth?

1 cm

Scientists found this meteorite in Antarctica in 1979. Where did it come from?

Differences Among Atoms

If different elements did not have distinctive atoms, then the information collected about the rock would have been limited. The mass spectrometer showed some interesting facts about the rock. A mass spectrometer does for elements what fingerprinting does for humans. ATOMIC MASS. **Atomic mass** is the sum of the masses of the protons and the neutrons in an atom. Remember, that each electron has so little mass, that its mass is not counted. Each proton and each neutron is said to have a mass of one atomic mass unit (a.m.u.). So, the atomic mass is equal to the number of protons *plus* the number of neutrons.

Let's look at an atom which has an atomic mass of 23. This atom has a total of 23 protons and neutrons in its nucleus. But, does it have twelve protons and eleven neutrons? Or does it have ten protons and thirteen neutrons? To find out, more information is needed.

ATOMIC NUMBER. The **atomic number** of an atom tells us the number of protons in its nucleus. If the atomic number is 11 then there are 11 protons. Suppose an element has an atomic mass of 23 and an atomic number of 11. The atomic number tells us there are 11 protons in the nucleus. How many neutrons are present? There are 23 − 11, or 12 neutrons. Remember that atomic mass is equal to the sum of the mass of the protons plus the mass of the neutrons.

Every element has its own atomic number! Sodium, for example, has an atomic number of 11. Sodium has an atomic mass of 23. So our atom would have to be a sodium atom. If the atomic number had been 10, the element would have been neon. Isn't it amazing that one proton can make the difference between the active metal sodium and the inert gas neon.

Review

I. Fill in each blank with the word (or words) that fits best. Choose from the words below.

mass   13   neutron   electrons   14   atomic mass   protons   27
atomic number

Aluminum, Al, has 13 protons. That means that an atom of Al has an atomic number of _____. If its _____ is 27 a.m.u., then it must have _____ neutrons. The atom is neutral so it must also have 13 _____.

II. Which statement seems more likely to be true?

A. _____ The atomic number tells what the element is.

B. _____ The atomic mass tells what the element is.

III. Fill in the blanks for each nucleus.

a	b	c
Atomic Number = 2	Atomic Number = 13	Atomic Number = 11
Atomic Mass = 4	Atomic Mass = 27	Atomic Mass = 23
_____ p	_____ p	_____ p
_____ n	_____ n	_____ n

_____ Which is the aluminum atom?

IV. Answer in sentences.

Why is it important to be able to analyze matter? Explain.

How Can Atoms of an Element Differ?

Exploring Science

A Critical Competition. In the 1940's, the world was at war. Chemists in many countries were working on a "super-bomb." A rare form of uranium, *U-235*, was needed as a fuel for the new bomb. A special kind of water was also needed. It was called heavy water. Heavy water contains atoms of a special type of hydrogen called **deuterium** (dyoo-TER-e-yum).

Chemists of many nations were working to produce heavy water. It was a slow and difficult job. French scientists had a small sample of heavy water. In 1940, it became clear that the German army was about to invade Paris. French scientists were afraid that if the Germans obtained their heavy water, it could help them win the war. Jean Joliot-Curie, a French chemist and Nobel prize winner, smuggled the heavy water out of France in time.

Norwegian scientists were not as lucky. Four hundred liters of their heavy water was captured by the Germans. Fortunately, this heavy water was destroyed by British commandos in 1942.

● Why didn't Jean Joliot-Curie destroy the French supply of heavy water?

Jean Frederic Joliet-Curie and his wife Irene working in their laboratory at the Paris Radium Institute.

Isotopes

Uranium-235 and deuterium are only two of many special forms of elements that have changed history. What was special about the uranium needed for the super-bomb? Why does deuterium make water "heavy?"

Uranium has an atomic number of 92. This means that every uranium atom has 92 protons. Most uranium atoms also have 146 neutrons in their nucleus. The atomic mass of these atoms is 238 a.m.u. Remember, the atomic mass is equal to the number of protons *plus* the number of neutrons.

Some uranium atoms have 92 protons but only 143 neutrons. The atomic mass of these atoms is only 235 a.m.u. They are still uranium atoms, because they have 92 protons. But the number of neutrons is different.

Atoms such as this are called isotopes. **Isotopes** (EYE-suh-topes) are atoms of the same element having the same atomic number but a different atomic mass. Therefore, an isotope of an element has the same number of protons, but a different number of neutrons.

A special shorthand notation is often used to write the names of isotopes. To indicate the isotope of an element, first write the element's chemical symbol. Then, write a hyphen after the symbol, followed by the atomic mass of the isotope.

Another name for deuterium is hydrogen-2. Deuterium can be written using the shorthand notation H-2. The most common isotope of hydrogen has one proton and zero neutrons. It has an atomic number of 1 and an atomic mass of 1 a.m.u. This is the only atom existing in nature that has no neutrons. Deuterium is a rare isotope of hydrogen. It has one neutron in its nucleus. Because of this neutron, a deuterium atom has a mass of 2 a.m.u.

Water is made from the elements hydrogen and oxygen. In the laboratory, water can be made using deuterium instead of common hydrogen (H-1). Water made with H-2 is called heavy water. Its mass is greater than that of regular water, which contains common hydrogen, H-1.

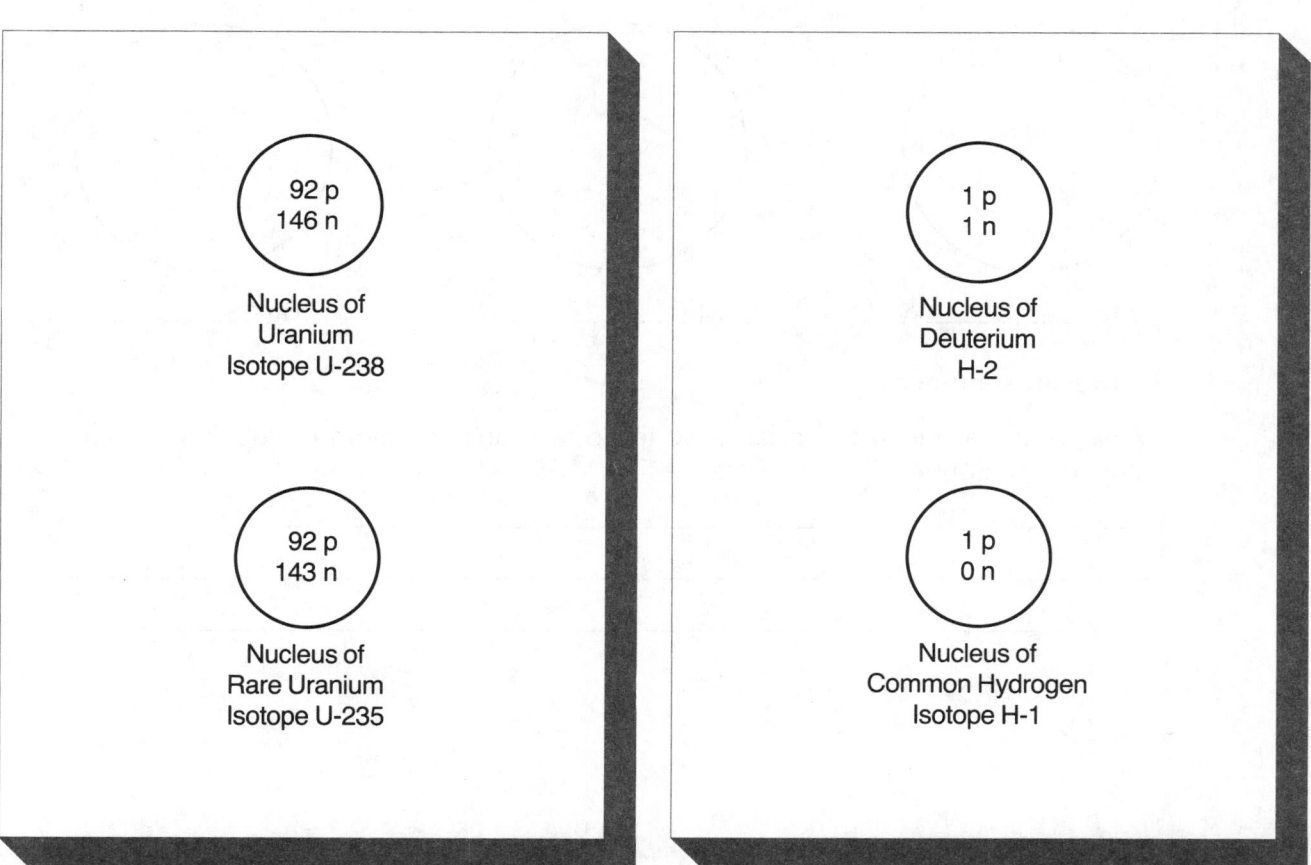

92 p
146 n

Nucleus of
Uranium
Isotope U-238

92 p
143 n

Nucleus of
Rare Uranium
Isotope U-235

1 p
1 n

Nucleus of
Deuterium
H-2

1 p
0 n

Nucleus of
Common Hydrogen
Isotope H-1

Review

I. Fill in each blank with the word (or words) that fits best. Choose from the words below.

protons   nature   atomic number   electron   isotope   atomic mass neutrons   element

Oxygen atoms have an atomic number of 8. There are 8

_____ in every oxygen nucleus. Oxygen atoms with an

atomic mass of 16 have 8 _____ . A rare

_____ of oxygen has an atomic mass of 17. These two

forms of oxygen have the same _____ . They have

a different _____ .

II. Which statement seems more likely to be true? _____

A. (17 p / 18 n) and (17 p / 20 n) are nuclei of isotopes of the same element.

B. (17 p / 18 n) and (18 p / 18 n) are nuclei of isotopes of the same element.

III. Draw nuclei of three isotopes that might exist for the element magnesium. Mg has an atomic number of 12.

Mg-_____   Mg-_____   Mg-_____

IV. Answer in sentences.

What would a scientist in a lab have to do to change an atom of cobalt-59 to an atom of cobalt-60?

_____

_____

_____

Lesson Six

How Are Electrons Arranged in an Atom?

Exploring Science

Yellow Streets. Towns across the country are changing their streetlights. These changes could save taxpayers millions of dollars in electric bills. The glaring white light of traditional streetlamps will be gone. It is being replaced by the golden-white glow of new sodium vapor lamps.

At dusk, when the lights first go on, a red glow appears in each lamp. This glow is caused by a small amount of neon in the lamps. The neon helps the sodium lamp start glowing.

The first sodium lamps had a definite yellow color. This yellow color is typical of the light produced when electricity is passed through sodium vapor. Unfortunately, the strong yellow color limited the uses for the lamps.

To solve this problem, mercury was added to the lamp. The temperature and pressure inside the lamp was also increased. The result is a lamp that glows with a golden-white light. This light is easy on the eyes. It is also economical, because sodium vapor lamps do not use a lot of electricity.

● What would the light be like if the tube only contained neon?

Electron Shells

You have learned that an atom has a small, heavy nucleus. Protons and neutrons are found in the nucleus. Electrons, however, are found in the space around the nucleus. Electrons are always moving around the nucleus in paths called orbits.

Electrons move very rapidly. Scientists believe that electrons move around the nucleus at different levels. Electrons with the smallest amount of energy are found closest to the nucleus. Those with more energy, are found farther away from the nucleus.

Niels Bohr developed a model to help explain the structure of the atom. He pictured different **energy levels** around the nucleus. Each electron had its place in a specific energy level.

When energy, such as heat, is supplied to an atom, the electrons farthest from the nucleus take in some of this energy. The electrons use this energy to move to the next higher energy level, jumping into a level farther from the nucleus. When the electrons fall back to the lower energy level, they give off the extra energy. This energy is given off in the form of light.

Atoms of different elements **emit,** of give off, light of different colors. Sodium atoms emit a bright yellow light. Neon atoms give off a bright red light. Mercury atoms emit a blue-white light.

Each energy level around the nucleus of an atom can hold only a certain number of electrons. The first level, closest to the nucleus, can hold only two electrons. The second level can hold eight electrons. Higher levels can hold even more electrons.

Let's look at a nucleus of a sodium atom. Sodium, Na, has an atomic number of 11 and an atomic mass of 23 a.m.u. Remember that each proton has a charge of +1. Neutrons carry no charge.

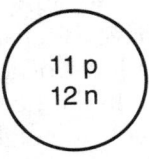

The nucleus of the sodium atom has a charge of +11. Recall that all atoms are electrically neutral. Thus, the Na atom must have 11 electrons. These electrons will constitute a charge of -11. But, where are they? Scientists have developed a way of showing the **electron configuration** (kun-fig-yuh-RAY-shun) of an atom. This configuration shows the probable locations of the electrons in an atom.

Look at the Bohr model of the sodium atom. The circles around the nucleus are the electron levels. The first level, nearest the nucleus, can hold two electrons. The second level can hold eight electrons. The third level has only one electron, even though it could hold more.

The single electron in the outer level is the one that gains energy in the sodium vapor

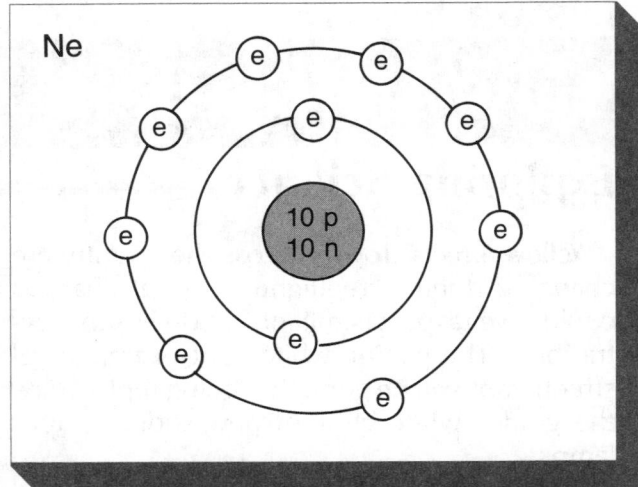

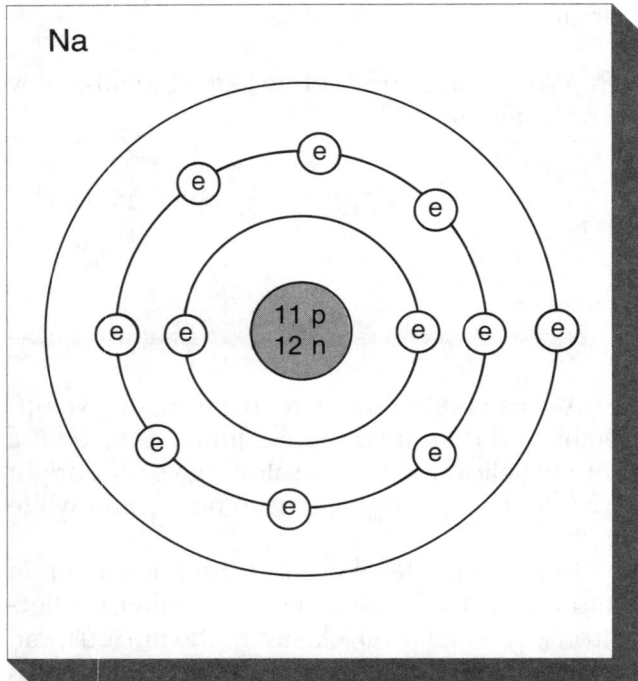

lamp. When this electron gains energy, it temporarily moves to the next higher energy level. As it drops back into its usual position in the lower energy level, it emits a yellow light.

We can make models of other elements also. Neon, NE, has an atomic number of 10 and an atomic mass of 20. Because neon has 10 protons, it must also have 10 electrons to be a neutral atom. The first level of any atom can hold only two electrons. In the neon atom, two electrons are in this level. The eight remaining electrons are in the second level. The diagram at the top of the page shows where the 10 electrons of the neon atom are found.

Carbon, C, is an atom with 6 protons and an atomic mass of 12. Because carbon is a neutral atom, it also has 6 electrons. There are two electrons in the first energy level. That is all that the first level can hold. The remaining 4 electrons are found in the second energy level. A diagram of the electron levels of a carbon atom would look like this:

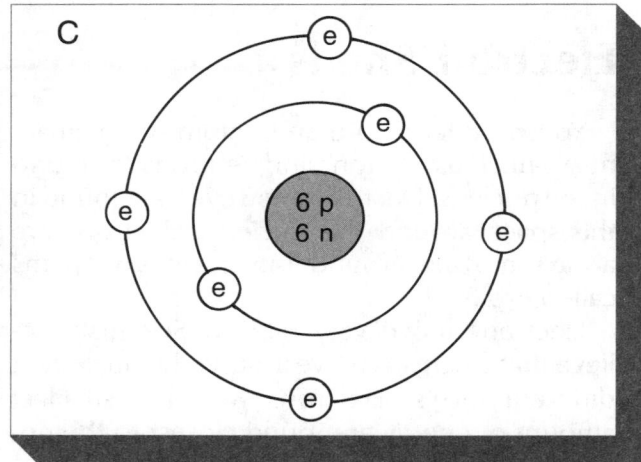

Let's look at an atom of chlorine. Chlorine, Cl, atoms have 17 protons, so they must also have 17 electrons. The atomic mass of chlorine is 35. Therefore, chlorine has 18 neutrons.

A model of chlorine shows three energy levels. Chlorine's first two electrons go in the first energy level. This level, remember, can only hold two electrons. Where do the remaining 15 electrons go?

The second energy level can hold a maximum of eight electrons. If eight of the remaining electrons go in the second energy level, the remaining 7 electrons must go into another energy level. Chlorine, therefore, has a third energy level. A model of the electron levels of the chlorine atom looks like this:

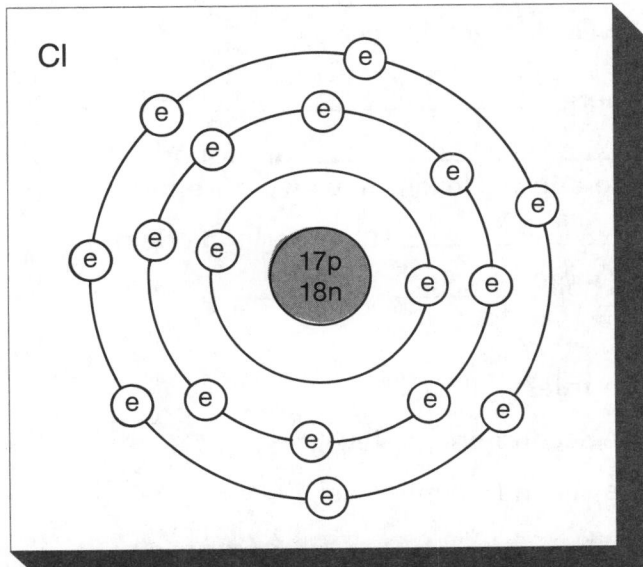

Remember, these are only simple models. There are several things to keep in mind about atoms when you construct these models. First, the electron orbits are not flat like those of the planets in our solar system. So, the shell that we draw as a circle actually represents a ball, or a *sphere*. You must imagine a three-dimensional shell.

Second, the electrons are moving in their orbits very rapidly. So, when we mark an electron on a shell, it only shows the possibility that the electron is in that spot for one instant in time.

Last, understand that this model of the atom is only a simple model showing the *possible* electron configuration of an atom. Today, scientists picture few orbits as spheres. Orbits are often pictured as pinwheels or other shapes. Any model of an atom only shows the *probability* of an electron's position at any given instant.

To Do Yourself **Constructing Atomic Models**

You will need:

Atomic model worksheet, colored paper dots (3 colors), glue

1. Decide what color will represent each subatomic particle.
2. Figure out how many protons, neutrons, and electrons make up a helium atom.
3. Construct a model of the helium atom by gluing the colored paper dots on the worksheet.
4. Repeat steps 2–4 for lithium and beryllium.

Questions

1. What is a model? _____

_____

2. How do scientists use models? _____

_____

Review

I. Fill in each blank with the word that fits best. Choose from the words below.

levels   energy   Bohr   eight   seven   light   two

Electrons move around a nucleus in energy _____ .

Electrons with the smallest amount of _____ are found in the first level. When electrons fall back to a lower level, they give off energy as _____ . _____ suggested a model for atoms. The second energy level can only hold _____ electrons.

II. Which statement seems more likely to be true?

A. _____ An atom has 9 electrons in the second energy level.

B. _____ An atom has 3 electrons in the second energy level.

III. Complete the nuclei and draw the energy levels with electrons for each atom.

Lithium:
(atomic number 3)
(atomic mass 7)

Sodium:
(atomic number 11)
(atomic mass 23)

IV. List all of the similarities between the atoms of lithium and sodium.

_____

_____

_____

_____

How Are Elements Organized?

Exploring Science

A Minute to a Lifetime. Once scientists had the ability to propel particles at atoms, new elements were discovered. In fact, some would argue that they were actually created.

In the 1980s, a few atoms of elements of atomic numbers 107, 108, and 109 were made in the laboratory. The 1990s brought atoms of elements 110 and 111. All these superheavy elements were unstable. The heaviest lasted for millionths of a second. Scientists continued to look for stable superheavy elements.

Element 106, sometimes called seaborgium, may be a breakthrough. In the 1990s, some isotopes of this artificial element were made that lasted a full minute. One of the new isotopes contained 159 neutrons. Another contained 160 neutrons.

Chemists now have a new hope. They are particularly interested in the isotopes of element 110. This superheavy element should have the properties of platinum.
● Draw a nucleus of a seaborgium isotope.

The Periodic Table

The list of elements continues to grow. Chemists have found the atomic number, atomic mass, and the electron configuration for each of the known elements. They have also found that each element has its own distinct properties. Many of these properties have been known for a long time.

As early as the 1860s, scientists saw that it would be helpful if all of this information could be placed in some kind of chart. By that time, more than 60 elements were known. Scientists believed that more elements probably existed. They hoped that if they put the known elements in a proper order that they would find some clues about other unknown elements.

Dimitri Mendeleev, a Russian chemist, did just this in 1869. He arranged the elements in order of their atomic masses. He also arranged them according to their properties. Mendeleev made a chart with columns of the known elements arranged in an up-and-down order. The elements with the lowest atomic masses were at the top of the columns. Those with higher atomic masses were at the bottom of the columns.

Mendeleev also had rows of elements going across his chart. The elements with similar properties were placed next to each other. When Mendeleev finished his chart, he saw

Mendeleev's Arrangement of the Known Elements
GROUP

	1	2	3	4	5	6	7	8
Period 1	?	Li	Be	B	C	N	O	F
Period 2	Ne	Na	Mg	Al	Si	P	S	Cl
Period 3	A	K	Ca	?	?	V	Se	Br
Period 4		Rb	Sr	In	Sn	Sb	Te	I
Period 5		Cs	Ba					

that there were three empty spaces. The three elements that would fit into these spaces were not yet known.

By comparing the information about the elements surrounding the three empty spaces, Mendeleev predicted the atomic masses and the properties of the three missing elements. Before Mendeleev died, each of these missing elements was found. The elements matched his predictions.

Mendeleev's table has been greatly changed and enlarged. Much new information has been added. Today's chemists use a newer table for quick reference. This new table is called the **Periodic Table of the Elements**.

Look at the modern table. Find gallium, Ga. The table tells you that the atomic number of gallium is 31. It also tells you that the atomic mass for gallium is 69.72 a.m.u. This is an average of the atomic masses for all of gallium's isotopes. An isotope, remember, is an atom of the same element that has the same atomic number but a different atomic mass.

You can find more information about gallium from the table. For example, gallium has properties similar to those of aluminum, Al, and indium, In. In the modern table, elements with similar properties are listed near each other in

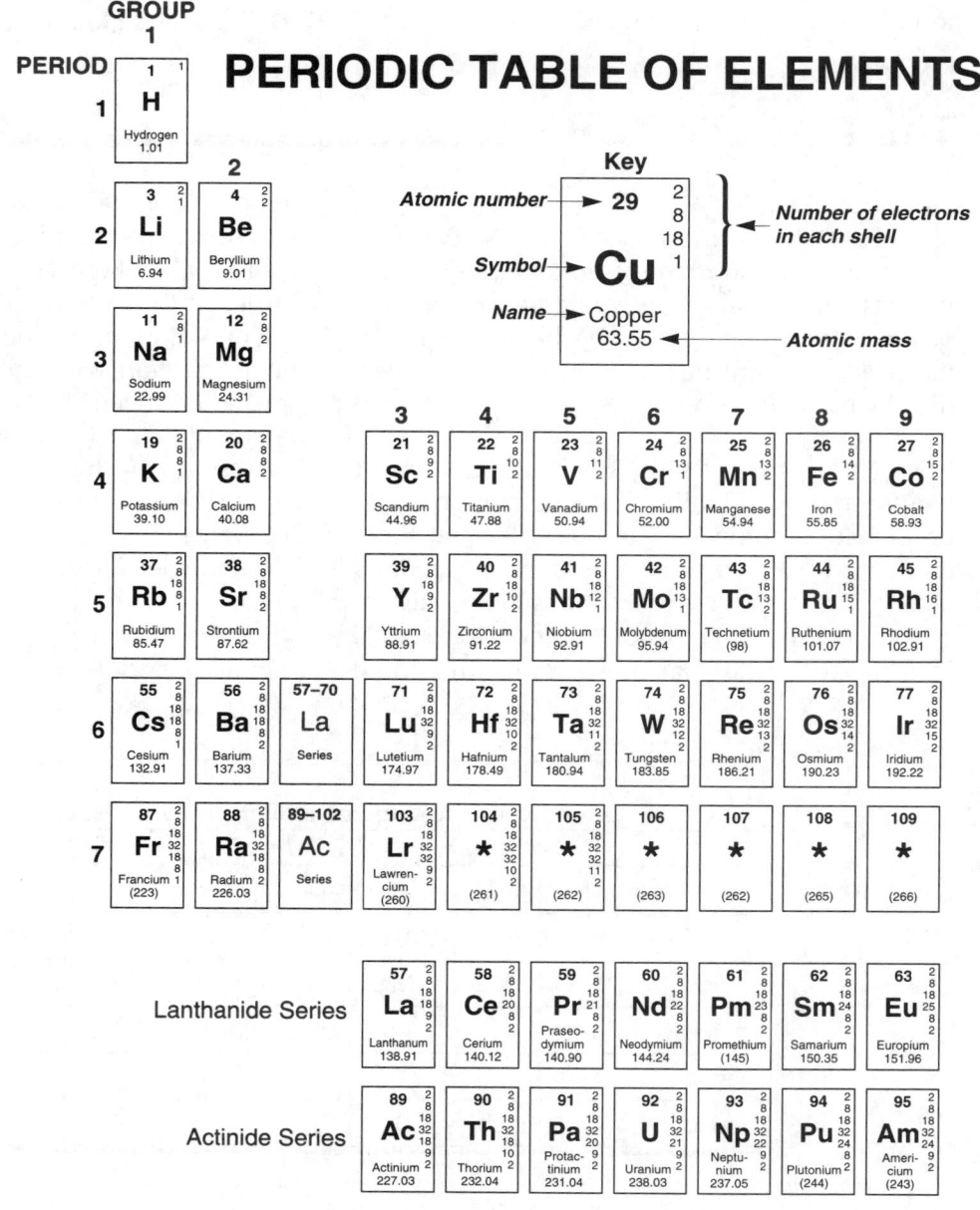

PERIODIC TABLE OF ELEMENTS

* Names for these elements have not been agreed upon.

vertical columns.

Each **vertical** column contains a **family of elements**. All of the elements in a family have similar properties. Elements in the **horizontal** rows of the periodic table are arranged in order of their atomic numbers. The atomic numbers get larger as you move from left to right.

Let's look a bit more closely at the properties that make elements similar.

METALS. The **metals** are generally elements that have either 1, 2, or 3 electrons in their **valence shells**, or outermost energy levels. Notice the dark black line that divides the Periodic Table into two sections. This line separates the metals from the other elements.

NONMETALS. The **nonmetals** are generally elements with 6 or 7 electrons in their valence shells. This family contains some very important elements.

METALLOIDS. The **metalloids** (MET-ul-oydz) have properties that lie in between those of the metals and the nonmetals. Many elements in this family have either 4 or 5 electrons in their valence shells.

NOBLE GASES. The **noble gases**, or **inert gases**, are those elements in the family marked *O*. Each member of this family has a complete set of electrons in its valence shell. All of these elements are gases at room temperature.

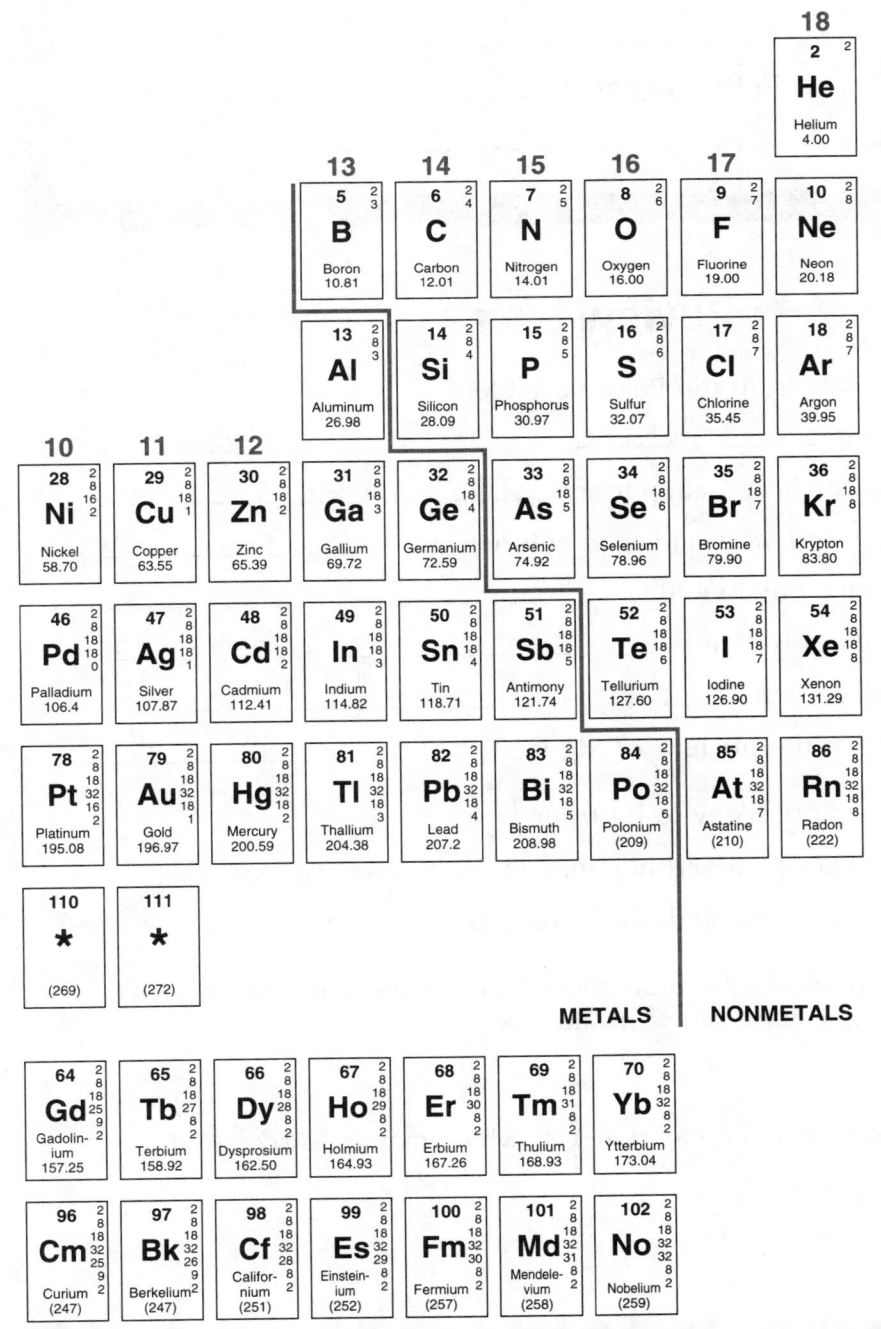

To Do Yourself — Classifying Elements

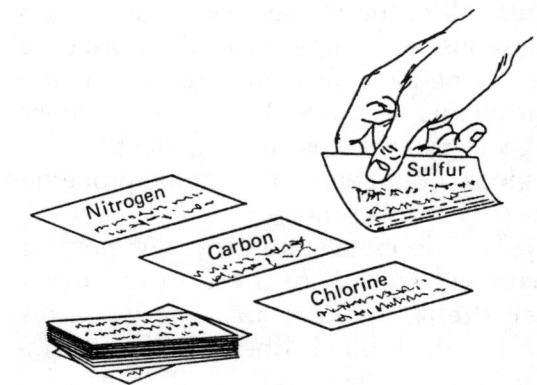

You will need:

15 index cards containing information about elements

1. Spread out the 15 element cards on your desk.
2. Examine each of the element cards for the information given.
3. Place the 15 element cards in any arrangement that contains *both* rows and columns.

Questions

1. How did you arrange the cards in the rows? _____

_____

2. How did you arrange the cards in the columns? _____

_____

Review

I. Fill in each blank with data from the Periodic Table.

The symbol for copper is _____.

The two elements most similar to copper are _____ and _____.

An element lighter than, but very similar to chlorine, is _____.

The noble gas with atomic number 10 is _____.

The atomic mass of sodium is almost _____.

K is a (metal, nonmetal) _____.

The number of protons in the nucleus of Mg is _____.

II. Which statement seems more likely to be true?

A. _____ The elements are arranged horizontally by atomic number.

B. _____ The elements are arranged horizontally by atomic mass.

III. Using data from the Periodic Table, draw a model of an atom of each of the following elements: magnesium, nitrogen, fluorine.

IV. Answer in sentences.

Using the Periodic Table, describe element 111, which may be discovered sometime in the future.

What Are Ions?

Exploring Science

21st-Century Rockets. For several decades, liquid fuels have been used for space flights. Although engines have improved, fuel efficiency for rockets is still very low. So much fuel has to be carried that the ship cannot transport very much cargo or travel very far.

A new fuel made of charged atoms is now being tested. Electrons can be removed from an atom. When this is done, the atom is no longer electrically neutral. It has more positive charges than negative charges. Electrically charged atoms can be made to move at very high speeds.

If one such atom is shot out of a rocket, the rocket would not move. If millions of these atoms were shot from the same rocket, the story would be different. The force of the atoms moving in one direction would "push" the rocket in the opposite direction. This idea is called **ion propulsion** (EYE-on pruh-PUL-shun), because each of the atoms with a positive charge is called an ion.

At this time, cesium, Cs, is the choice for use as an ion-propellant. Cesium gives up an electron easily. Ion propulsion of cesium requires only one-tenth the weight of liquid fuels.

● Why is sodium also being considered as an ion propellant? (Use the Periodic Table to help with this question.)

Ion propulsion boosts this rocket off the launch pad.

Ions

Sodium, Na, costs 20¢ per kilogram. Cesium, Cs, costs $300 per kilogram. It is easy to see why researchers would rather use sodium as an ion-propellant than cesium. However, it is much harder to strip an electron from a sodium atom than it is from a cesium atom.

You have learned that atoms are neutral. This means that they have the same number of positive and negative charges.

Sometimes an atom can lose an electron. When this happens, the atom is no longer neutral. It becomes positively charged, because it now has more protons than electrons.

Sometimes an atom gains an electron. When this happens, the atom is again no longer neutral. In this case, there are more electrons than protons, so the atom becomes negatively charged.

When an atom is no longer neutral in charge, it is called an **ion** (EYE-on).

METALS. Metals usually have one, two, or three electrons in their valence shells. Ions of metals have a positive charge. This means that metal atoms give up, or lose, electrons.

Let's look at a sodium atom. The sodium atom has one electron in its valence shell. Suppose the atom loses this electron. When this happens, the sodium atom becomes a sodium ion with a positive charge.

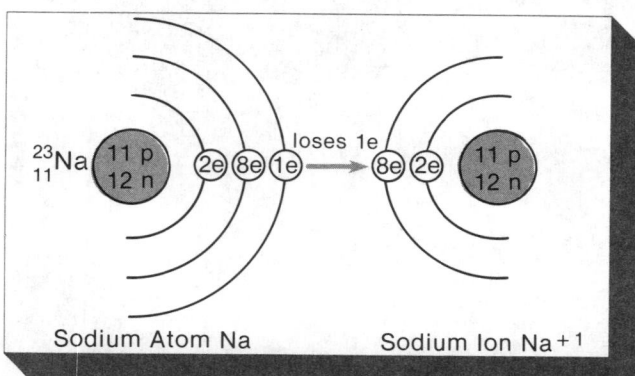

Sodium Atom Na   Sodium Ion Na^{+1}

The outer level of the sodium ion now has eight electrons. Having eight electrons in an outer shell is a very stable condition. This is true even for atoms whose outer energy level can hold more than eight electrons. Once an outer energy level has eight electrons, it does not gain or lose any more electrons easily.

Some metals, like magnesium, Mg, and Calcium, Ca, have two electrons in their valence shells. The atoms of these elements give up 2 electrons fairly easily. Their ions have a +2 charge.

NONMETALS. Ions of nonmetals have a negative charge. This means that the atoms of nonmetals gain electrons. Let's look at an atom of fluorine.

Flourine, F, has an atomic number of 9. Therefore, fluorine has 9 protons and 9 electrons. Seven of fluorine's electrons are in its outer valence shell. If one more electron is added, the valence shell will have 8 electrons, making the fluorine ion very stable.

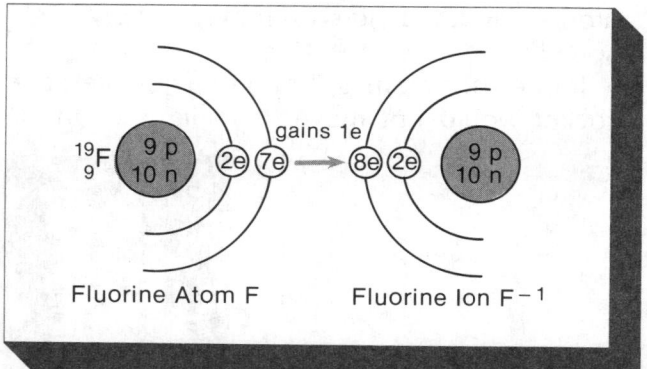

Fluorine Atom F   Fluorine Ion F^{-1}

The fluorine atom accepts an electron easily and becomes a fluorine ion. This ion has a negative charge. The number of electrons is greater than the number of protons.

Some nonmetals, like oxygen, O, and sulfur, S, gain 2 electrons. Their ions have a charge of −2. Look at the Periodic Table. What other elements do you think might have ions with a −2 charge?

NOBLE GASES. The noble gases are sometimes called *inert* gases. These elements do not usually form ions. Why? All these gases, except helium, He, have 8 electrons in their valence shells. The noble gases, therefore, are very stable. Because they are already stable, atoms of the noble gases do not gain or lose electrons easily. This explains why the chemicals in this family are chemically inactive.

Let's look at an argon atom. Argon, Ar, has an atomic number of 18. So it has 18 protons and 18 electrons. Look at the diagram on page 47.

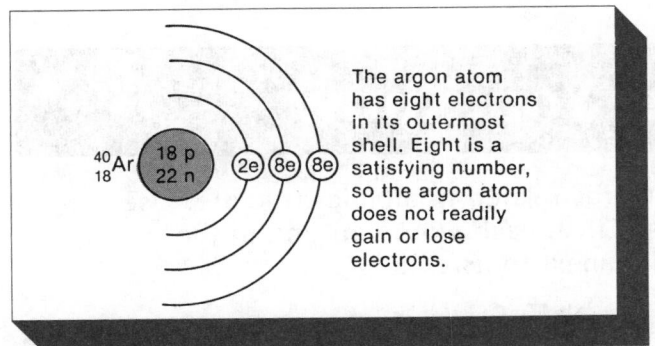

The argon atom has eight electrons in its outermost shell. Eight is a satisfying number, so the argon atom does not readily gain or lose electrons.

As you can see, argon has 3 energy levels. The valence shell of argon has eight electrons. Remember, eight electrons in the valence shell is a very stable number. So, although its valence shell is not filled, the argon atom is stable.

To understand ions better, it is important to remember these points.

● Atoms are electrically neutral.

● *Ions* are atoms that have become charged by gaining or losing electrons.

● Atoms that lose electrons become positive ions.

● Atoms that gain electrons become negative ions.

● Metal ions are positive.

● Nonmetal ions are negative.

● Eight is a "satisfying number" for the valence shell. Atoms with eight electrons in their valence shells are stable and do not gain or lose electrons easily.

● The nucleus is not affected when an atom becomes an ion.

Review

I. Fill in each blank with the word that fits best. Choose from the words below.

ion   stable   metals   nonmetals   positive   negative   noble gases

When a sodium atom loses one electron, it becomes a sodium

_____ . It has a _____ charge. An atom

with eight electrons in its valence shell is _____ . When the

fluorine atom adds one electron, it gets a _____ charge.

_____ do not usually form ions.

II. Which statement is more likely to be true?_____

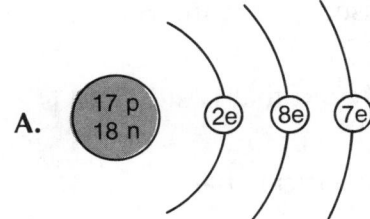

A.

This atom will gain an electron.

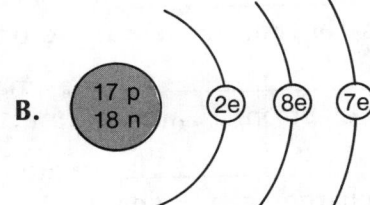

B.

This atom will lose an electron.

III. Draw an atom of cesium. Use the Periodic Table. Then show how it becomes an ion.

IV. Answer in sentences.
Why is it unlikely that a neon atom would become a neon ion? Explain.

Review What You Know

A. Hidden in the puzzle below are eight words related to atomic structure. Use the clues to help you find these words. Circle each word you find in the puzzle. Then write each word on the line next to its clue.

```
L P R O T O N E Q
I I L L E V E L U
S N U C L E U S I
N U M B E R T O S
D U N I C L R P O
O R B I T Q O G T
L O O N R E N E O
C D Z J O C N A P
M A S S N T E T E
```

Clues:

1. It has a charge of +1. _____

2. It has a charge of −1. _____

3. A subatomic particle with no electrical charge. _____

4. Where protons and neutrons are found. _____

5. The path of a subatomic particle. _____

6. Another term for valence shell is energy _____ .

7. The *m* in amu. _____

8. An atom of an element with the same atomic number but a different atomic mass. _____

B. Write the word (or words) that best completes each statement.

1. _____ An example of a chemical symbol is **a.** + **b.** p **c.** N

2. _____ Many symbols are based on **a.** the number of protons **b.** Latin names **c.** isotopes

3. _____ The smallest piece of gold that is still gold is a(n) **a.** electron **b.** atom **c.** nucleus

4. _____ An atom has **a.** no electrical charge **b.** a + charge **c.** a − charge

5. _____ Two positive charges will **a.** attract **b.** form a bond **c.** repel

6. _____ The nucleus of an atom is **a.** electrically neutral **b.** positively charged **c.** negatively charged

7. _____ If an atom has 6 protons, 6 is its **a.** atomic number **b.** atomic mass **c.** amu

8. _____ Isotopes of hydrogen have different atomic
   a. numbers   **b.** charges   **c.** masses

9. _____ The electrons of an atom are located in   **a.** the
   nucleus   **b.** energy levels   **c.** different phases

10. _____ An atom that has gained or lost electrons is
    called a(n)   **a.** isotope   **b.** proton   **c.** ion

11. _____ An atom with one electron in its valence shell is
    a   **a.** metal   **b.** nonmetal   **c.** noble gas

12. _____ An atom with eight electrons in its valence shell
    is a   **a.** metal   **b.** nonmetal   **c.** noble gas

13. _____ A metal ion has   **a.** no charge   **b.** a + charge
    c. a − charge

14. _____ The Periodic Table lists elements according to
    their atomic numbers and   **a.** properties   **b.** phases   **c.** chemical symbols

15. _____ The atomic number of an element is equal to its
    a. number of electrons   **b.** number of protons   **c.** number of neutrons

16. _____ The atomic mass of an element is equal to its
    number of   **a.** protons   **b.** protons plus its electrons   **c.** protons plus its
    neutrons

17. _____ The particles of an element in the gas phase
    are _____ than the particles of the same element in the liquid phase
    a. farther apart   **b.** the same as   **c.** closer together

18. _____ Elements with properties that lie between the
    metals and nonmetals are called   **a.** noble gases   **b.** metalloids   **c.** plasma

19. _____ An element with eight electrons in its valence
    shell is   **a.** stable   **b.** likely to form an ion   **c.** a metal

20. _____ The mass of one atom containing 17 protons and
    18 neutrons is   **a.** 17 amu   **b.** 18 amu   **c.** 35 amu

C. Apply What You Know

1. Using the Periodic Table of the Elements, complete each of the following
   statements.

   a. The symbol for iron is _____ .

   b. Ag is the symbol for _____

   c. The atomic number of sodium is _____ .

   d. The atomic mass of sodium is _____ .

   e. Sodium has _____ electron(s).

   f. Sodium has _____ electron(s) in its outermost energy
   level.

   g. Calcium is in group _____ .

h. An element in the same family as calcium is _____ .

i. Carbon has _____ neutron(s).

j. The elements in the last column of the Periodic Table

are _____ .

2. Study each of the diagrams. Then match each diagram with its description.

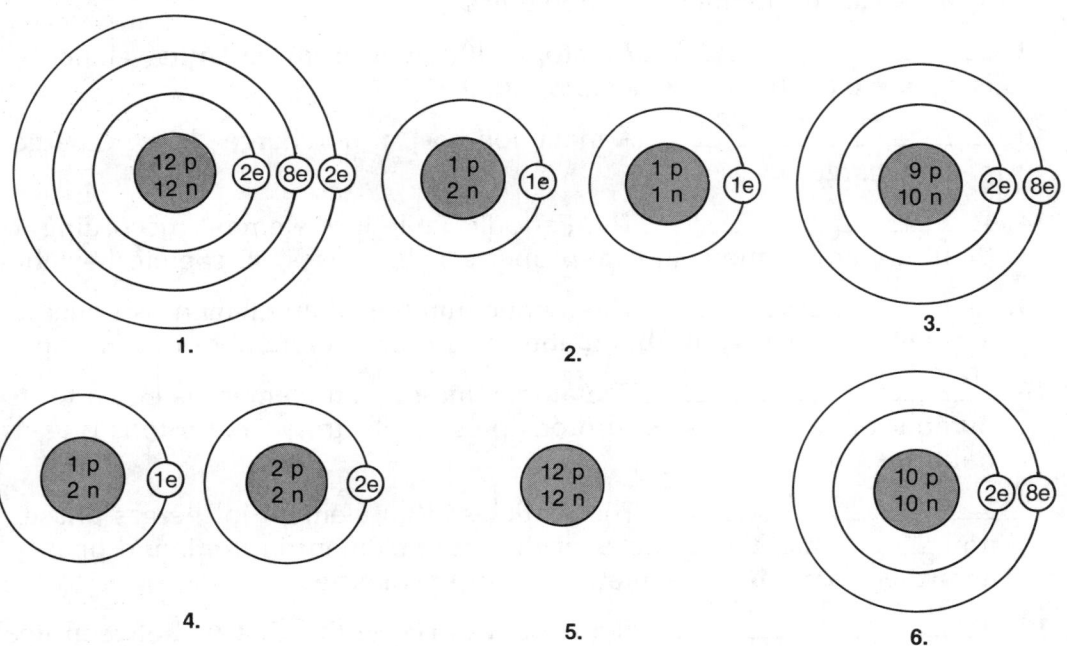

a. The nucleus of an atom is shown in diagram _____.

b. Isotopes of the same element are shown in diagram _____.

c. An atom of a noble gas is shown in diagram _____.

d. An atom of a metal is shown in diagram _____.

e. An ion of a nonmetal is shown in diagram _____.

f. The magnesium atom is shown in diagram _____.

g. The magnesium atom is shown in diagram _____.

D. Find Out More

1. Construct a three-dimensional model of an atom. Pick an element with an atomic number between 3 and 10. Draw the nucleus of the atom. Then draw its valence shells. Use wire, beads, and glue to make a model of the atom you have drawn.

2. Using an encyclopedia, identify three useful isotopes. Find the chemical symbols for the three isotopes. Draw a diagram of each isotope. Under each diagram, write the name of the isotope and a brief description of how it is useful.

Careers in Physical Science

New Products Require New Uses of Matter. Do you enjoy experimenting with new things? Do you like to test materials to see if they will bend or snap? Do you measure carefully and take pride in your accuracy? If so, you might consider a career in materials science.

Materials Engineer. A materials engineer is a problem-solver. This scientist looks for new ways to use matter. Materials engineers may work to develop new substances, or may find new uses for existing matter.

A materials engineer attends at least five years of college. This scientist must learn about the special properties of different materials. Special course-work helps the scientist gain this knowledge. Many of these courses involve chemistry and physics.

Materials engineers have specialties just as doctors do. For example, a metallurgist (MET-uhl-uhr-jist) works only with metals. If a new plastic is needed, a plastics engineer is given the problem. If a new type of glass is desired, a ceramics engineer is hired. Materials engineers develop many new products to meet the needs of society.

Materials Technicians. Most materials engineers do not work alone. Materials technicians (tek-NIH-shunz) are needed to put the engineer's ideas to work.

Materials technicians usually work in a laboratory. Most of these technicians have very special skills. For example, one metallurgy technician may run an electron microscope. Another may be responsible for testing the strength of a new metal. A third technician may test the metal's resistance to corrosion.

A ceramics technician may grind new lenses. Another ceramics technician may test the heat conductivity of a new material. Another technician may measure the resistance of a new ceramic glaze.

The training a technician receives is very specialized. Some technicians get their training in special high school courses. Others attend two-year colleges with programs in a specific field. Many companies also provide courses. These courses help to keep their technicians up-to-date.

The work done in the materials sciences is very important. Our society uses the technology of materials science every day. Perhaps there is a place for you in this exciting field.

This materials engineer is examining the metal used in this instrument before it is launched into the atmosphere.

A materials technician puts the ideas of the materials engineer to work.

UNIT

3

COMPOUNDS

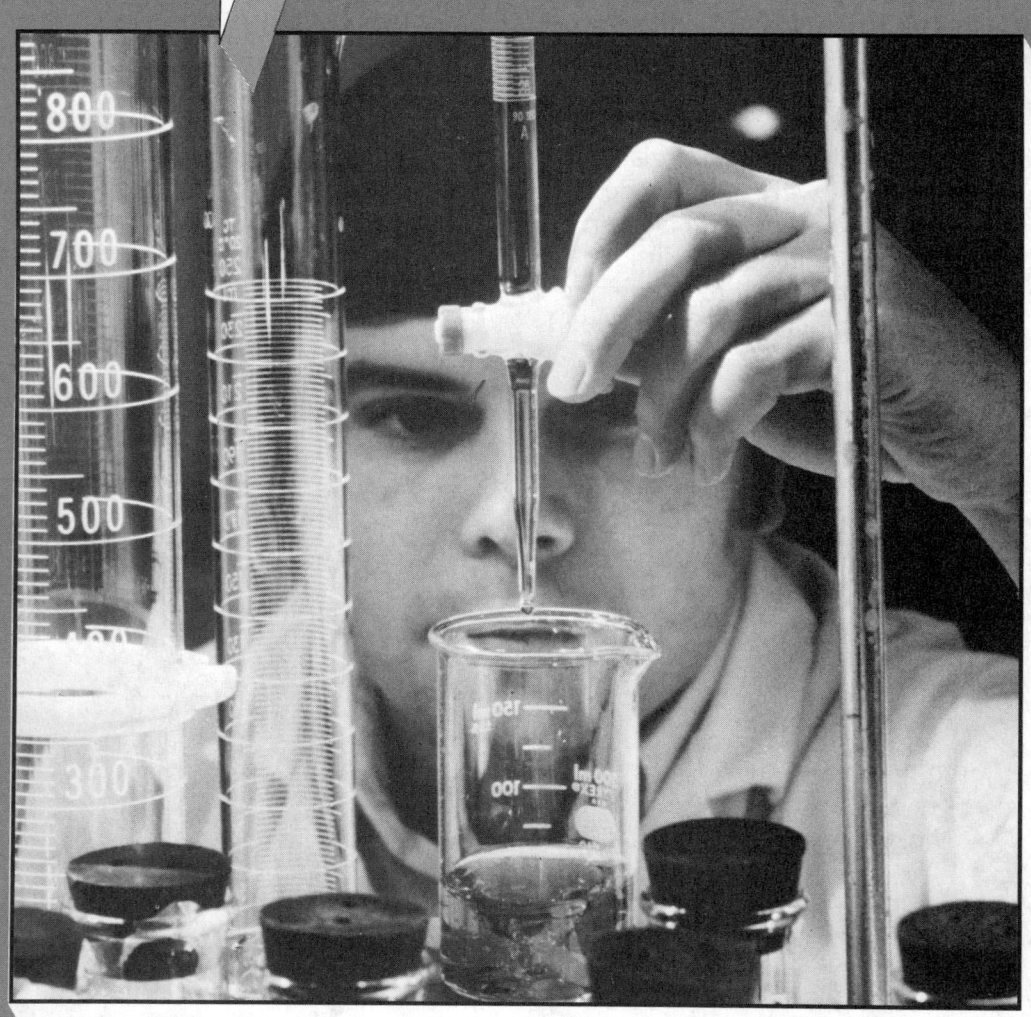

What Is a Compound?

Exploring Science

"Send This to the Lab!" History books report many cases of poisonings of famous people. Modern lab tests show that some deaths of famous people may not have been from "natural causes."

For example, Napoleon was convinced that he was being poisoned. He even mentioned this fear in his will. When he died, in 1821, doctors stated that Napoleon had died from a stomach ailment.

Many years later, scientists studied a sample of Napoleon's hair. It contained more than ten times the normal amount of deadly arsenic. History may have been changed if lab tests had been available in 1821.

Now, let's look at a more recent case. Mrs. X was a wealthy and suspicious woman. She was also a keen observer. One morning, as she was about to fix her coffee, she noticed that the sugar looked a bit odd. She sent some of the crystals from the sugar bowl to a local lab for analysis. The results showed the presence of the elements carbon, hydrogen, oxygen, and nitrogen. These elements seemed harmless.

However, the bottom line showed the conclusion: $C_{21}H_{22}N_2O_2$—strychnine (STRIK-nyn), a deadly poison.

● Why would the elements C, H, N and O seem harmless?

Scientists now believe that Napoleon died from arsenic poisoning.

Compounds

Mrs. X was correct not to fear the elements carbon, hydrogen, oxygen, and nitrogen. Carbon is a harmless black substance. The other elements are always present in the air we breathe.

Yet, when these elements unite in special ways, dangerous substances, such as strychnine, may be formed. Strychnine is a compound. A **compound** is a substance made up of two or more elements which have combined chemically.

When elements combine, they often lose their special properties. In strychnine, the one black solid (carbon) and the three colorless gases (hydrogen, nitrogen, and oxygen) combine to form white crystals. These four common elements combine and form a dangerous poison.

Sometimes the situation is reversed. This is the case with sodium chloride. This compound is table salt, a substance we consume all the time. Yet, the metal sodium is a dangerous element. The gas chlorine is equally harmful.

Fortunately, when they form the compound sodium chloride, they lose their deadly properties. The silvery solid (sodium) combines with the greenish gas (chlorine) to form the white crystals of sodium chloride.

How do scientists name compounds? If two elements are combined, the metal is named first. Then, the nonmetal is given with the ending -*ide*. A compound made from magnesium and oxygen, for example, becomes magnesium oxide. A compound formed from calcium and chlorine is named calcium chloride.

Nineteenth-century scientists working to form compounds were often surprised. Some elements form compounds very easily. Sodium and fluorine, for example, form sodium fluoride as soon as they contact each other. Other elements require heating to combine. Some need an electrical spark. And some elements won't combine under any conditions.

In the 18th century, an English chemist named John Dalton made a remarkable discovery. Dalton found that elements combine in definite proportions (pruh-POR-shunz) by

mass. For example, it is known that 40 g of calcium, Ca, will combine with 32 g of sulfur, S, to form the compound calcium sulfide, CaS. The ratio, then, of calcium to sulfur in this compound is 40:32 (read 40 to 32). This means that twice as much calcium (80 g) will combine with twice as much sulfur (64 g).

If you tried to unite 40 g of calcium with 35 g of sulfur, 3 g of the sulfur would be left over. These three grams would remain as the element sulfur.

Consider the compound water. The ratio of hydrogen to oxygen is 2 to 16 by mass. This means that 20 g of hydrogen combines with 160 g of oxygen. 200 g of hydrogen would combine with 1600 g of oxygen.

To form the compound water, how much oxygen would be needed to combine with 2000 g of hydrogen? The ratio must be 2 to 16. So, the answer must be 16,000 g of oxygen.

To Do Yourself   How Can the Formation of a Compound Be Shown?

You will need:

Black and white paper squares

1. Put together a checkerboard pattern of black and white squares that measures 6 squares long by 6 squares wide.
2. Put together another 6 by 6 pattern of squares alternating 2 black rows with 1 white row.
3. Put 4 squares of the same color together to form a diamond shape. Place two black, and two white diamonds together in an alternating fashion to form a larger diamond.
4. Now randomly build a 4 by 4 square without looking.

Questions

1. What is the ratio of black to white squares in step **1**? _____

   Step **2**? _____

   Step **3**? _____
2. Find the ratio for step **4**. Will atoms combine this way in a compound?

   _____

How much oxygen would be needed to combine with 1 g of hydrogen to form the compound water? The proportion must be 2 to 16. Since the first number is cut in half, the second number must be treated in the same way. 1 g of hydrogen unites with 8 g of oxygen.

Try these problems:

1. Magnesium oxide can form from 24 g of magnesium, Mg, and 16 g of oxygen, O. How much oxygen is needed to unite with 48 g of magnesium?

Magnesium:Oxygen

24:16
48:?

The answer must be 32. Since the first number is doubled, the second number must also be doubled.

2. Calcium fluoride forms when 40 g of calcium, Ca, combine with 38 g of fluorine, F. How many grams of calcium are needed to unite with 80 g of fluorine to make calcium fluoride?

Calcium:Fluorine

40:38
?:380

The answer must be 400 g of calcium. Since the second number is multiplied by 10, the first number must also be multiplied by 10.

Dalton's rule shows the orderly way that compounds form. Elements cannot combine any which way. Every compound is made up of specific elements in definite proportions.

Review

I. Fill in each blank with the word that fits best. Choose from the words below.

elements   32g   sulfide   compound   64g   metal

14 g of lithium combine with 32 g of sulfur to form the _____ called lithium _____ . Lithium is named first because it is a _____ . If 28 g of lithium are used, then you need _____ of sulfur. The _____ change properties when they combine.

II. Which statement seems more likely to be true?

A. _____ All compound names end in -ide.

B. _____ All compounds are made of more than one element.

III. 24 g of magnesium combine with 16 g of oxygen to form a compound. This compound is called _____ . 12 g of magnesium would unite with _____ of oxygen.

IV. Answer in sentences.

Why do you think elements are called the "building blocks" of matter? Explain.

What Are Molecules?

Exploring Science

Computer Assistance. The pictures shown are special computer graphics. They certainly don't resemble Pac-Man or other computer game characters. They look more like Christmas ornaments than chemical figures. But, they actually represent molecules.

Molecules (MAHL-uh-kyoolz) are the smallest natural units of many elements and all compounds. By using computer pictures, scientists can look at chemicals in a new way.

Each element in a compound is assigned a color. Each atom is shown as a sphere. So, by looking at the graphics, we can easily see how a molecule is built. We can even see the angles between the atoms.

Computer graphics provide another important feature. Parts of some complex molecules can rotate. They can change positions. By using a computer program, the graphics can rotate as well. A scientist can see many three-dimensional views just by giving the computer commands.

Since over ten million chemicals are known, computer help is welcome. Perhaps such graphics will be part of your chemistry course in the near future.

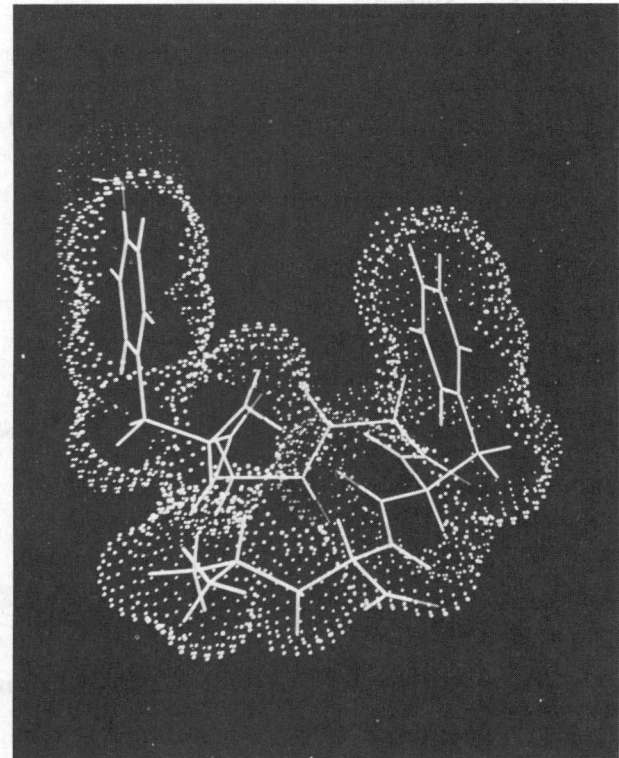

How might looking at molecules on a computer screen be helpful to scientists?

● Compare

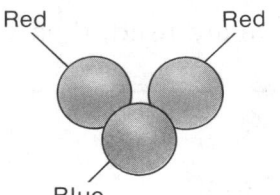

and

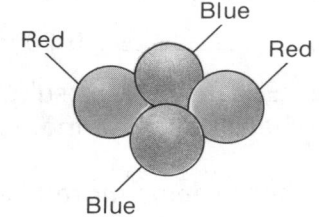

Molecules and Formulas

Molecules can be described without using a computer. Scientists use a **formula** (FOR-myu-la) to describe a molecule. A chemical formula tells us the kind and number of atoms of each element found in a compound. For example, O_2 is the formula for a molecule of oxygen. There are two atoms in this molecule. But, there is only one type of atom. Oxygen is one of the seven diatomic elements. *Di-* means two. So,

diatomic means two identical atoms tied together in a molecule.

The seven diatomic elements are oxygen, hydrogen, nitrogen, fluorine, chlorine, bromine, and iodine. Their formulas are O_2, H_2, N_2, F_2, Cl_2, Br_2 and I_2. The little number, or **subscript**, tells us the number of atoms in the compound.

Atoms in a molecule are held together by a

chemical tie. This chemical tie is called a **bond**. A bond is the force which holds the atoms of a molecule together.

Molecules are the smallest units of compounds. Let's look at a molecule of hydrogen sulfide. Hydrogen sulfide has the formula H_2S. The subscript 2 shows that there are 2 atoms of hydrogen. No subscript follows the S. When no number follows the symbol of an element, the number 1 is understood. So there are 3 atoms in one molecule of H_2S. A molecule of hydrogen sulfide always has this formula. That's why Dalton's Rule works so well. The proportion must be the same every time the compound hydrogen sulfide is formed. It is always H_2S.

Consider the compound Al_2O_3. This compound has 2 elements—aluminum, Al, and oxygen, O. But this molecule has five atoms—2 aluminum atoms + 3 oxygen atoms.

Simple sugar has the formula $C_6H_{12}O_6$. One molecule of sugar has 24 atoms. There are 6 atoms of carbon, C, + 12 atoms of hydrogen, H, + 6 atoms of oxygen, O.

Sometimes a formula has no subscripts. For example, HCl represents a molecule of hydrogen chloride. This molecule has only two atoms, 1 of H + 1 of Cl.

Can you describe the molecule that has the formula H_2CO_3? There are 6 atoms in this molecule. There are 2 hydrogen atoms + 1 carbon atom + 3 oxygen atoms.

Review

I. Fill in each blank with the word that fits best. Choose from the words below.

diatomic   **CO**   **H_2SO_4**   **molecule**   **F_2**   **AlF_3**

A formula describes a _____ . Sometimes, a molecule is

of an element like _____ . Fluorine is called a

_____ element. An example of a molecule of a compound

with 4 atoms is _____ . An example of a formula showing

three kinds of atoms is _____ .

II. Which statement seems more likely to be true?

A. _____ Every formula has a subscript.

B. _____ Co and CO are different substances.

III. Fill in the blanks.

The names of the elements in $MgCl_2$ are _____ and

_____ .

The name of the compound $MgCl_2$ is _____ .

The number of atoms in one molecule of $MgCl_2$ is _____ .

The number of atoms in Na_2CO_3 is _____ .

IV. Answer in sentences.

How is a molecule of Na_2S different from a molecule of NaF?

_____

_____

Lesson Three

What Are Ionic Bonds?

Exploring Science

Encyclopedia Crystals. Superman enters the Fortress of Solitude. He chooses a long, clear crystal. He places this crystal in a control panel. Visions of the past appear. Other crystals contain pre-recorded messages of lessons in history and science.

This scene was from *Superman II*. Fans loved the special effects. Others claimed that the scene was too ridiculous to be believed. These critics may be surprised by the findings of some recent research.

Physicist Lynn A. Boatner is making crystals which can store data. It is hoped that thousands of images might be stored in one crystal. Surprisingly, the crystal is only about the size of a thumb. The images are fixed in the crystal with laser light.

The crystals are built from a mixture of compounds: tantalum oxide, niobium oxide, and potassium carbonate (poh-TAS-ee-uhm CAR-buh-nayt). The best combination of these compounds is still being sought. When the best combination is found, the images stored in the crystal will not fade.

● What metals are in the compounds being tested by Dr. Boatner?

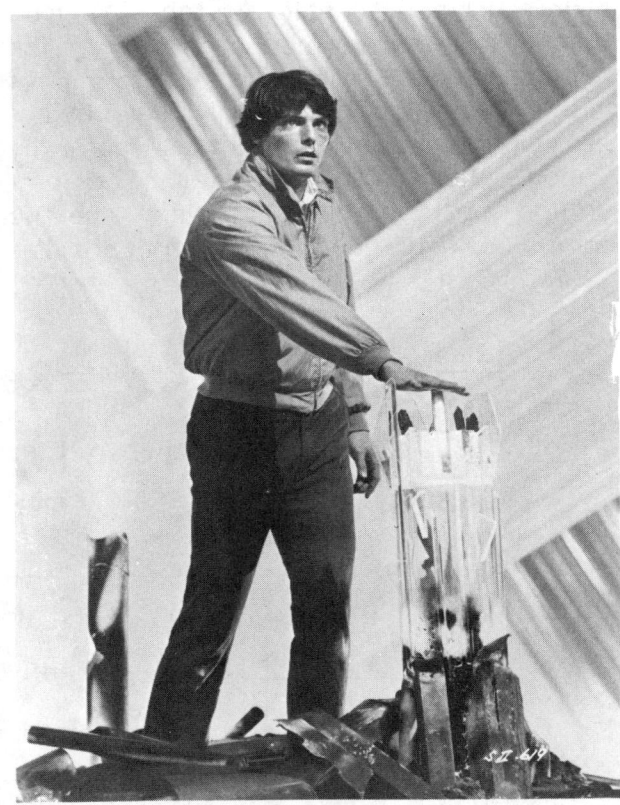

Christopher Reeve as Superman in *Superman II*.

Ionic Bonds

Many substances used for crystal building are ionic compounds. **Ionic compounds** form when metal ions combine with nonmetal ions. Metal ions have a positive charge. Nonmetal ions have a negative charge. Opposite charges attract, or pull toward, each other. When this happens, an **ionic bond** forms. An ionic bond is created by the force of attraction between + charges and − charges.

Let's look at how an atom of sodium, Na, forms an ionic bond with an atom of chlorine, Cl.

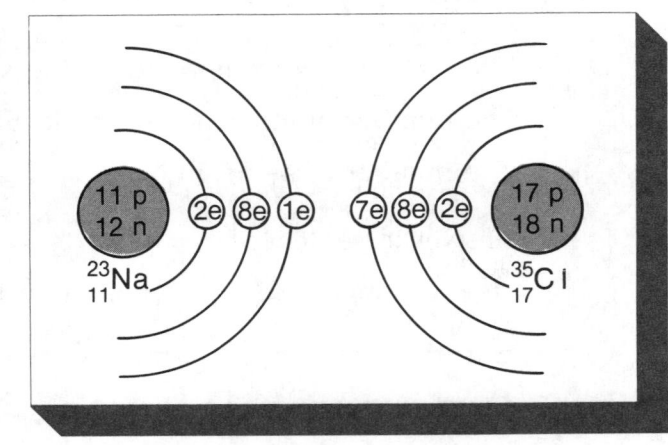

Both atoms are electrically neutral. The positives and negatives are equal in each atom. The valence shell of sodium has only one electron. The valence shell of chlorine has seven electrons. Therefore, the chlorine atom only needs one electron to reach the magic number of eight.

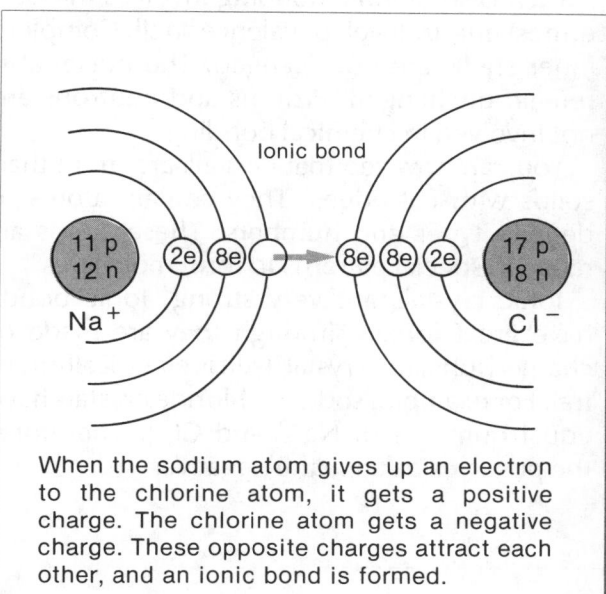

When the sodium atom gives up an electron to the chlorine atom, it gets a positive charge. The chlorine atom gets a negative charge. These opposite charges attract each other, and an ionic bond is formed.

If the sodium atom gives up one electron to the chlorine atom, the sodium ion gets a charge because it has lost one electron.

$$\begin{array}{r} +11 \\ -10 \\ \hline +\,1 \end{array}$$

The chlorine ion has a charge because it has gained one electron.

$$\begin{array}{r} +17 \\ -18 \\ \hline -\,1 \end{array}$$

An ionic bond now exists between the positive ion and the negative ion. An ionic crystal is building. NaCl involves one ion of sodium and one ion of chlorine.

Let us look at another ionic bond forming. This time we will see how a sodium, Na, atom bonds with a sulfur, S, atom.

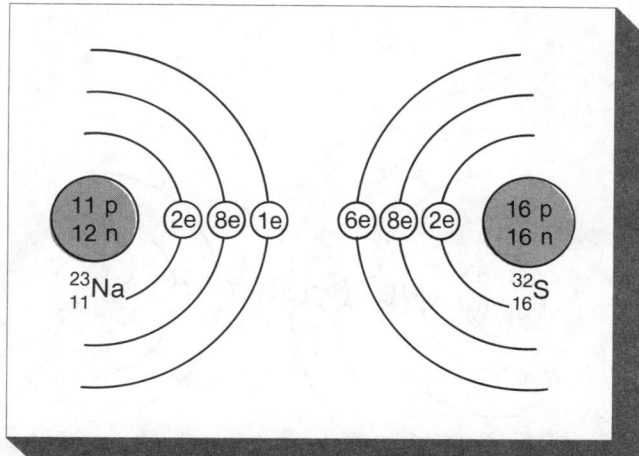

The outer energy level of sulfur needs 2 electrons to reach the magic number of 8. Sodium can only donate, or give, 1 electron. Therefore, every sulfur atom needs two sodium atoms to fill its outer level. The bonding can be represented like this:

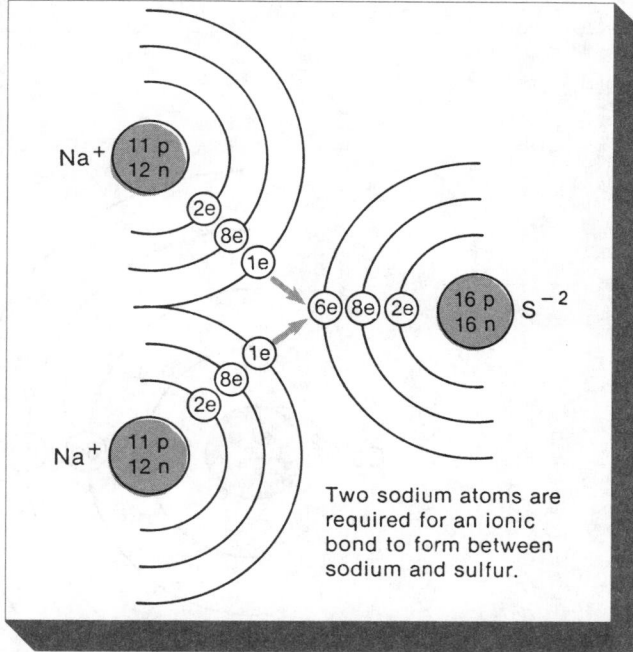

Two sodium atoms are required for an ionic bond to form between sodium and sulfur.

Each sodium ion has a charge of +1. The sulfur ion has a charge of −2 since it has gained the 2 electrons.

Now you can see why the formula for sodium sulfide is Na_2S. This ionic crystal will have twice as many metal ions as nonmetal ions.

Let us look at the bonding between calcium and fluorine.

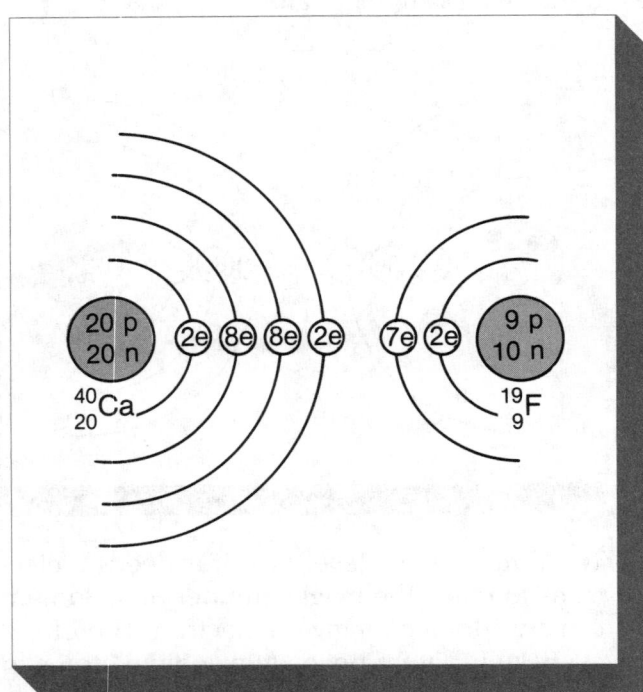

The calcium, Ca, atom has 2 electrons to donate. But fluorine can only take 1 electron to have a complete shell. Therefore, 2 atoms of fluorine are needed to make a successful ionic bond. The bonding can be shown like this:

The calcium ion has a +2 charge since it donated 2 electrons. Each fluorine ion has a −1 charge. The formula for calcium fluoride is CaF_2.

Each case of ionic bonding involves the outermost energy level, or valence shell. Complete inner shells are not changed. The nuclei also remain unchanged. Protons and neutrons are not involved in chemical bonding.

You can now see that crystals are more than solids with flat edges. They contain atoms of definite types and numbers. These atoms arrange in special patterns to form molecules.

Ionic crystals are very strong. Ionic bonds have great force. Although they are made of charged ions, the crystal itself is electrically neutral. For example, sodium chloride crystals have equal numbers of Na^{+1} and Cl^{-1}. Therefore, the positives balance the negatives.

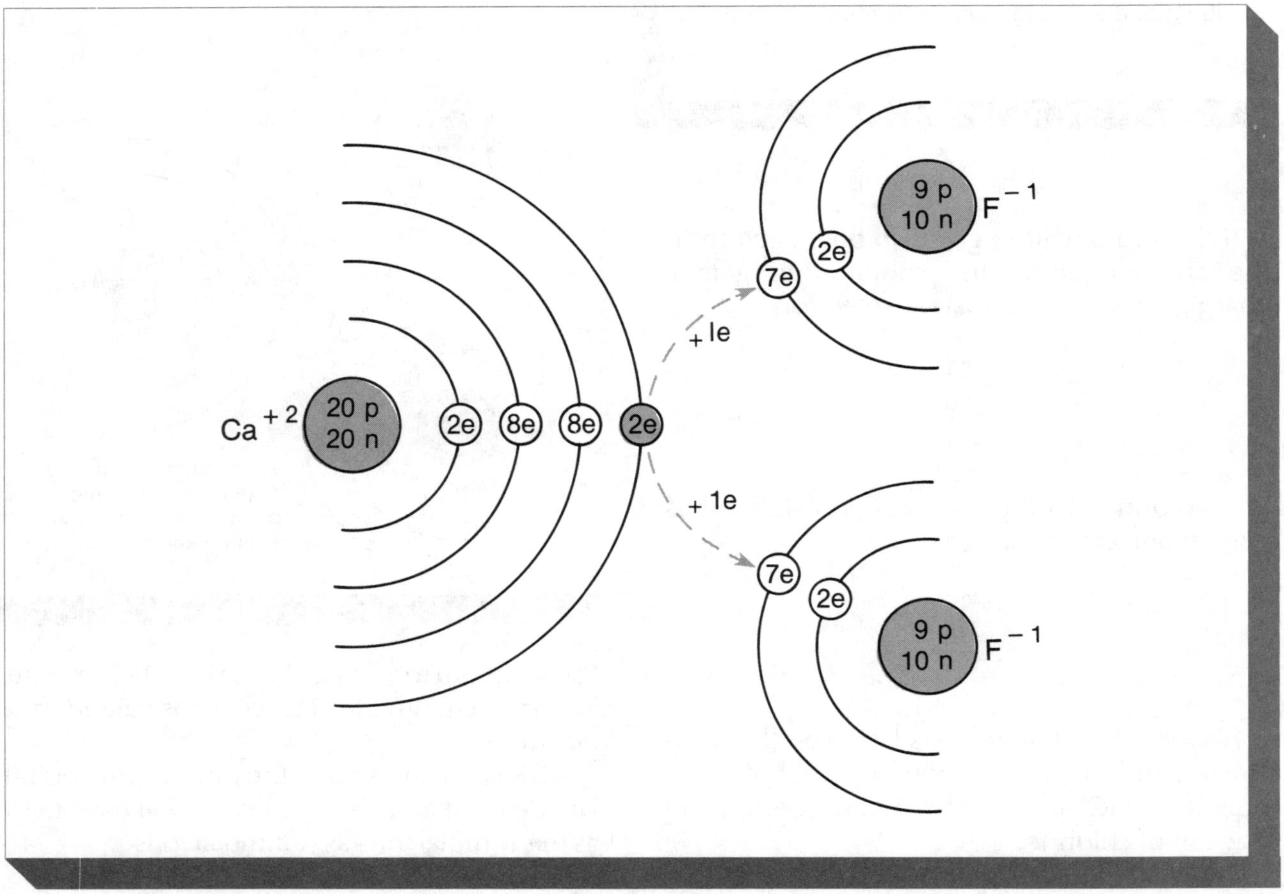

I. Fill in each blank with the word that fits best. Choose from the words below.

attraction   **metal**   **ionic**   **AgF**   **electron**   **negative**   **AgF₂**

Silver fluoride is an _____ crystal. It has equal numbers

of _____ and nonmetal ions. The formula for silver

fluoride is _____ . An _____ moves from

Ag to F. The bond is formed from the _____ between the

positive and _____ ions.

II. Which statement seems more likely to be true?

A. _____ Metal ions are positively charged.

B. _____ Crystals must have even numbers of ions.

III. Calculate the charge on this zinc ion.

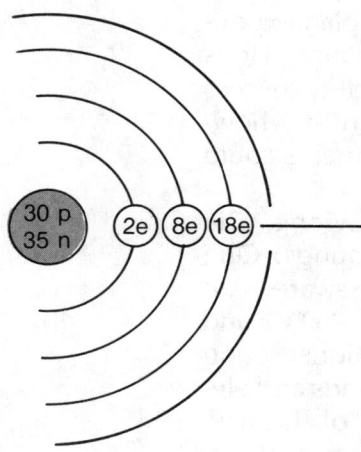

IV. Answer in sentences.

Why are metals called electron-donors?

_____

_____

_____

Lesson Four

What Is an Element's Valence?

Exploring Science

The Danger of Labels. A high school football player with the nickname "Fumbles" usually stays on the bench. "Strikeout Jimmy" won't be called in to pinch-hit in the bottom of the ninth inning. When the prediction is for poor performance, a person usually isn't even given a chance.

This was the case with the elements which were called inert in the late 1800's. **Inert** (IN-uhrt) means lacking power to move. These elements were listed in the last column of the Periodic Table.

Scientists predicted that helium, neon, argon, krypton, xenon, and radon would not form compounds. So, they didn't plan experiments using these elements. In 1932, Linus Pauling, a famous chemist, predicted success for xenon compounds. Most scientists, however, still believed that the inert elements could not form compounds.

In 1962, this label was proved wrong. The inert elements could form compounds. One experiment involved unusual glassware. Another involved temperatures of 400°C and −195°C. But, under the right conditions, xenon could combine with other compounds and elements. $XePtF_6$ and XeF_4 were two of the new compounds.

Since the achievement of forming compounds of xenon in 1964, similar advances have been made with other inert gases. Fluorine forms compounds with krypton and radon as well as with xenon. Many stable xenon compounds have been made. Today, the "no-longer-inert" gases are usually called the noble gases, although the name inert is still used by some chemists. But the joke is on the people who insist on saying *noble* instead of *inert.* One dictionary gives the following for the fifth meaning of the word *noble:* 5. *Chemistry.* Inactive or inert.

● How many atoms are in one molecule of XeF_4?

NOBLE GASES

O

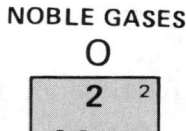

2	2
He	
Helium	
4.00	

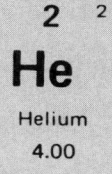

10	2 8
Ne	
Neon	
20.18	

18	2 8 8
Ar	
Argon	
39.94	

36	2 8 18 8
Kr	
Krypton	
83.80	

54	2 8 18 18 8
Xe	
Xenon	
131.30	

86	2 8 18 32 18 8
Rn	
Radon	
(222)	

Valence

The coach has a good idea of a player's potential batting average. **Potential** (poh-TEN-chul) refers to what can be done if conditions are right. A batter's potential refers to the number of times the ball will be hit when he goes to bat.

The **valence number** of an element refers to its potential combining power. For example, if it is likely to give an electron, its valence is +1. If it is likely to give two electrons, its valence is +2.

Xenon, with 8 electrons in its valence shell, is not likely to combine with other atoms. So, its valence number is 0. We learned from the story that xenon can enter into a compound. If the conditions are very special, xenon can outreach its potential.

Nonmetals have negative valence numbers. If an atom is likely to take one electron, its valence is −1. If it is likely to take two electrons, its valence is −2.

Some elements have more than one valence. They will be discussed later. For now, let us look at the new power valence numbers provide.

We can now predict formulas more easily. When the valence numbers are added, the total must be zero. For example, sodium has a valence of +1. Chlorine has a valence of −1. Therefore, the formula for sodium chloride must be NaCl since the sum of the valence numbers, (+1) + (-1), equals 0.

Calcium, Ca, has a valence of +2. Sulfur, S, has a valence of −2. So calcium sulfide must be CaS, since +2 cancels −2. Again, we have a neutral combination.

Barium, Ba, has a valence of +2. Magnesium, Mg, has a valence of +2. These elements do not have the potential to form a compound. Because +2 and +2 equals +4, we cannot get a neutral combination using these two elements.

Zinc, Zn, has a valence of +2. Fluorine, F, has a valence of −1. The formula for zinc fluoride must include two atoms of fluorine to get a total of zero. +2 and −1 and −1 equal zero. Therefore, the formula for zinc fluoride is ZnF_2.

A long time ago, a chemistry student figured out a criss-crossing system for writing formulas. To use this system, first write the symbols for each element. Let's use Zinc and Fluorine.

Start by writing the chemical symbols for the two elements.

$$Zn \qquad F$$

Next, write the valence number for each element above its symbol.

$$Zn^{+2} \qquad F^{-1}$$

Cross out the + and − signs of each valence number.

$$Zn^{\cancel{+}2} \qquad F^{\cancel{-}1}$$

After you have done this, criss-cross the valence numbers, and use them as subscripts.

$$Zn^2 \diagdown \diagup F^1$$
$$Zn \diagup \diagdown F_2$$

Remember, the subscript 1 is never written. The formula for our molecule is therefore written ZnF_2.

Let's try this again. This time we will combine aluminum with chlorine. First, write the symbols for each element.

$$Al \qquad Cl$$

Now, add the valence number of each element.

$$Al^{+3} \qquad Cl^{-1}$$

Cross out the + and − next to each valence number. Then, criss-cross the numbers.

$$Al^3 \diagdown \diagup Cl^1$$
$$Al \diagup \diagdown Cl_3$$

The formula for aluminum chloride is $AlCl_3$.

Try using the criss-crossing system yourself for each of the following. The valence numbers for the atoms have already been written for you.

$$Ag^{+1} \diagdown \diagup O^{-2} \qquad\qquad Al^{+3} \diagdown \diagup O^{-2}$$
$$Ag_? \diagup \diagdown O_? \qquad\qquad Al_? \diagup \diagdown O_?$$

Silver oxide   Aluminum oxide

What formula did you get for silver oxide? There will be two atoms of silver for every atom of oxygen in silver oxide. Its formula should be Ag_2O.

What formula did you get for aluminum oxide? Every molecule of aluminum oxide requires 2 atoms of aluminum and 3 atoms of oxygen. Therefore, its formula is Al_2O_3.

Review

I. Fill in each blank with the word that fits best. Choose from the words below.

potential   + 2   valence PbI_2 zero   − 2   inert   PbI

The noble gases were called _____ elements. They had a valence number of _____ . The valence number of an element refers to its _____ combining power. Lead can give two electrons, so, its valence is _____ .Iodine's _____ is −1. The formula for lead iodide is _____ .

II. Which statement seems more likely to be true?

A. _____ An element can have only one valence.

B. _____ Metals have + valence numbers.

III. Write formulas for each of the following:

$K^{+1}Cl^{-1}$ _____

$Ca^{+2}O^{-2}$ _____

$Na^{+1}S^{-2}$ _____

IV. Answer in sentences.

Why do all the elements in column IA on the Periodic Table have a valence of +1?

_____

_____

_____

What Is a Covalent Bond

Exploring Science

Tinkertoys and Daydreams. A chemistry textbook with new kinds of diagrams appeared in 1861. The diagrams looked like tinkertoy blueprints. Water was not written as H_2O, it was written as H—O—H. Methane was not CH_4, it was

$$H—\overset{\displaystyle H}{\underset{\displaystyle H}{C}}—H.$$

Friedrich August Kekulé (kuh-COOL-ay) was the author of that textbook. He believed that such diagrams could solve some of the chemical puzzles of that time.

Benzene was one of those mysteries. Benzene was discovered in 1825. A molecule of benzene had six carbon atoms and six hydrogen atoms. Kekulé guessed that a molecule of benzene might look like this

$$H—C{=}C—C{=}C—C{=}C—H.$$

But, he knew that this diagram must be wrong. A benzene molecule acts as if it is **symmetrical** (suh-ME-tri-kuhl), or balanced. This diagram did not show a balanced molecule.

After several years of studying this problem, the solution came to Kekulé in a daydream. The chains of his diagram came alive and danced. One end of the chain began to coil itself like a snake. Its head bit its tail. Kekulé became alert. Benzene is a ring not a chain! A beautifully balanced molecule was drawn.

A benzene molecule.

● What is true about every carbon in Kekulé's drawings?

Covalent Bonds

Kekulé's diagrams are called **structural formulas.** They show how the atoms are bonded as well as what atoms are present in a molecule. The dashes in the diagram represent a different kind of bond.

They are not ionic bonds. No electrons move from one atom to another. Instead electrons are shared. A **covalent** (KO-VAY-lent) **bond** is formed by a shared pair of electrons. Each dash in the diagram represents one covalent bond.

Carbon has four electrons in its valence shell. It is not likely to give them away, as a metal atom might. It is also not likely to gain four electrons, as a nonmetal might. Instead, it shares those electrons.

Methane consists of one carbon atom and four hydrogen atoms. Its formula is CH_4. Each of the four hydrogen atoms has one electron to share. Hydrogen has a valence of 1. The carbon atom has four electrons to share. Therefore, the valence of carbon is 4.

If the 4 electrons of the carbon atom combine in covalent bonds with the single electrons of 4 hydrogen atoms, a molecule of methane is formed.

The structural formula for methane is drawn as

$$H—C—H.$$
(with H above and H below the central C)

Each dash shows that a pair of electrons is being shared. Thus, each dash represents a covalent bond.

An **x** and **o** diagram may make this clearer.

$$H \overset{x}{\underset{o}{\times}} C \overset{o}{\underset{x}{\times}} H$$
(with H above showing ox and H below showing xo)

x = one carbon electron
o = one hydrogen electron

Now carbon is satisfied with eight electrons in its valence shell. Hydrogen's valence shell can only hold 2 electrons. So, each hydrogen also has a complete valence shell.

To Do Yourself   How Can a Model Show Covalent Bonding?

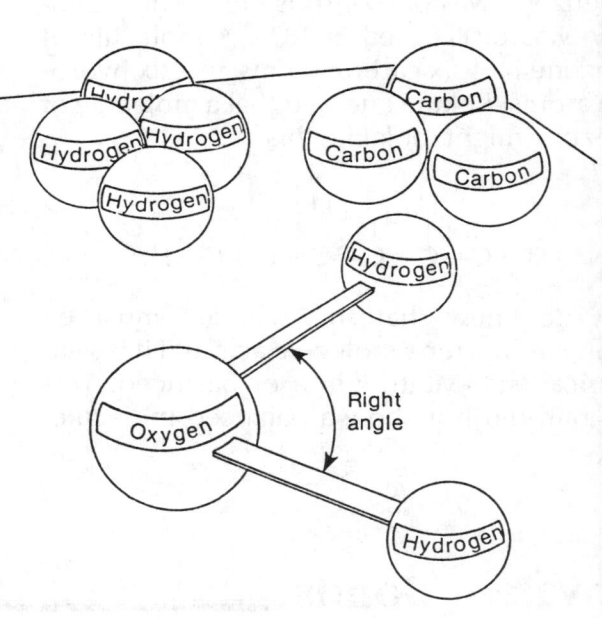

You will need:

Styrofoam spheres of 3 different sizes, connecting sticks, small self-sticking labels, a pen or pencil

1. Make six labels which read hydrogen. Make one label which reads oxygen. Make three labels which read carbon.
2. Label two spheres hydrogen. Label one sphere oxygen. Using the connecting sticks, attach these three spheres so that the hydrogens are at a right angle to each other.
3. Label one sphere carbon. Label four spheres hydrogen. Using the connecting sticks, attach the hydrogens to the carbon.
4. Label two spheres carbon. Now connect three hydrogen spheres to each of the carbon spheres.

Questions

1. What molecule did you build in step **2**? _____

2. What is the formula for the molecule you built in step **3**? _____
3. How many shared (covalent) bonds are in the molecule you built in step **4**?

_____

4. What is the formula for the molecule you built in step **4**? _____

_____

Let's look at Kekulé's structural formula for water. It was H—O—H. Oxygen, O, has six electrons in its valence shell (x). Hydrogen, H, has one electron (o).

$$H \, {}^{o}_{x} \, O \, {}^{x}_{o} \, H$$

with xx above and xx below the O

There is one shared pair of electrons between the oxygen and each hydrogen atom. Oxygen has the magic number of 8. Again, each hydrogen has a complete shell with 2 electrons.

Let's look at ammonia. Kekulé drew it as

H—N—H.
|
H

Nitrogen, N, has five electrons in its valence shell (x). Hydrogen, H, has one electron, (o).

$$H \, {}^{x}_{o} \, N \, {}^{o}_{x} \, H$$

with xx above and xo below the N, then H below

Again, covalent bonds result in complete valence shells for both nitrogen and hydrogen.

Review

I. Fill in each blank with the word that fits best. Choose from the words below.

C   covalent   shared   H   structural   ionic   N   electron

If an _____ leaves one atom, an _____ bond is forming. But if a pair of electrons is _____ then a _____ bond has formed. A _____ atom must have four bonds around it in the _____ formula. A _____ atom only needs three.

II. Which statement seems more likely to be true?

A. _____ Ionic and covalent bonds are the same.

B. _____ All bonds hold atoms together.

III. Write a structural formula for a diatomic molecule of hydrogen.

IV. Answer in sentences.

Why does the nucleus of carbon remain unchanged when it forms methane? Explain.

_____

_____

_____

What Is Organic Chemistry?

Exploring Science

Chocolate Fever. Nutrition experts caution adults and children, "Chocolate is to be treated as a poison." This is excellent advice.

The average American eats eight pounds of chocolate a year. Chocolate boutiques have opened all over the country. A magazine devoted to chocolate has been published. A laboratory group at Pennsylvania State University is studying the essence of chocolate—cocoa butter. Using new microscopic techniques, this laboratory group is making interesting discoveries about chocolate.

Every cocoa butter molecule has four parts: three units of one type attached to one unit of another type. These units may differ slightly from chocolate to chocolate. So, they line up differently when they form crystals.

We can begin to see the difference between chocolates which are $1.29 a pound and $20.00 a pound by looking at how these crystals form. Some crystals form feathers, bow ties, or spheres that look like fancy flowers. The difference in how the crystals form may explain why some chocolate is gritty, some snaps, and some shines.

Someday maybe scientists can invent a chocolate which is both nutritious and delicious.

● How would you describe the shapes shown on the right?

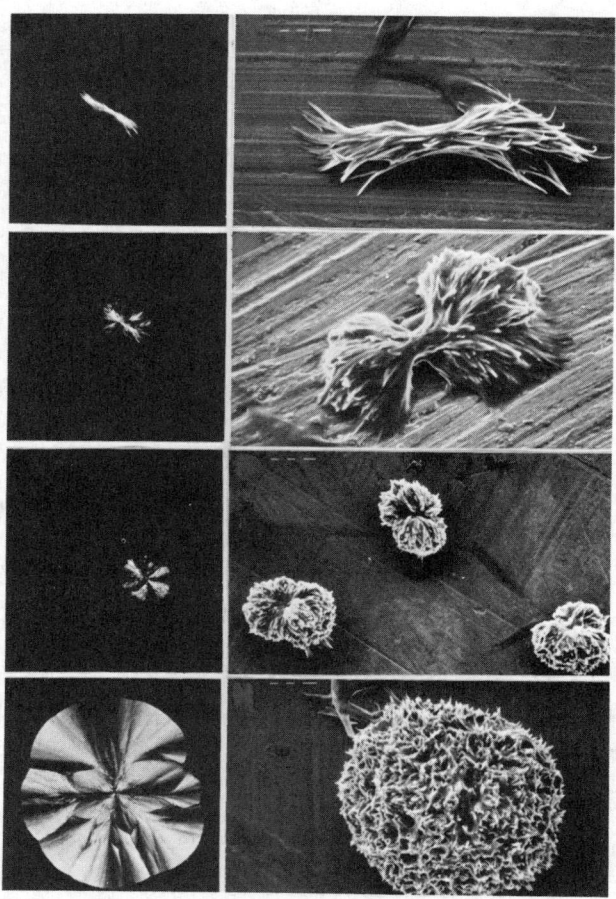

These are just a few of the shapes that cocoa butter crystals have.

Organic Chemistry

Cocoa butter is an organic compound. All **organic** (or-GAN-ik) **compounds** contain the element carbon. They are present in chocolate, plastic, human cells, and thousands of other unexpected places.

One of the simplest carbon substances is also the most beautiful: the diamond. A diamond is not a compound. It is a crystal of carbon. Each carbon atom is bonded to four other carbon atoms. It is like a giant three-dimensional jigsaw puzzle made of nothing but carbon atoms.

Organic compounds have covalent bonds. So, structural formulas are best used. As Kekulé showed us, some molecules are not simple.

CH_3COOH is the formula for the acid found in vinegar. Its structural formula is

Between the carbon and the lone oxygen is a double bond. A **double bond** is formed when two pairs of electrons are shared by two atoms. In this case, two electron pairs are shared by the carbon atom and the oxygen atom. To show a double bond in a structural formula, a double dash (=) is used.

C_2H_2 is acetylene. It is better shown as $H—C \equiv C—H$. You have probably guessed that the $\equiv$ stands for a **triple bond.** In acetylene, three electron pairs are shared by the two carbon atoms.

Carbon's ability to form single, double, and triple bonds adds to the variety of organic compounds. Carbon compounds can be straight chains. They can be branched chains. They can be rings. They can also be a limited number of bonded large units as in cocoa butter.

Some organic substances have repeated large units without limit. These large units are called **polymers** (POL-ee-merz). The prefix poly- means many. Polymers are very important compounds. They form the proteins in humans and fish. Polymers form the threads in a polyester suit. Polymers also form the plastics which are changing our lives.

Review

I. Fill in each blank with the word that fits best. Choose from the words below.

polymers   **electrons**   **organic**   **double**   **covalent**

Compounds with carbon are called _____ . If atoms share two pairs of _____ there is a _____ bond. All carbon bonds are _____ . Giant organic molecules are usually _____ .

II. Which statement seems more likely to be true?

A. _____ Each carbon atom can have 4 covalent bonds.

B. _____ Diamonds are organic compounds.

III. How are these molecules alike? How are they different?

$$
\begin{array}{ccccccc}
& H & H & H & H & & \\
& | & | & | & | & & \\
H{-} & C & {-}C & {-}C & {-}C & {-}H & \\
& | & | & | & | & & \\
& H & H & H & H & &
\end{array}
\quad \text{and} \quad
\begin{array}{c}
\begin{array}{ccccc}
H & H & H \\
| & | & | \\
H{-}C & {-}C & {-}C{-}H \\
| & | & | \\
H & & H \\
\end{array} \\
H{-}C{-}H \\
| \\
H
\end{array}
$$

_____

_____

IV. Answer in sentences.

How might two chains of compounds containing only carbon atoms and hydrogen atoms be different?

What Are Radicals?

Exploring Science

Invisible Danger. In New York, six people were hospitalized with heart ailments. They had eaten sausage with nitrites. In New Orleans, ten children became seriously ill. They had eaten bologna and frankfurters treated with nitrites. In Florida, a three year old boy died. He had eaten a hot dog with added nitrites.

Why did they eat these chemicals? They probably did not know these chemicals were in their food. Sodium nitrate and sodium nitrite are used in meat and fish processing. They are present in most packages of bacon and hot dogs. Sodium nitrite helps to keep meat an attractive pink color for a long period of time.

Why does the government allow the use of such chemicals? It is a question of proving the danger. The cases above all involved meats with more nitrite than is legally allowed. But there is other evidence. For the past two decades, scientists have collected new data about nitrates and nitrites. It shows that the nitrite group joins with other chemicals as a probable cause of cancer.

Now, labels showing the presence of nitrites and nitrates must be placed on these foods. Some companies are packing products which are nitrite free. It is now up to the consumer to choose the right package.

● Why do meat-packers continue to use sodium nitrite?

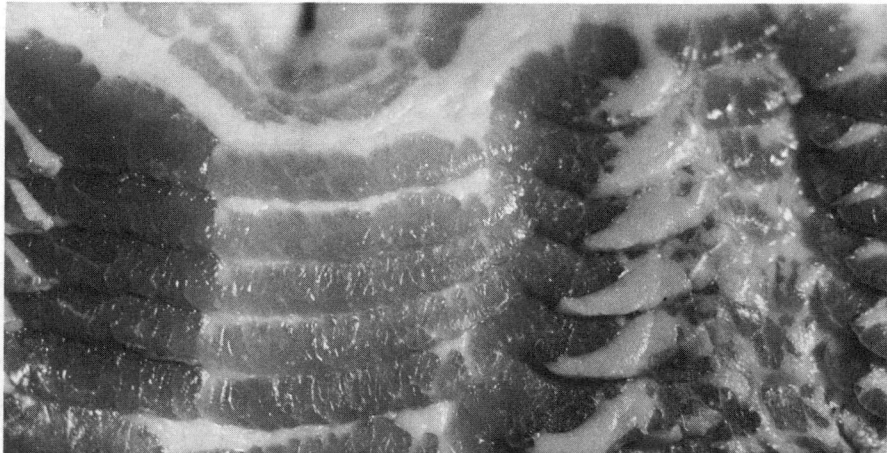

Cured with Water, Salt, Sodium Phosphates, Suga
Sodium Erythorbate, Sodium Nitrite, Flavorings.

What can labels tell you about food products?

Radicals

Sodium nitrate and sodium nitrite are compounds that have special groups of atoms. You should know that there are more than two elements in sodium nitrate. It doesn't end in -ide.

The formula for sodium nitrate is $NaNO_3$. The sodium ion is Na^{+1}. The nitrate group is NO_3^{-1}. A group of atoms that acts as a charged unit is called a **radical.** The nitrite group is a slightly different radical. It is NO_2^{-1}. Thus, sodium nitrite is $NaNO_2$.

This is a list of some other common radicals:

bicarbonate	HCO_3^{-1}
carbonate	CO_3^{-2}
hydroxide	OH^{-1}
phosphate	PO_4^{-3}
sulfate	SO_4^{-2}
sulfite	SO_3^{-2}
acetate	$C_2H_3O_2^{-1}$
ammonium	NH_4^{+1}

Let's use a criss-crossing trick. You must remember to treat the radical as a unit. So, if you need two of the radicals, just put the radical in parentheses. Then, add a subscript. For example:

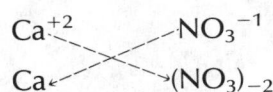

The formula for calcium carbonate is $Ca(NO_3)_2$. This is read as Ca-NO_3 taken twice. Try another.

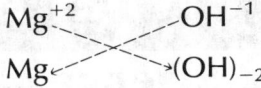

The formula for magnesium hydroxide is $Mg(OH)_2$. You must remember to use parentheses for hydroxide. You need two groups, not two hydrogens.

On your list of common radicals there is only one which acts as a metal. That is NH_4^{+1}, ammonium. So you can have a compound with two radicals. Ammonium nitrate is NH_4NO_3.

Radicals are also called *polyatomic ions*. The prefix *poly-* means "many." Can you explain why this new term is logical?

Review

I. Fill in each blank with the word that fits best. Choose from the words below.

CO_3^{-2} radical   element $C_2H_3O_2^{-1}$ SO_4^{-2} charge

A _____ acts as a group in a compound. It always has a

_____ . Also, it is made up of more than one

_____ . _____ is a radical formed

from three elements. _____ has four atoms in the
radical.

II. Which statement seems more likely to be true?

A. _____ A compound with a radical may end in -ide.

B. _____ Ammonia (NH_3) is the same as ammonium (NH_4^{+1}).

III. Circle the radical in each compound.

$AlPO_4$ NH_4Cl $Ca(NO_3)_2$
$NaHCO_3$ $KC_2H_3O_2$ Li_2SO_4

IV. Answer in sentences.

Should compounds involving the acetate radical be called "organic compounds?"

_____

_____

Lesson Eight

What Are Acids and Bases?

Exploring Science

Acid problems. One hot June day, a strange thing happened in Baltimore, Maryland. Several women working in a power plant noticed tiny holes in their panty hose. The holes got bigger and bigger. Finally, the fabric fell apart.

Researchers suggested an explanation for the holes. Many power plants give off sulfur compounds. These gases rise into the air. They can then combine with water vapor to form sulfuric acid.

The humidity was very high on that June day. So, the acid mist explanation was reasonable. Sulfuric acid can damage much more than thin fabric.

This small case is an example of a bigger problem. For several decades, scientists have warned careless manufacturers about the dangers of sulfur in the air. Some plants release many sulfur and nitrogen compounds into the air. These compounds combine with water and return to earth as acid rain and snow.

Because of this acid rain and snow, lakes have been polluted. Fish have died. Crops have been

How do you think acid rain has changed this statue?

damaged. Forests have been destroyed. Even concrete and steel buildings have been damaged.

● How do you think acid rain might be harmful to humans?

Acids and Bases

Sulfuric (suhl-FYOOR-ik) acid, H_2SO_4, is just one of the group of compounds called acids. Some are strong. Others are weak. But, all should be treated with caution.

An **acid** is a substance which gives off hydrogen ions when placed in water. Hydrogen ions are H^{+1}. In the discussion of ionic bonds it was mentioned that ions can separate in water. So, acids must have hydrogen ionically bonded to a nonmetal or a radical. In sulfuric acid the H^+ is bonded to the sulfate group SO_4^{-2}.

These are some common acids.

Sulfuric Acid	H_2SO_4
Hydrochloric Acid	HCl
Nitric Acid	HNO_3
Carbonic Acid	H_2CO_3
Boric Acid	H_3BO_3
Acetic Acid	$H_2C_4H_4O_6$

Hydrochloric (HY-dro-KLOR-ik) acid is one of the strong acids. When placed in water, almost all of its hydrogen ions are released into the water. Acetic (uh-SEET-ik) acid is a weak acid. Most of its H^+ ions stay bonded to the acetate groups. That is why acetic acid can be used in vinegar.

All acids are sour, like vinegar and the acid in lemons, citric acid. But, this property cannot be used to identify acids. **You must never taste an unknown liquid!**

Chemists have developed an easy method for identifying acids. Special chemicals change color when certain ions are present. These substances are called **indicators** (in-di-CAY-torz). **Blue litmus** (LIT-mus) is one of the indicators of an acid. It indicates the presence of H^+ ions by turning red.

Sodium hydroxide (HY-drox-yd) is one of a

group of compounds called bases. A **base** is a substance that gives off hydroxide ions when placed in water. The radical for the hydroxide group is OH^{-1}. Bases have their hydroxide group ionically bonded to a metal.

These are some common bases.

Sodium Hydroxide	$NaOH$
Potassium Hydroxide	KOH
Magnesium Hydroxide	$Mg(OH)_2$
Calcium Hydroxide	$Ca(OH)_2$
Aluminum Hydroxide	$Al(OH)_3$
Ammonium Hydroxide	NH_4OH

If a base releases a large number of OH^{-1} ions into water, it is a strong base. If few OH^{-1} ions are released, then it is a weak base.

A property of all bases is that they are slippery to the touch. They also have a bitter taste. But, remember, you must never taste or touch unknown liquids.

Once again, we can turn to the indicator shelf. This time **red litmus** is used. If red litmus paper comes in contact with a base, it turns blue.

So far, we have encountered strong acids, weak acids, strong bases, and weak bases. Now what happens when we put two of these together?

If we put a strong acid in water, it releases many H^+ ions. If we put a strong base in the same test tube, it releases many OH^- ions. The number of H^+ and OH^- ions might be the same. When this happens, it is as if neither the acid nor the base was there. A neutral condition is created. The number of H^+ ions is equal to the number of OH^- ions.

When we test the neutral liquid with blue litmus, it stays blue. If we try red litmus paper, it stays red. This perfect balancing of acid and base is called **neutralization** (NEW-trul-ih-ZAY-shun).

Suppose we add equal amounts of an unknown acid to an unknown base. If, when we test the mixed liquids with blue litmus, it turns red, what have we discovered?

The unknown acid must be a stronger acid. The unknown base must be a weaker base. The number of H^+ ions is still greater than the OH^- ions. We have not yet reached a neutral point.

So the terms **acidic** (more H^+ ions) and **basic** (more OH^- ions) are not useful when describing how acidic or how basic a substance is. The litmus is a general test. Litmus turns pink whether it is testing a very strong acid or a weak acid. A scale has been created to solve this problem of indicating strength. On this scale, an exact number (a **pH number**) is assigned to a liquid to show how acidic or basic it is. Pure water has the pH value of 7. It has equal numbers of H^+ and OH^- ions.

Any liquid with a ph of less than 7 is acidic. A

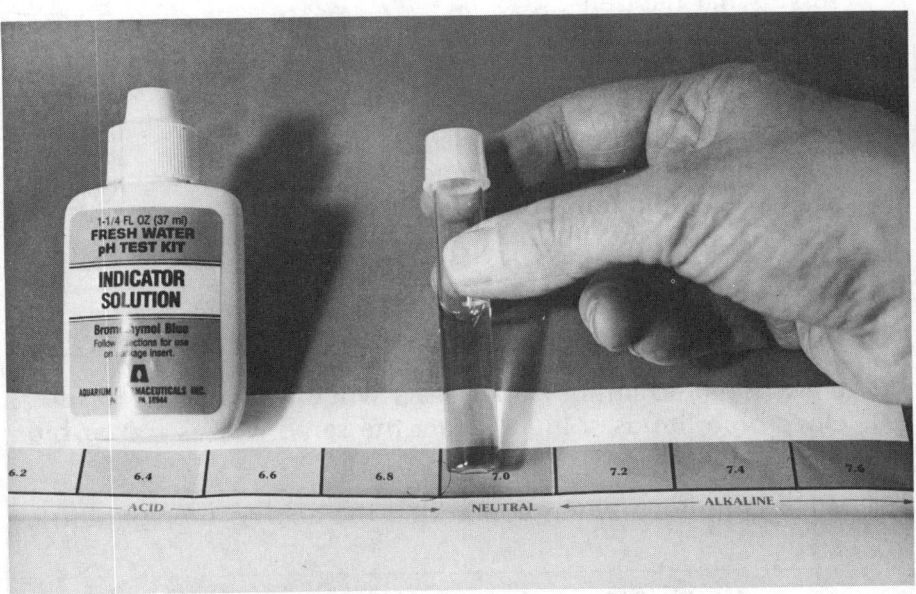

Many scientists use indicator solutions instead of pH paper. These solutions are also compared to a color chart.

pH of 1 would be a strong acid, very acidic. A pH of 4 would show a weak acid.

Any liquid with a pH of more than 7 is basic. A pH of 13 would be a strong base, very basic. A pH of 10 would be a weak base.

Many pH indicators have been developed. Most are more accurate than litmus. With pH paper, for example, we can actually match a color change to a pH number. Other indicators can add to this accuracy, to give us values from 6.5 to 3.0.

6.5 is the pH of milk. It is close to the neutral value of 7.0. The pH value of 3.0 is for soft drinks. So, both milk and soft drinks could be called acidic. But, there is really quite a difference between their pH values.

To Do Yourself   How Are Acids and Bases Identified?

You will need:

3 acids, 3 bases, 7 small beakers, blue litmus paper, red litmus paper, blue litmus solution, safety glasses

1. Put on your safety glasses.
2. Carefully pour 10mL of each acid into each of 3 beakers. Pour about 10mL of base into each of the 3 remaining beakers. **CAUTION: Use great care when working with chemicals. Pour the solutions very slowly so that they do not splash onto your skin or your clothing.**
3. Test each acid by touching a piece of blue litmus paper to the solution. Observe what happens.
4. Test each base by touching a piece of red litmus paper to the solution. Observe what happens.
5. Slowly pour about 10mL of base into the last beaker. Add some blue litmus solution to the beaker (just enough to make the liquid change color).
6. Carefully add about 15 mL of acid to the base and blue litmus solution. Observe what happens.

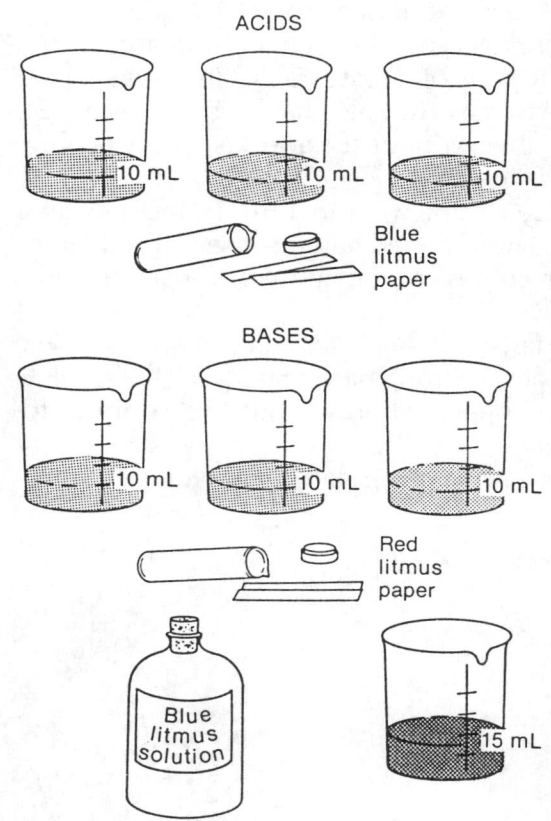

Questions

1. How does blue litmus react when it touches an acid? _____
2. How does red litmus paper react when it touches a base? _____
3. Does blue litmus solution react the same way as red or blue litmus paper?

_____

4. What happened when you added the acid to the solution in step **6?**

_____

Approximate pH Values for Some Common Substances		
	Substance	**pH Value**
Acidic Substances	lemon juice	2.3
	vinegar	2.8
	soft drinks	3.0
	milk	6.5
Neutral Substances	pure water	7.0
Basic Substances	blood	7.3
	baking soda	10.0
	ammonia	11.1
	lye	14.0

Review

I. Fill in each blank with the word that fits best. Choose from the words below.

acid   pink   O   base   blue H^+ water   7 OH^-

NaOH is a _____ . It would turn litmus

_____ . It releases many _____ ions

in _____ . A neutral liquid has a pH of

_____ . NaOH can be neutralized with a strong

_____ . It would release many _____

ions.

II. Which statement seems more likely to be true?

A. _____ A weak acid neutralized a weak base.

B. _____ Litmus is the best indicator.

III. Match the terms on the following lists:

A	B
_____ pH = 7	acid-base scale
_____ pH	strong acid
_____ pH = 1	slightly basic
_____ pH = 8	pure water

IV. Answer in sentences.

Compare the properties of lemon juice (pH = 2) and soap (pH = 10) to the general properties of acids and bases.

What Is a Salt?

Exploring Science

Salt B.C. and A.D. In ancient times, salt was used to purchase slaves. Salt was used to pay Roman Soldiers. Salt supplies were targets during battles. Salt smugglers were even sent to prison.

It is difficult to think of table salt, sodium chloride, as being a valuable substance. After all, a kilogram of salt can be bought in a supermarket for about 30¢. Tons of salt are used to dump on icy streets.

But this was not always the case. Man has used sodium chloride for at least 5,000 years. As a food preservative, it became a necessity for life. When there was no refrigeration, salt was used to preserve foods. Without salt, food would spoil.

● People exposed to high heat are often told to take salt pills. What does this tell you about the body's need for salt?

Salts

Sodium chloride is called salt around the dinner table. We don't say, "Please pass the NaCl." But the term salt can be misleading. Sodium nitrate is a salt. Potassium chloride is also a salt.

A **salt** is a compound formed from the positive ion of a base and the negative ion of an acid. Let's see how a salt is formed. We need to choose a base—NaOH. Then, we need to choose an acid—HCl.

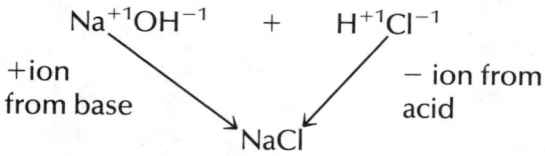

Let's choose a different base—KOH

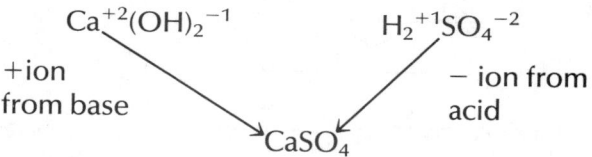

By changing the base, we form KCl, potassium chloride. Potassium chloride is a salt. Let's make a third salt. This time we'll use CaOH as our base and H_2SO_4 as our acid.

$$Ca^{+2}(OH)_2{}^{-1} \qquad H_2{}^{+1}SO_4{}^{-2}$$

+ion
from base

− ion from
acid

$CaSO_4$

Calcium sulfate is our new salt.

Many salts can be built in this way. Pick a common acid and a common base from the last lesson. Try to write a formula for your new salt. Remember to check the valences.

Imagine placing an acid and a base in the same test tube. We've already discussed this combination. These two substances can result in neutralization. Whenever we neutralize an acid with a base, we form a salt and water.

An acid + a base = a salt + water

This chemical sentence is called a **word equation**. You will learn about equations in a later unit.

The attraction between the positive ion and the negative ion holds the solid salt together. This is an ionic bond. Most salts are ionic crystals. Sodium chloride crystals, for example, are cubes. Each crystal contains the same number of sodium ions and chlorine ions. Each sodium ion is bonded to 6 chlorines and each chlorine ion is also bonded to 6 sodiums. So it always comes out even.

Salts are very different from each other. Some are bitter. Some are salty. Some dissolve easily in water. Some do not dissolve at all.

The variety of salts adds to their usefulness. Look at the content label from any food, cleanser, or medicine. You will probably find at least one salt listed.

Review

I. Fill in each blank with the word that fits best. Choose from the words below.

covalent   +   acid   salt   ionic   −   base

A salt forms from the _____ ion of a base and the

_____ ion of an acid. They attract and form an

_____ bond. If NH_4OH is the _____

and HCl is the _____ then the _____
NH_4Cl will form.

II. Which statement seems more likely to be true?

A. _____ There is only one true salt.

B. _____ A neutralization always produces a salt.

III. Next to each formula write acid, base, or salt.

A. _____ HCl   E. _____ NaCl

B. _____ NaOH   F. _____ $MgCl_2$

C. _____ H_2SO_4 G. _____ HNO_3

D. _____ $CaSO_4$ H. _____ KOH

IV. Answer in sentences.

How might the salt $MgSO_4$ be formed?

_____

_____

_____

How Are Molecules Affected By Heat and Cold

Exploring Science

A Cold Grave. For ten days they chipped away at ice and gravel. At four feet, they saw the ice-covered coffin. For 138 years, this box and its contents remained frozen. It was buried by a group of explorers in 1846.

The team, led by Owen Beattie was a research group from the University of Alberta. They had been given permission to investigate this burial site. They were also to conduct an autopsy, an examination of the body.

It was hoped that the data would help to solve the mysteries about the expedition of 1846. No one had survived, and rumors of terrible events had spread. 138 years later, families still wanted answers.

The corpse was that of Chief Petty Officer John Torrington. It was in excellent condition.

The examination showed that he had died of natural causes.

One group was very interested in the condition of the body. People in this group believed that a person's body could be frozen at the time of death. The body should be kept at super-cold, or cryogenic (KRY-uh-jen-ik), temperatures. At some later time, advances in medical science might make it possible to bring the person back of life.

Examination of Torrington's body showed that all of the bone marrow cells had been destroyed. This fact was strong evidence that the cryogenic process would not work.

● Why was the cryogenics group interested in this investigation?

This is the gravesite of Chief Petty Officer, John Torrington.

Motion of Molecules

The account of the Beattie Team is a true story. Many people believe that frozen bodies might one day be revived. Why do people think that super cold conditions can preserve human cells?

The key is in our knowledge of molecules and their movement. Many molecules in human cells are very big and complex.

Let's look at water to learn about **molecular** (mol-eh-KYOOL-ar) **motion,** or the movement of molecules. The diagram shows four H_2O molecules.

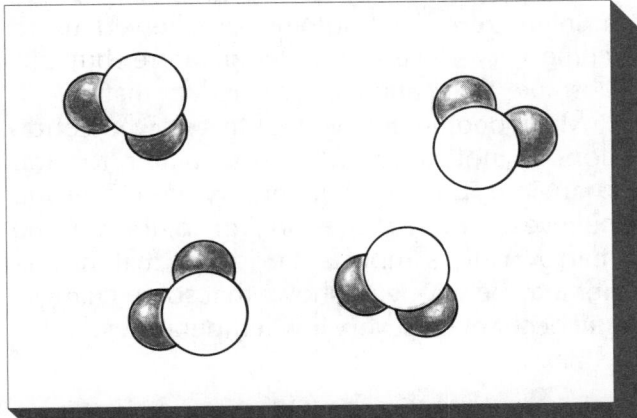

Our diagram is like a photograph. What would we see in a moving picture? That would depend on the amount of heat. If we add a lot of heat, the movement of the molecules would speed up. They would speed forward. They would speed backward. They would speed up in every direction, showing **random motion.**

The molecules would crash into one another or **collide.** The force of the collision would cause the molecules to change direction. They would collide again. If they were not contained, they would spread further and further apart. Eventually, the liquid water would become a gas. **Evaporation** (EE-vap-or-AY-shun) would take place.

Suppose we took heat away. The molecules

To Do Yourself   How Does Heating and Cooling Affect a Balloon?

You will need:

Erlenmeyer flask, balloon, ringstand, ring, bunsen burner, tongs, heat-resistant mitt, heat-resistant pad, safety glasses

1. Stretch the opening of the balloon over the opening in the flask. Check to make sure that the balloon is secure.
2. Put on your safety glasses.
3. Place the flask on the ring over the bunsen burner. Heat the flask for a few minutes. **CAUTION: Use great care when heating anything. Always wear a heat-resistant mitt when handling hot objects. Never set hot objects directly on a desk or table top. Place them on a heat-resistant pad.** Observe what happens to the balloon.
4. Turn off the flame. Using the tongs, carefully remove the flask from the heat source. Place the flask on a heat-resistant pad. Allow it to cool for a few minutes. Observe what happens.

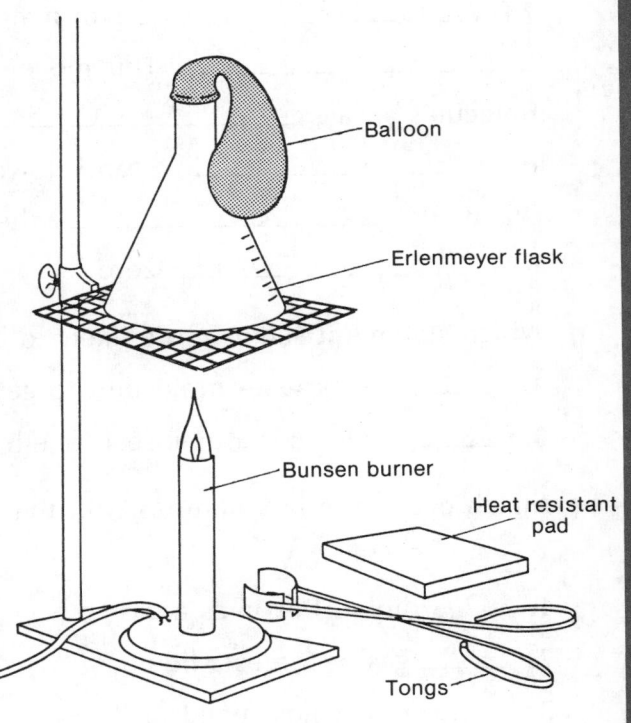

Questions

1. What happened to the balloon when it was being heated? _____

2. What happened to the balloon as it cooled? _____

3. Where is the force that made the balloon react this way coming from? _____

would slow down. They would not collide as much. The force of the collisions would be less. The molecules would come closer together. **Condensation** (con-DEN-SAY-shun) would occur. The gas would turn back to a liquid.

If we remove even more heat, the molecules are slowed even more. The collisions are fewer. The molecules become closer. The liquid becomes a solid. **Freezing** changes water to ice.

This idea of molecules moving randomly at speeds determined by heat is called the **kinetic-molecular theory** (ki-NET-ik THEE-uh-ree). **Kinetic** refers to motion. What happens when so much heat is removed that the molecules stop moving? The ideal temperature at which this can happen is −273° C. Scientists call −273°C **absolute zero.** Technology has allowed us to come very close to this temperature, but true absolute zero cannot be reached for matter.

Many people believe that this slowed condition of molecules will cause matter to stop changing. This is what the cryogenics group believes. The changes in Torrington's bone marrow cells is important evidence that this belief may be wrong. It shows that some changes still occur even at very low temperatures.

Review

I. Fill in each blank with the word that fits best. Choose from the words below.

random	collide	solid	kinetic	molecules	liquid	motion
absolute	gas					

The _____ of matter are in constant

_____ . This movement is _____ so

molecules always _____ . Motion of molecules is fastest

in a _____ and slowest in a _____ .

This is the _____ theory. Movement of molecules will stop

at _____ zero.

II. Which statement seems more likely to be true?

A. _____ Heat causes molecules to get further apart.

B. _____ Molecules are closest together in a liquid.

III. Match each term in column **A** with the term it relates to in column **B**.

A	**B**
1. _____ liquid becomes a gas	a. absolute zero
2. _____ gas becomes a liquid	b. heat added
3. _____ molecules speed up	c. heat taken away
4. _____ molecules slow down	d. condensation
5. _____ molecules stop moving	e. evaporation

IV. Answer in sentences.

If a sealed container which is filled with water is heated, what might happen to the molecules?

Review What You Know

A. Unscramble the groups of letters to form science words. Write the words in the blanks.

1. TKECIIN (theory of molecular motion) _____

2. SAOBLTEU ROZE (−273°C) _____

3. CDICIA (pH of less than 7) _____

4. CIAMOTID (molecules of paired atoms) _____

5. CTDINIAOR (chemical used to find pH) _____

6. LSAT (one product of an acid + a base) _____

7. ZIITOETUNLNAAR (way to get a pH of 7) _____

8. IOGARNC (any substance containing carbon) _____

9. PCOONUDM (chemically combined elements) _____

10. GYOCCINRE (super-freezing temperatures) _____

B. Write the word (or words) that best completes each statement.

1. _____ Two or more elements which have chemically combined form a(n)   **a.** metal   **b.** compound   **c.** indicator

2. _____ The force that holds atoms together is a(n)   **a.** element   **b.** bond   **c.** ion

3. _____ In ionic compounds, metals combine with   **a.** metals   **b.** noble gases   **c.** nonmetals

4. _____ The type of bond formed by the attraction of a positive charge and a negative charge is a(n)   **a.** ionic bond   **b.** covalent bond   **c.** triple bond

5. _____ The type of bond formed by atoms sharing electrons is a(n)   **a.** ionic bond   **b.** covalent bond   **c.** compound

6. _____ Elements always combine   **a.** at high temperatures   **b.** to give a pH of 7   **c.** in definite proportions

7. _____ Hydrogen and oxygen are   **a.** diatomic elements   **b.** noble gases   **c.** metals

8. _____ The number of electrons in the outer shell of an atom determines its   **a.** valence   **b.** pH   **c.** atomic number

9. _____ Organic compounds always contain   **a.** oxygen   **b.** hydrogen   **c.** carbon

10. _____ Compounds that release H^+ ions in water are   **a.** acids   **b.** bases   **c.** salts

11. _____ Compounds that release OH⁻ ions in water are
   a. acids **b.** bases **c.** salts

12. _____ A compound is neutral if its pH is **a.** 0 **b.** 14
   c. 7

13. _____ A base is most likely to have a pH of **a.** 0 **b.** 14
   c. 7

14. _____ A group of atoms that act as a charged unit is
   called a(n) **a.** ion **b.** polymer **c.** radical

15. _____ Absolute zero is **a.** 0°C **b.** −100°C **c.** −273°C

16. _____ The phase of matter in which molecules are
   closest together is a **a.** gas **b.** liquid **c.** solid

17. _____ One molecule of H_2SO_4 contains **a.** 3 atoms
   b. 6 atoms **c.** 7 atoms

18. _____ Neutralization of an acid and a base results in
   the formation of **a.** a salt **b.** water **c.** a salt and water

19. _____ The phase of matter in which molecules move
   fastest is a **a.** gas **b.** solid **c.** liquid

20. _____ In the formula Na_2S, the number 2 is the
   a. coefficient **b.** ion **c.** subscript

C. Apply What You Know

   1. Circle the choice that makes each statement true.

   a. An example of a chemical formula is (Sn , HF).

   b. A diatomic molecule is shown in the formula (H_2, H_2O).

   c. Na_2S is a molecule with (2 atoms, 3 atoms).

   d. A formula showing a radical is ($CaCO_3$, CO_2).

   e. The formula for a compound ending in -ide is (NaCl, H_2SO_4).

   f. K^{+1} can combine with F^{-1} to produce (KF, KF_2).

   g. Ca^{+2} can combine with S^{-2} to form (CaS, CaS_2).

   h. Al^{+3} can combine with Cl^{-1} to form (Al_3Cl, $AlCl_3$).

   i. H_2SO_4 is (an acid, a base).

   j. Two molecules of H_2SO_4 contain (14 atoms, 12 atoms).

   k. One molecule of $Ca(OH)_2$ contains (4 atoms, 5 atoms).

   l. F^{-1} shows (a radical, an ion).

2. Study the diagram. Then use the words below to fill in the blanks.

sodium fluoride   **sodium fluorine**   **Na^{+1}**   **F^{-1}**   **NaF**   **Na$_2$F$_2$**
ionic bond   **covalent bond**   **valence**   **electron**

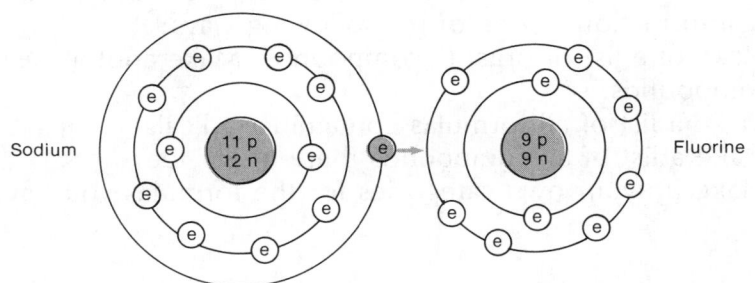

a. The _____ moves from the sodium atom to the fluorine atom.

b. the metal ion is written as _____ .

c. The nonmetal ion is written as _____ .

d. These atoms are held together by _____ bonds.

e. The name of the compound formed by these atoms is _____ .

f. The formula for this compound is _____ .

3. Study the diagram. Then use the words below to fill in the blanks.

ionic   **covalent**   **hydrogen**   **nitrogen**   **oxygen**   **NH$_3$**   **H$_2$O**
1   **6**   **8**   **16**   **18**   **2**   **shares**   **gives up**

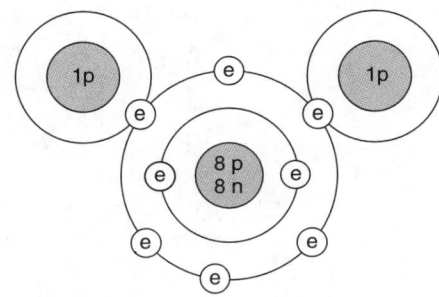

a. The small atoms in this diagram are _____ .

b. The large atom is _____ .

c. The formula for the compound is _____ .

d. These atoms are held together by a _____ bond.

e. The atomic number of the large atom is _____ .

f. The atomic mass of the large atom is _____ .

g. Each of the small atoms _____ one electron with the large atom.

D. Find Out More

1. Collect labels and copy ingredients from foods, cleaning products, and cosmetics. Find the chemical formulas for the ingredients on the labels. You may be able to get a book from your have found the formulas, organize your information in one of the following ways:

   a. Make one list of organic compounds. Make another list of inorganic compounds.

   b. Make a list of all formulas containing radicals.

   c. Make a list of all compounds that end in -ide.

   d. Make up your own categories for the formulas you have collected.

Summing Up
Review What You Have Learned So Far

A. Study the illustrations. Then write the word (or words) that best completes each statement.

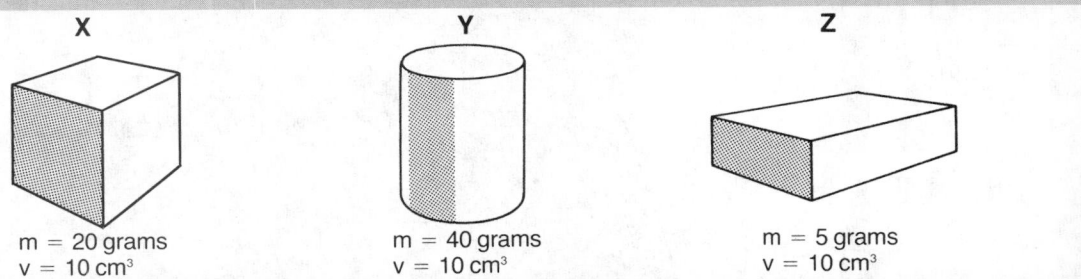

X	Y	Z
m = 20 grams	m = 40 grams	m = 5 grams
v = 10 cm³	v = 10 cm³	v = 10 cm³

1. _____ The density of object Y is **a.** 4 g/cm³ **b.** 40 g/cm³ **c.** .4 g/cm³

2. _____ The density of object X is **a.** .2 g/cm³ **b.** 20 g/cm³ **c.** 2 g/cm³

3. _____ The density of object Z is **a.** .5 g/cm³ **b.** 5 g/cm³ **c.** 50 g/cm³

4. _____ The object that will float in water is **a.** object X **b.** object Y
   c. object Z

5. _____ The object with the greatest density is **a.** object X **b.** object Y
   c. object Z

A portion of the Periodic Table of Elements is shown. Use the Periodic Table to answer the questions. Circle the underlined word or phrase that makes each statement true.

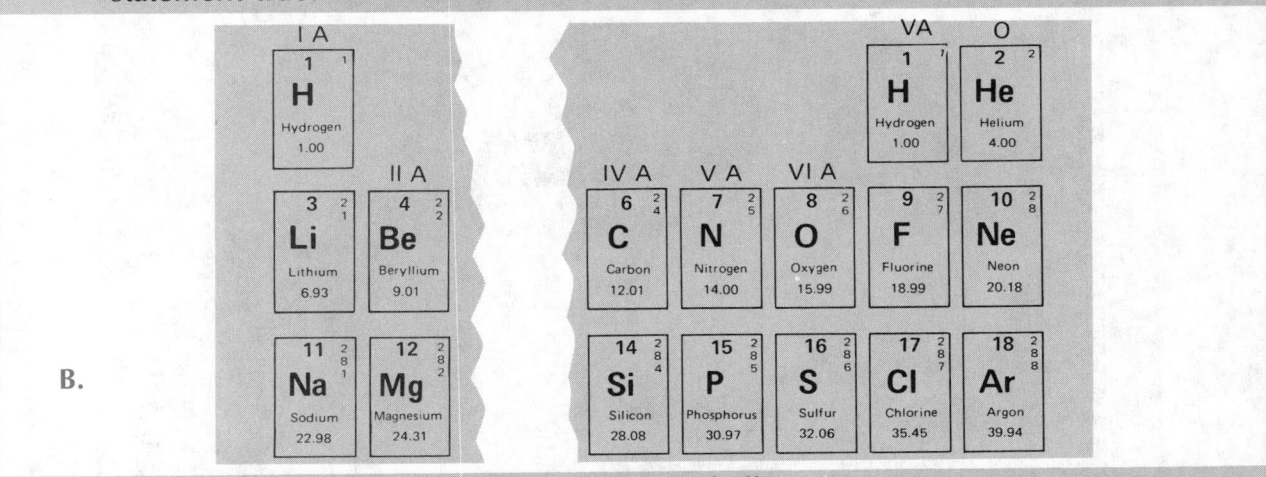

B.

1. Argon has (2/8) electrons in its valence shell.
2. The atomic mass of magnesium is (12/24.31) amu.
3. Phosphorus is in the same family as (nitrogen/sulfur).
4. The element with properties most similar to those of fluorine is (chlorine/neon).
5. All of the elements in family O are (gases/solids).
6. Sodium and lithium are (metals/nonmetals).
7. The atomic number of helium is (4/2).
8. The symbol for silicon is (S/Si).

UNIT
4

MIXTURES

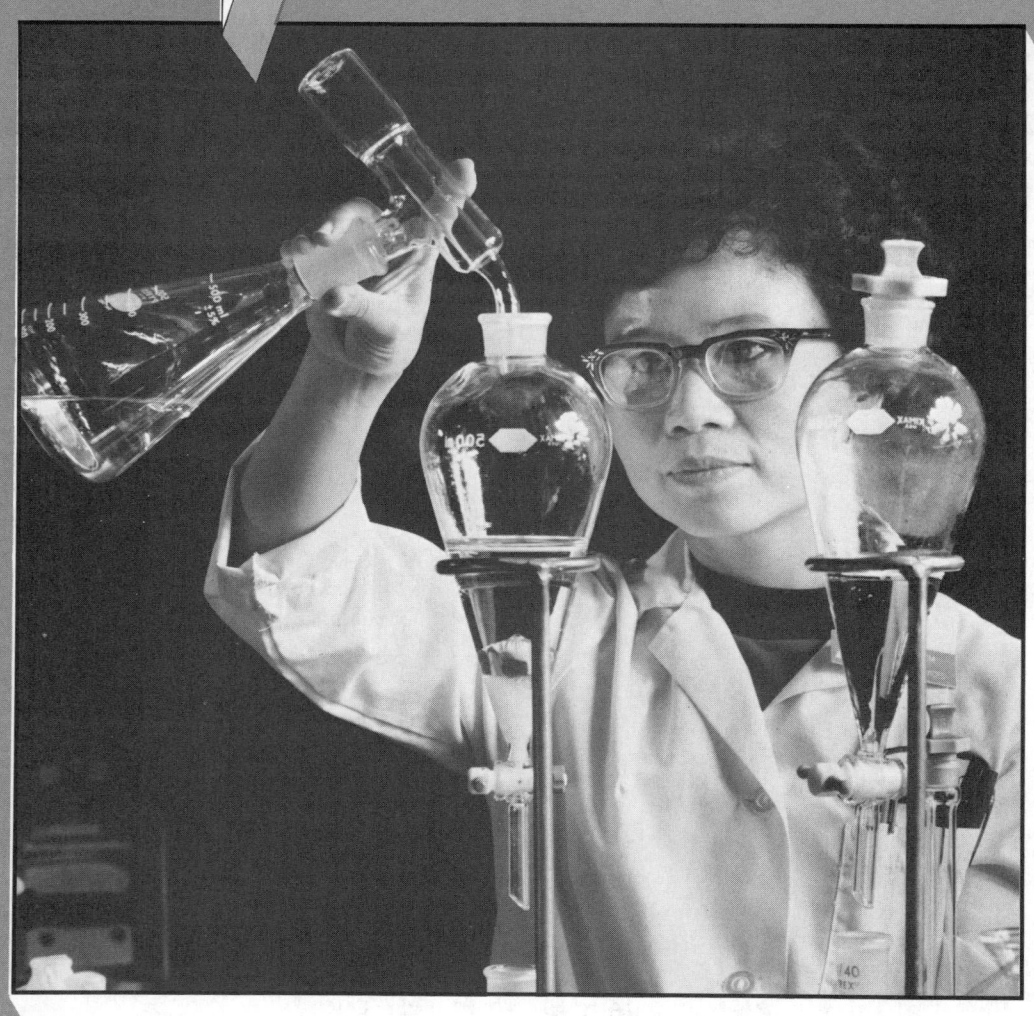

Lesson One

What Are Mixtures?

Exploring Science

New Recipes. A restaurant owner sees that dessert sales have dropped. His customers have decided that his pastries are too high in calories. He changes his recipes. The menu now claims, "All desserts made with artificial sweetner." They don't taste quite as good, but dessert sales are terrific.

A soap maker wanted to make a product that dissolves the fuzzy lint that builds up on cotton socks after several washings. They found a plant chemical that plants use to dissolve cellulose and added it to their detergent.

An airplane manufacturer knows that lighter planes require less fuel. He asks his metal supplier, Alcoa, to make a change. Scientists at Alcoa develop Alithalite (uh-LITH-uh-lyt). This new metal is a mixture of aluminum, lithium, and traces of several other metals. Alithalite is stiffer, but the aircraft is 18% lighter.

A jet engine producer reads about the political problems of Zambia and Zaire. These small countries provide cobalt to the U.S. The hard metals needed in jet manufacturing contain cobalt. The jet engine producer fears that supplies of this element may be cut off, so he seeks help.

Scientists at NASA are trying to replace the cobalt with nickel. Their new recipe metal is not strong enough. A team at Los Alamos Laboratory has invented a tungsten substitute for cobalt, but tungsten is also imported. The metal *molybdenum* (muh-LIB-duh-nuhm) is another possibility. This metal is mined in New Mexico, so it should be easier to get. The search goes on!

Changing a recipe for an industrial metal is harder than changing cake ingredients. But scientists continue working to create new metals. The rewards are worth the effort.

● Should Alithalite be listed on the Periodic Table of the elements? Explain your answer.

This aluminum can be mixed with lithium and other metals to produce alithalite.

Mixtures

Suppose that a stew is being cooked for supper. You might add an extra potato for an unexpected guest. You might add extra pepper for greater flavor. You might add flour for thicker gravy. But, the result would still be stew.

A stew is an obvious mixture. Other mixtures, however, are not as easy to identify. A **mixture** is matter containing two or more substances which are not chemically combined.

Air is a mixture. Ocean water is a mixture. Soil is also a mixture. In fact, most of the substances found in nature are mixtures. We rarely find a pure compound or a pure element.

Air is a mixture of gases. Milk is a mixture of solids and liquid. Alithalite, mentioned in the story, is a mixture of solids. A mixture of metals is called an **alloy** (AL-oy). Alithalite is an alloy. It contains aluminum, lithium, and several other metals.

When elements or compounds are mixed, they keep their own properties. Remember, this is not the case with a compound. When lithium is added to aluminum, it lends its properties to the alloy. This is how pepper adds its properties to a stew.

Let's consider the alloy called 14 karat gold.

To Do Yourself   How Are Mixtures Prepared?

You will need:

Sodium chloride, ammonium dichromate, graph paper, magnifying lens, two test tubes, stopper.

1. Place equal amounts of sodium chloride and ammonium dichromate in a test tube. Place a stopper in the test tube and shake.
2. Spread out some of the mixture on a sheet of graph paper.
3. Choose three boxes. Using the magnifying lens, count the number of sodium chloride and ammonium dichromate crystals in each box. Record your results in the table below. **Use caution when working with chemicals.**

	Box 1	Box 2	Box 3
sodium chloride			
ammonium dichromate			

Questions

1. Are the crystals in each box equal? _____

2. Is there a mixture in each box? _____

3. Does the same proportion exist in every part of the mixture? _____

Pure gold is much too soft to be made into a useful ring. Copper is more durable, but it is not as attractive. So copper is added to gold to make an alloy that is both pretty and durable. A recipe for a mixture has been invented.

Suppose we want to change the recipe. We could add another metal. We could also change the amounts of each of the metals. In this way, we change the ratio or proportion of the metals. We can do this with an alloy because a mixture has no definite ratio. There is no definite formula.

If we added more gold to the copper, the alloy would be prettier. It might be 18 karat gold. If we added more copper, we might make 10 karat gold. This alloy would be more durable.

When we work with mixtures, we are concerned with substances and their amounts. Salt water might be 10 g of NaCl and 100 mL of H_2O. Or it might be 10 g of NaCl and 1000 mL of H_2O. Both mixtures are called salt water, but they have very different tastes.

100 g of soil might contain 1 g of iron, 2 g of magnesium, and 97 g of other substances. Another sample might have 10 g of iron and 6 g of magnesium. Both mixtures are soil; but they will support the growth of very different plants.

Now the work at Los Alamos Laboratories should have more meaning. Scientists invented a recipe for a hard metal containing tungsten. Now, they are trying to replace the tungsten with molybdenum.

Look on the Periodic Table. Tungsten, W, is right under molybdenum, Mo. The two elements have similar properties. Alloys made with these elements should also be similar.

Review

I. Fill in each blank with the word that fits best. Choose from the words below.

metal   alloy   amounts   compound   properties   mixture

A _____ of copper and zinc forms bronze. Bronze is

called an _____ . Each metal keeps its

_____ . Small _____ of other metals

may be added.

II. Which statement seems more likely to be true?

A. _____ Water from a kitchen faucet is a mixture.

B. _____ There are more types of elements than mixtures.

III. Label the following substances as element, compound or mixture.

_____ air   _____ sea water

_____ sodium chloride   _____ 14 karat gold

_____ tungsten   _____ $C_6H_{12}O_6$

IV. Answer in sentences.

Describe three different alloys you might create using silver and copper.

_____

_____

_____

How Can Mixtures Be Separated?

Exploring Science

Cooperation Offers Hope. A woman and her child walk six miles along a dirt road. They are carrying large empty containers. They are looking for water. The woman and child come from a fishing village by the ocean, but the ocean is salt water. Drinking salt water will make them very sick.

The woman and child are going to a deep well. Doctors say that drinking from this well may make them sick. But, they must have water, so they take the risk.

Wealthier nations can solve this problem. They can have an expensive plant built. Salt water can be boiled. Then the water vapor can be cooled and pure water can be collected. Poor villages cannot afford this type of solution.

Students from Canada and Saudi Arabia have joined together to attack this problem. By studying salt-water fish, they have formed a plan.

Salt-water fish can take in water molecules from their surroundings and leave salt molecules behind. Scientists are hoping to build plants that can operate on the same principle. Some of these plants will be built on ships. Then the plants can move from village to village leaving supplies of pure water at every stop.

Since 80% of the world's illness is caused by poor water supplies, this is an important project. Saudi Arabia is nearly all desert, so it needs more sources of fresh water. Citizens of Canada and the U.S. are concerned about fresh water because of industrial pollution. It is possible that plants like the desalinization plant could reclaim pure water from a polluted source.

●How can a water supply become polluted?

A desalinization plant removes the salt from salt water to produce fresh water.

Separation of mixtures

In the case of seawater, scientists only wish to recover one part of a mixture—water. Sometimes, we want to separate a mixture and recover all of its parts. The method used would depend upon the mixture and our reason for separating it.

During the gold-rush days of 1849, people knew that many streams carried small gold particles. Gold was a part of that water mixture. The miners didn't care about the water. They wanted the gold. So they used fine wire mesh to screen out the tiny gold pieces.

This method, called **filtration** (fill-TRAY-shun) separates the parts of some mixtures. If the problem of the villagers had been sand in their water, filtration could have solved it. They could pour water through paper or cloth and collect only the water part of the mixture. But salt water cannot be separated in this way. Particles of salt are so small, they pass through the finest filter.

Some mixtures can be separated because of the special properties of one of its parts. Remember, the substances in a mixture keep their own properties. A mixture of pepper and iron powder, for example, could be separated by using a magnet. The iron would be attracted to the magnet, but the pepper would not.

The plant in the story uses a process called distillation. **Distillation** (dis-tuh-LAY-shun) uses evaporation and condensation to separate a mixture. When the liquid mixture is heated, the water part evaporates. The water vapor is then cooled, causing condensation. Every drop of

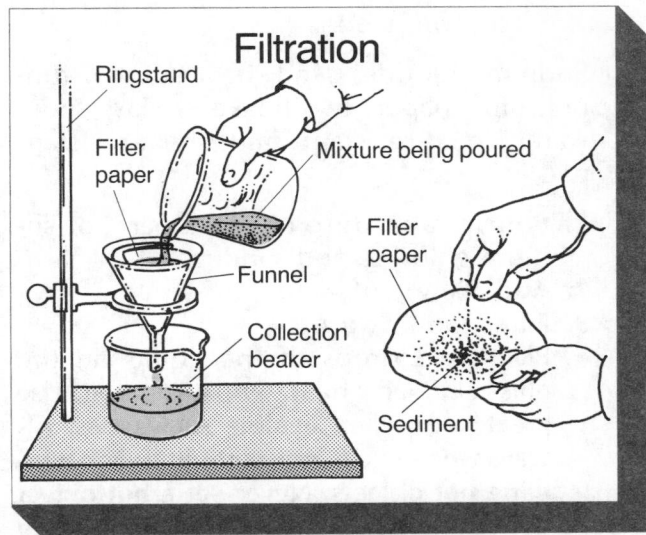

Filtration

Ringstand
Filter paper
Mixture being poured
Funnel
Filter paper
Collection beaker
Sediment

liquid which condenses is pure H_2O. Salt and pollutants are left behind, usually as a solid.

Distillation is very useful if we want to save all of the parts of a mixture. Suppose a chemist was analyzing a 50 mL liquid sample. He would not want to lose any part of it. Distillation would be a valuable procedure.

All of these techniques are called **physical** (FIH-zih-kul), or **mechanical** (muh-CAN-ik-ul), methods. If you filter gold particles from water, it's physical. If you use a magnet to pull out iron pieces, it's physical. If you evaporate water, it's physical. Mixtures can be separated using physical methods. Compounds cannot be separated using these methods.

Distillation

1. A solution is heated in a flask and the water turns to steam. Solids or liquids that have not reached their evaporation point remain in the flask.
2. The steam enters the condenser and is cooled. As it cools, it changes back to a liquid.
3. The condensed liquid comes out of the condenser and enters the receiving flask.

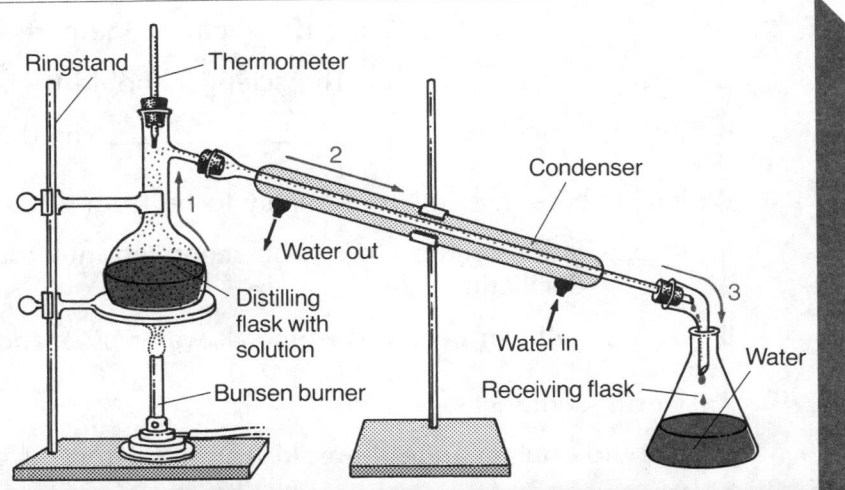

Ringstand
Thermometer
Condenser
Water out
Distilling flask with solution
Water in
Bunsen burner
Receiving flask
Water

To Do Yourself   How Are Mixtures Separated?

You will need:

Sodium chloride, sand, heat source, funnel, filter paper, two beakers, glass slide, water, heat-resistant mitt, heat-resistant pad.

1. Prepare a mixture of equal parts of sodium chloride and sand.
2. Add 100 mL of water and stir.
3. Filter the mixture.
4. Place two drops of the liquid on the glass slide and heat gently. **Caution: Use great care when heating substances. Always wear a heat-resistant mitt when handling hot objects. Never set a hot object directly on a desk or table top. Place them on a heat-resistant pad.**

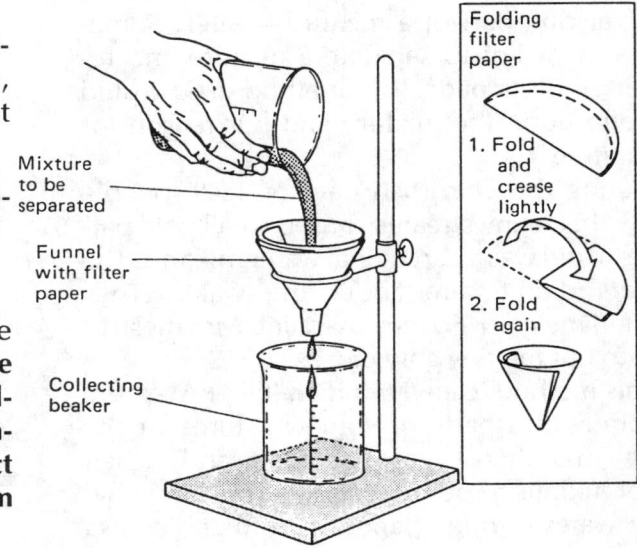

Questions

1. Did all substances pass through the filter? _____

2. What substances passed through the filter? _____

3. What happened on the glass slide? _____

Review

I. Fill in each blank with the word that fits best. Choose from the words below.

liquid   **filtration**   **distillation**   **physical**   **compound**   **filter**   **mixture**

Sugar water is poured through _____ paper. The

_____ which is collected is still sweet.

_____ did not separate the parts of the

_____ . The scientist should try _____ .

Both techniques are _____ methods.

II. Which statement seems more likely to be true?

    A. _____ Sodium chloride can be separated into sodium and chlorine by distillation.

    B. _____ Salt can be saved from salt water by evaporation.

III. Answer in sentences.

    How could you separate the gold from a mixture of gold, iron powder, and salt?

What Are Solutions?

Exploring Science

Art in Peril. For thousands of years, ground water dissolved the limestone and a cave formed. About 30,000 years ago, people entered the cave and left their hand prints. Then, during the next 10,000 years, people painted animal figures on the cave walls—cave bears, rhinoceroses with long hair, mammoths, lions, and buffalo-like bison. Where was this cave?

Three explorers, led by Jean-Marie Chauvet, found the cave and its paintings in southern France on December 18, 1994. The three became the first humans to enter the cave since the paintings were made. Today it is called Chauvet cave.

There are many such caves in southern France and Spain. The first cave paintings known to modern humans was found by a little girl exploring a cave in Spain with her father in 1879. The bulls she saw on the ceiling of Altamira cave became famous the world over. Other famous cave paintings were found in Lascaux in France in 1940.

Altamira and Lascaux became great tourist attractions, but problems arose. The colors, bright for 10,000 years or more, began to fade. Whitish crystals grew on, and over, the art.

Researchers were called in. The caves were closed while they collected data. The scientists found that each person entering the cave gives off 50 grams of water and 17 liters of CO_2 in an hour's visit. The water and gas from breathing mixed with the natural cave water. The mixture dissolves colors and limestone. Tourists were destroying the caves just by breathing.

● What plan would you suggest to preserve Chauvet cave, Altamira, and Lascaux?

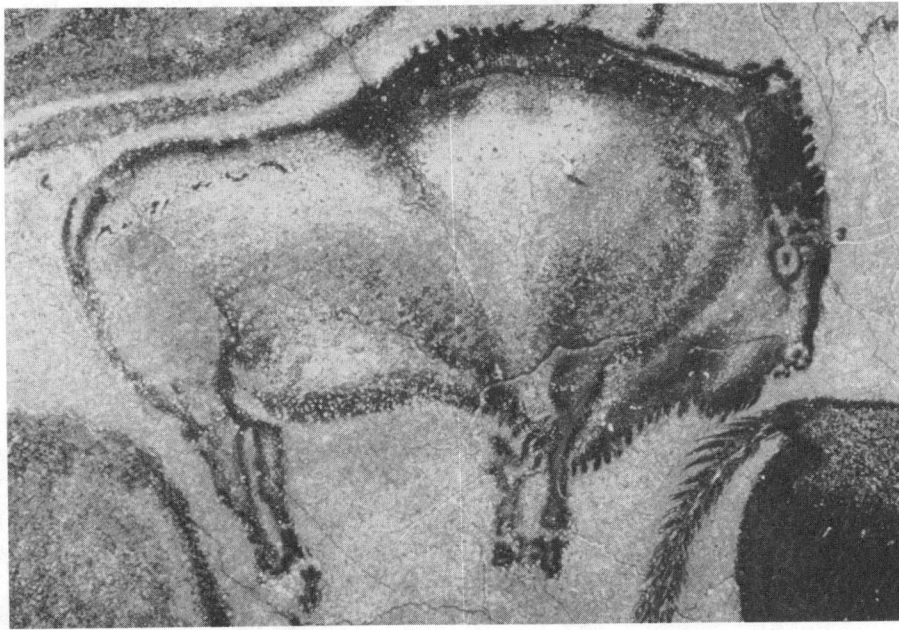

This bison was painted on the ceiling of the Altamira Cave in Spain almost 15,000 years ago.

Solution

If we put a drop of water on limestone, we would see no change. But it is possible that a change took place. Two or three molecules might have left the solid and entered the water drop. The molecules would have the formula $CaCO_3$, calcium carbonate. Limestone is calcium carbonate.

Suppose we continued to drop water on the

limestone. This is how the caves formed. Every drop of water would remove more molecules. After thousands of drops of water, the rock would seem to disappear. It dissolves into the water. The result is water containing millions of molecules of $CaCO_3$. The water became a mixture.

This is a special kind of mixture. It is called a **solution** (suh-LOO-shun). A solution is a uniform mixture. This means that every milliliter of the mixture is just like every other milliliter. Suppose we collected the drops of liquid in a test tube. The solution at the top would be exactly the same as the solution at the bottom. We could seal the test tube and let it stand for two weeks. Nothing would change. The $CaCO_3$ would not settle out. This is another property of solutions that makes them special.

If we tried to filter the solution, nothing would remain on the filter paper. If a solution is poured in, the same solution filters through.

Solutions cannot be separated by usual filtration methods.

This should remind you of the salt water problem in the last lesson. Salt water is also a solution. Remember, it could not be filtered. Salt water had to be separated by distillation.

In each of the cases discussed, water was involved. Water was the solvent. A **solvent** (sahl-VENT) does the dissolving. The substance which is dissolved is called the **solute** (sahl-YOUT). In forming the caves, the solute was calcium carbonate. It was dissolved in water, the solvent.

SOLID IN LIQUID. Each of the solutions mentioned so far involves a solid solute and a liquid solvent. Water dissolves more kinds of solids than any other liquid. In fact, we use water as a solvent every day.

LIQUID IN LIQUID. When liquids mix to form a solution, they are **miscible** (MISS-ih-bull). Alcohol and water are miscible. When liquids do not

To Do Yourself   What Are Some Properties of Solutions?

You will need:

Copper sulfate, water, four test tubes, funnel, filter paper, beaker, safety gloves

1. Dissolve a small amount of copper sulfate in 50mL of water. CAUTION: **Handle all chemicals with great care. Avoid direct contact with the skin.**
2. Pour equal amounts of the solution into four test tubes.
3. Set one test tube aside. Check it every 10 minutes to see if any of the copper sulfate settles out.
4. Filter the contents of a second test tube. Observe what happens.
5. Observe the remaining test tubes for similarities and differences.

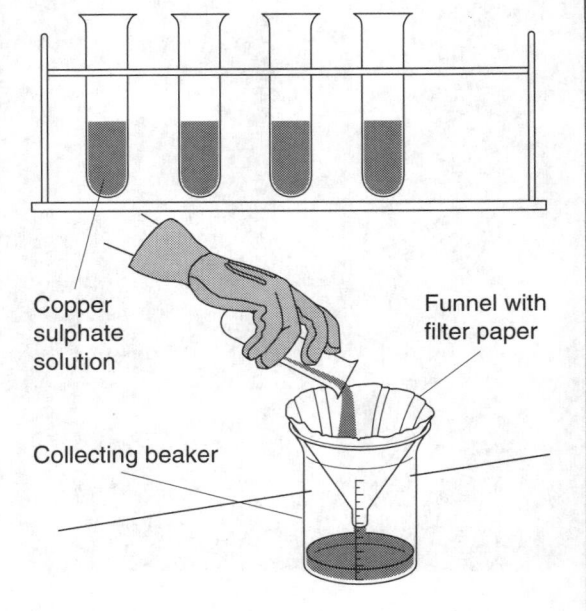

Copper sulphate solution

Funnel with filter paper

Collecting beaker

Questions

1. Did any copper sulfate settle out? _____

2. Did filtering separate the solution? _____

3. How did the remaining test tubes compare with one another? _____

mix, they are **immiscible** (ih-MISS-ih-bull). When oil is put into water, it forms a layer. The molecules do not mix. Oil and water are immiscible.

SOLID IN SOLID. Remember the mixtures called alloys. Zinc and copper are melted together to form brass. Every piece of this mixture is exactly like every other piece. It is uniform. It is a solution.

GAS IN LIQUID. Club soda is little more than carbon dioxide (CO_2) dissolved in water. You will read more about this solution later. Ocean water has many dissolved gases, such as carbon dioxide and oxygen. Fish could not live in water without dissolved oxygen.

GAS IN GAS. Oxygen and nitrogen are found in air. Air is a solution, too. It is a solution of many gases. Air is perhaps the most important solution on our planet.

Think about the air in the Altamira Caves. For thousands of years, the caves were unchanged. The air contained a steady level of gases. Then people came. If five people stayed one hour, they would add 85 liters of CO_2 to the air. The pure cave water would dissolve some of this

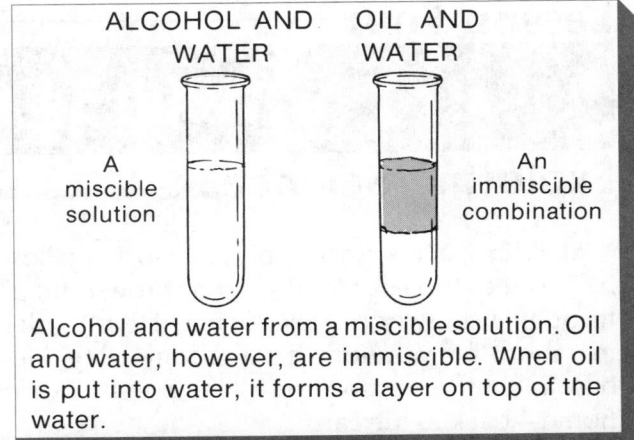

ALCOHOL AND WATER   OIL AND WATER

A miscible solution

An immiscible combination

Alcohol and water from a miscible solution. Oil and water, however, are immiscible. When oil is put into water, it forms a layer on top of the water.

extra CO_2. This new solution is what damaged the art.

Pure water did not affect the color of the paintings. But the new solution of CO_2 in water dissolved the color. Drop by drop, year after year, the molecules were dissolved. The colors faded. If action is not taken, the dissolving will continue. The walls of Altamira could look the way they did fifteen thousand years ago—before the paintings were made.

Review

I. Fill in each blank with the word that fits best. Choose from the words below.

solvent   can   mixture   clear   compound   cannot   dissolves

Copper sulfate is a blue _____ . It _____

in water. The blue liquid is _____ . It _____

be filtered to remove the solute. Water is the _____ . This

_____ is called a solution.

II. Which statement seems more likely to be true?

A. _____ All mixtures are solutions.

B. _____ All solutions are mixtures.

III. Circle the letter of the examples of solutions.

    a. milk   **c.** salt water   **e.** zinc

    b. tea   **d.** brass   **f.** air

IV. Answer in sentences.

How could you test to see if a green liquid is a solution?

Lesson Four

What Is Saturation?

Exploring Science

Made in Space. An assembly line worker takes his coffee break. He drinks a brown fluid through a straw from a plastic bag. He is thinking about his vacation. It starts tomorrow. He has a shuttle reservation—two weeks back home—back to the Earth.

Why would any company make a product in a plant in space? It would certainly be more expensive. It would also involve different technology. A gravity-free factory would be entirely new. The products produced there would also be new.

The first commercial products from space may be super-crystals. Crystals are important in our high-tech world. For example, silicon chips are cut from crystals. The microcomputers in many schools depend upon these small chips. Medical and business systems use them, too.

But there are problems. Crystals grown on Earth look perfect. They appear uniform and pure. But when the chips are cut and used, defects are found. The atoms in the crystal do not arrange themselves perfectly because of gravity.

Experiments in space offer new hope for the future. Crystals, far more complex than silicon, have been grown there. They are more perfect than Earth crystals. They are also 10 to 100 times larger.

Space crystals do not grow at the bottom of a test tube as they do on Earth. They grow in the center of the tube. The crystals float in the center of the tube. This freedom may explain why the crystals are larger and more perfect.

● Why do crystals on Earth grow in the bottom of the test tube?

SPACE CRYSTALS

EARTH CRYSTALS

Do you notice any differences between the space crystals and the Earth crystals?

Saturation

Let's explore how Earth crystals were formed before the high-tech age. Then, we can appreciate this new technology. It all begins with solutions. Remember a solution forms by dissolving a solute in a solvent.

Imagine a solution with sugar as the solute

and water as the solvent. A sample of water will dissolve just so much sugar. Then it will become saturated. A **saturated** (sa-choor-AY-tud) solution can hold no more solute. If we add one more tiny crystal of sugar, it will fall to the bottom.

Suppose the sample was 100 mL of water at 20°C. We add 10 g of sugar. It dissolves. The solution can hold more solute; it is **unsaturated.** So we add another 10 g of sugar. Again, it dissolves. In fact, we can add a total of 179 g of sugar to 100 mL of water. Then it becomes saturated.

Let us now raise the temperature to 50°C. At a higher temperature, the solution can hold more sugar. It is unsaturated again. We can add 81 g more of sugar before reaching the new saturation point.

By raising the temperature, we increased the amount of solute which can be held. This is true of most solid-in-liquid solutions. What happens to our saturated solution when the temperature drops again?

Let's return the temperature to 20°C. The 260 g of sugar stay in solution. It is **supersaturated.** It holds more solute than usual at that temperature. Now let's add one more sugar crystal. This is called a seed crystal. It drops to the bottom of the container. Slowly, it begins to grow. Sugar molecules come out of solution—one, two, three at a time. They begin building a large sugar crystal.

Even pioneers in the early West knew how to make large sugar crystals. First they made supersaturated sugar solutions. Then, a seed crystal tied to a string was placed in the solution. Sugar molecules came out of solution and combined with the seed crystal. The large sugar crystal that formed was called rock candy.

These crystals were pretty, just as laboratory crystals are pretty. But crystals formed on a string are far from perfect. Our high-tech world demands perfect crystals.

The gravity-free space process eliminates "the string." Now large perfect crystals can grow in the middle of the solution. Someday, perhaps you will be the worker who checks their growth.

To Do Yourself   How Do You Know When a Solution is Saturated?

You will need:

Test tube, beaker, water, table salt, small jar, graduated cylinder, balance

1. Place 5 g of salt in a beaker. Find the mass of 5 g of salt and the beaker together.
2. Put 10mL of water in the test tube.
3. Add the water to the beaker a little at a time. Shake vigorously. Do this until no more of the salt will dissolve.
4. Find the new mass of the beaker and salt.

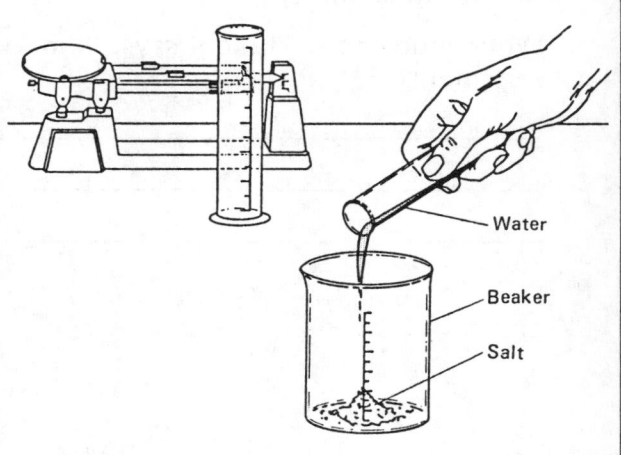

Questions

1. How much salt was needed to form a saturated solution? _____
   _____

2. How did you know the solution was saturated? _____
   _____

3. What could you do to dissolve more salt? _____
   _____

Review

I. Fill in each blank with the word that fits best. Choose from the words below.

solute   **unsaturated**   **temperature**   **saturated**   **more**   **supersaturated**

A salt solution can dissolve no more _____. It is

_____. If you raise the _____, it could

hold _____ salt. It has become _____

again. By adding salt and cooling, a _____ solution could form.

Which statement seems more likely to be true?

A. _____ Crystals can grow in a supersaturated solution.

II. B. _____ An unsaturated solution of salt can hold no more salt at that temperature.

21 g of copper sulfate ($CuSO_4$) can dissolve in 100 mL of H_2O at 20°C. Mark each of the following statements as true or false.

III. A. _____ Copper sulfate is the solvent.

B. _____ 42 g of $CuSO_4$ can dissolve in 200 mL H_2O at 20°C

C. _____ 21 g of $CuSO_4$ can dissolve in 300 mL H_2O at 20°C

D. _____ 100 mL of H_2O at 20°C is saturated with 21 g of $CuSO_4$.

Answer in sentences.

IV. When studying, a student says, "I'm saturated!" What does he mean by this statement?

_____

_____

_____

Lesson Five

What Is Solubility?

Exploring Science

Oceans at Risk. Ocean water is more than H_2O. Any Pacific surfer can tell you of the taste of salt. Any observer of sea life can tell you of dissolved oxygen. Oceanographers also study the amount of dissolved carbon dioxide, nitrate, and phosphate in our oceans.

But in the 20th century a new substance has found its way into our oceans: oil. Every year, oil spills threaten our planet. Oil cannot dissolve in water. It forms a slick on the ocean surface. It smears the coastline.

On March 24, 1989, the *Exxon Valdez* struck a reef in Alaska's Prince William Sound. About 240,000 barrels of crude oil spread out over water filled with marine life. The oil affected more than 2,600 square kilometers of coastline. The slick destroyed sea otters, fish, nesting birds, and bald eagles.

Although $4 billion was spent on clean-up, the region may never recover. The oil had to be skimmed or sponged off the water surface and the rocks. Thousands of volunteers and hired workers struggled to save as much wildlife as possible. Because of their efforts, no species has

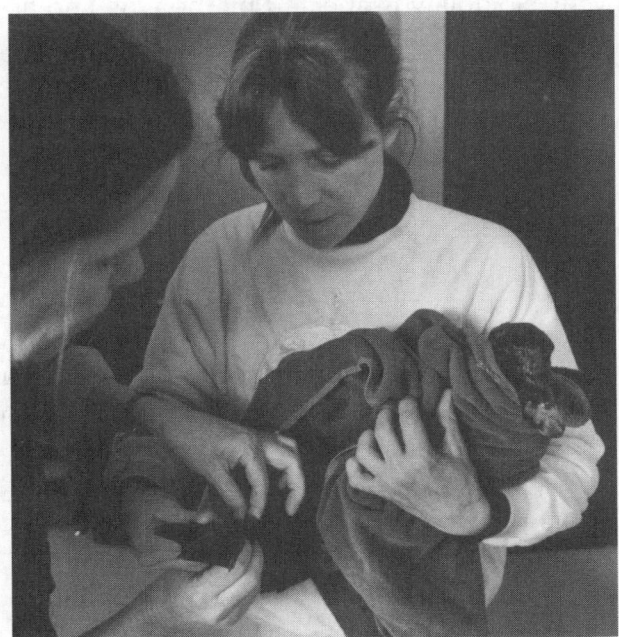

become lost completely to the Sound region, but the cost has been great.
● Oil floats on the water's surface. Why do you think that fish, who live underwater, are affected?

Solubility

The clean-up of any oil spill is a serious problem because oil is **insoluble** in water—it does not dissolve well in water.

Soluble materials are those that will dissolve in a given solvent. For example, carbon dioxide is soluble in water. Potassium phosphate is soluble in water. Both of these compounds are found in ocean water.

Sometimes, a solute is insoluble in only one solvent. For example, iodine dissolves easily in alcohol, but is does not dissolve well in water. Iodine, therefore is soluble in alcohol, but insoluble in water.

Remember, water is the best solvent. So tables showing solubility in water are helpful. Here is a portion of such a table.

s = soluble i = nearly insoluble ss = slightly soluble	carbonate	chloride	iodide	nitrate	oxide	phosphate	sulfate
barium	i	s	s	s	s	i	i
calcium	i	s	s	s	ss	i	ss
lead	i	ss	ss	s	i	i	i
potassium	s	s	s	s	s	s	s
silver	i	i	i	s	i	i	ss

Notice that i = nearly insoluble, not totally insoluble. If lead phosphate is put into water, a few molecules do enter solution. But the change is too small to observe. Thus, we usually say that lead phosphate is insoluble in water.

Scientists do not fully understand why a given substance will or will not dissolve in a given solvent. Solubility is affected by the type and arrangment of the molecules of both the solute and the solvent.

Even when a substance is soluble, there are limits. Each mL of solvent will dissolve just so much solute. When we can measure volume, this is easy to calculate. We can find out how close we are to saturation.

The volume of the oceans cannot be measured. The temperature is different at different depths. The water is moving. That is why the work of the *Knorr* is so complex. The data is very important, so the years of research are well spent.

Review

I. Fill in each blank with the word that fits best. Choose from the words below.

soluble   dissolves   more   solute   molecules   less   solvent

BaO easily _____ in water. CaO is only slightly

_____ . So, BaO is _____ soluble than

CaO. In both cases, water is the _____ . The

_____ of water are _____ effective in dis-

solving CaO.

II. Which statement seems more likely to be true?

A. _____ Most potassium compounds are soluble.

B. _____ Water can only dissolve solids.

III. By using the solubility table, label the following compounds as s, ss, or i.

A. _____ barium carbonate

B. _____ barium oxide

C. _____ potassium chloride

D. _____ silver sulfate

E. _____ lead phosphate

F. _____ calcium chloride

G. _____ lead nitrate

s = soluble i = nearly insoluble ss = slightly soluble	carbonate	chloride	iodide	nitrate	oxide	phosphate	sulfate
barium	i	s	s	s	s	i	i
calcium	i	s	s	s	ss	i	ss
lead	i	ss	ss	s	i	i	i
potassium	s	s	s	s	s	s	s
silver	i	i	i	s	i	i	ss

IV. Answer in sentences.

You have two unlabeled samples of chemicals, silver chloride and potassium chloride. How might you find which one is silver chloride?

How Can Solubility and Rate of Solution Be Affected?

Exploring Science

The Pop in Soda Pop. Rising dough makes its own bubbles. Soda pop, on the other hand, does not have natural bubbles. The bottler has to force a gas into the liquid.

The gas is CO_2. The liquid is usually flavored water. Anyone who has tasted soda after it has lost its bubbles knows that such flavored water has little appeal.

If the bottler just bubbled CO_2 through the liquid, it would quickly become saturated. The liquid can hold very little CO_2. If a bottle of this liquid were opened, there would be no hiss. There would be no bubbles.

Instead, the bottler pours water into a large sealed chamber. The temperature is between 0°C and 4°C. The pressure in the chamber is four times greater than in a regular room. The water can now dissolve more and more CO_2 as it moves through the chamber.

This carbonated water is then mixed with flavorings. It is kept under pressure until it is capped. When the temperature rises, CO_2 molecules leave the liquid. The space between the liquid and the cap fills with CO_2. But the space is small. The pressure under the cap builds to two or three times room pressure.

The CO_2 molecules rapidly leave the solution when you pop the top. The pressure is released before you hear a sound. You see the bubbles. The trapped gas molecules escape.

When the pressure in a bottle of soda or seltzer is released, it can actually blow up a balloon.

● Why might a bottle of soda explode if it's left in the trunk of a car in July?

Changing Solubility and Rate of Solution

When working with solutions, there are two things to think about. One is: *How much* solvent will dissolve? The second thing is: *How fast* will the solvent dissolve? The first question has to do with *solubility*. The second question has to do with *rate of solution*. Let's look at solubility first.

In the soda-pop example, we were dealing with a gas dissolving in a liquid. Gases are sort of special. Because their molecules are so far apart and moving so fast, gases behave differently than solids (or liquids).

Changes in pressure have a great affect on gases. So do changes in temperature. If you increase the pressure on a gas, you squeeze its molecules closer together. If you lower the temperature, gas molecules move slower.

Suppose you want to dissolve a gas in a liquid. The greater the pressure and the colder the temperature, the more gas will dissolve. This is just the opposite of what happens with a solid. Higher temperatures help increase the solubility of a solid in a liquid. (Solids are not affected much by changes in pressure.)

Let's look at copper sulfate again. Scientists have collected data on how much $CuSO_4$ can dissolve in water. At 20°C, 21 g of $CuSO_4$ saturates 100 mL of water. At 60°C, 40 g of $CuSO_4$ saturates 100 mL of water. Scientists conduct experiments and record the amounts in a table.

If you were preparing a $CuSO_4$ solution, this data would be important. You might heat the water to 60°C and dissolve 40 g of $CuSO_4$. But if you put it aside and it cools, the crystals might come out. It will be supersaturated, and your solution might be ruined.

This data is so important that scientists often graph it. Look at the graph for copper sulfate. The line on the graph shows how temperature affects solubility. The large dots show the data we used. At 20°C, 21 g of $CuSO_4$ saturates 100 mL of water. At 60°C, 40 g of $CuSO_4$ saturates 100 mL of water. By using such a graph, scientists can find the saturation amounts for any temperature. For example, how many grams of $CuSO_4$ will dissolve in 100 mL of water at 30°C?

Because every solute behaves differently, a graph is needed for each solid. Look at the graph for table salt, NaCl. The curve is different. Heat affects the solubility of NaCl less than that of $CuSO_4$. How much NaCl can dissolve at 20°C in 100 mL of water? How much can dissolve at 50°C?

The second thing to think about when working with solutions is rate of dissolving. Imagine that you have a 1-gram lump of sugar. You want to dissolve this sugar in water as quickly as pos-

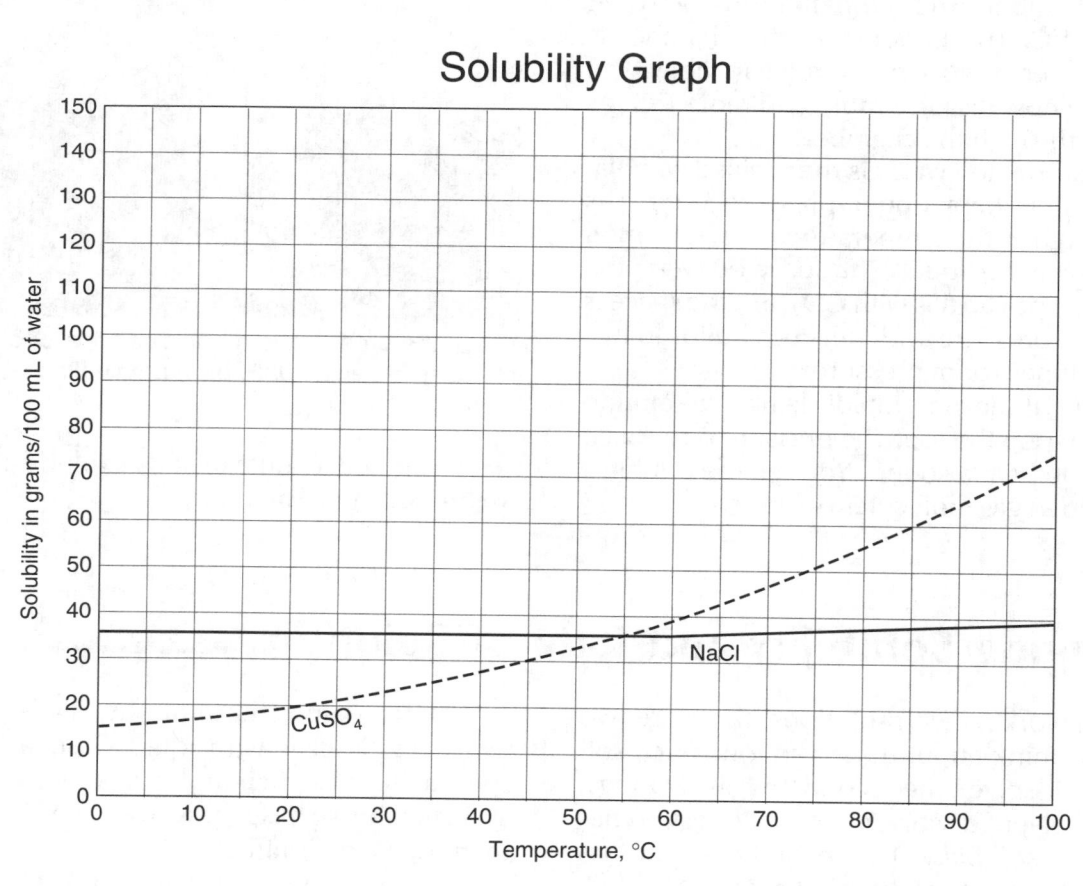

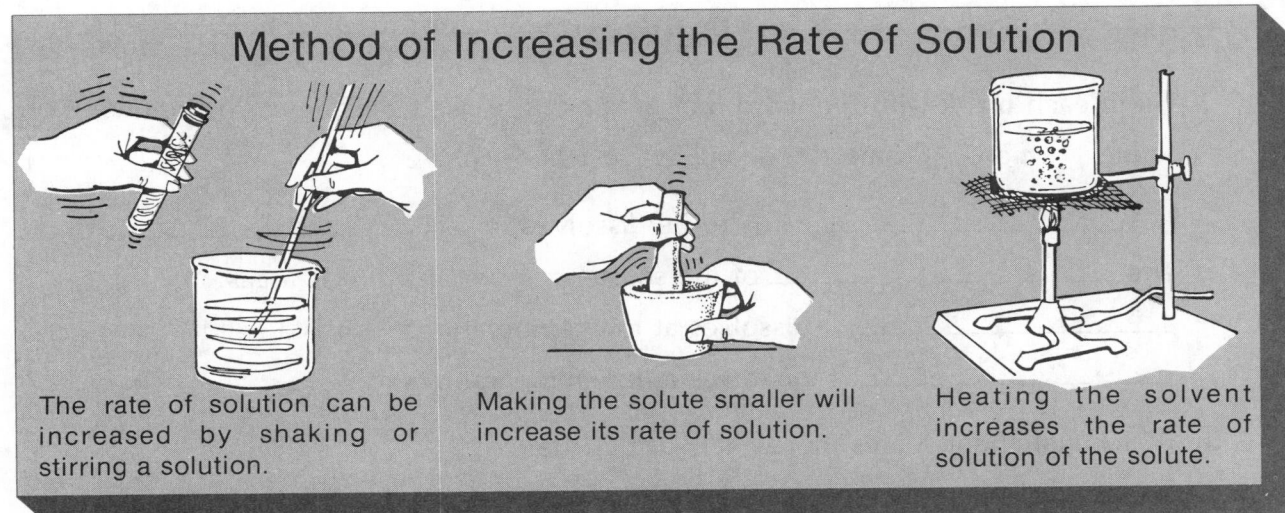

Method of Increasing the Rate of Solution

The rate of solution can be increased by shaking or stirring a solution.

Making the solute smaller will increase its rate of solution.

Heating the solvent increases the rate of solution of the solute.

sible. Can you think of any ways you could speed up this process?

First, you could grind the sugar lump into powder. This would increase the amount of sugar that is in contact with the water at one time.

Next, you could stir the mixture. You might even shake it. This would also increase the contact between the sugar and the water.

A third thing you could do would be to heat the solvent. Heating causes molecules to speed up. This would increase the number of collisions between sugar molecules and water molecules. Thus, heating will increase the rate of dissolving. In fact, all of these actions—grinding, stirring, and heating—will speed up the rate of dissolving a solid in a liquid.

To Do Yourself   How Does Heat Affect Dissolving Rate?

You will need:

Two salt tablets, two sugar cubes, two aspirin tablets, hot and cold water, two beakers.

1. Pour 100 mL of hot water into one beaker and 100 mL of cold water into the other.
2. Drop a salt tablet at the same time into each beaker. Observe carefully.
3. Repeat steps 1 and 2 for two sugar cubes.
4. Repeat steps 1 and 2 for two aspirin tablets.

Questions

1. Did each experiment give the same result? _____

2. What do you conclude? _____
3. What could you have done to make the substances dissolve even faster?

_____

Review

I. Fill in each blank with the word that fits best. Choose from the words below.

solute   **more**   **water**   **solubility**   **less**   **phase**

A _____ is to be dissolved in _____ .

The _____ of the solute is important. If it is a gas,

_____ dissolves at low temperatures. If it is a solid,

_____ dissolves at low temperatures.

II. Which statement seems more likely to be true?

A. _____ Heating is the only way to speed up the dissolving of salt in water.

B. _____ An insoluble material is still insoluble when it's powdered.

III. Use the table to fill in the data.

100 mL of H_2O at 50° C can hold _____ g of KBr.

200 mL of H_2O at 50° C can hold _____ g of KBr.

100 mL of H_2O at 70° C can hold _____ g of KBr.

100 mL of H_2O at 20° C can hold _____ g of NaCl.

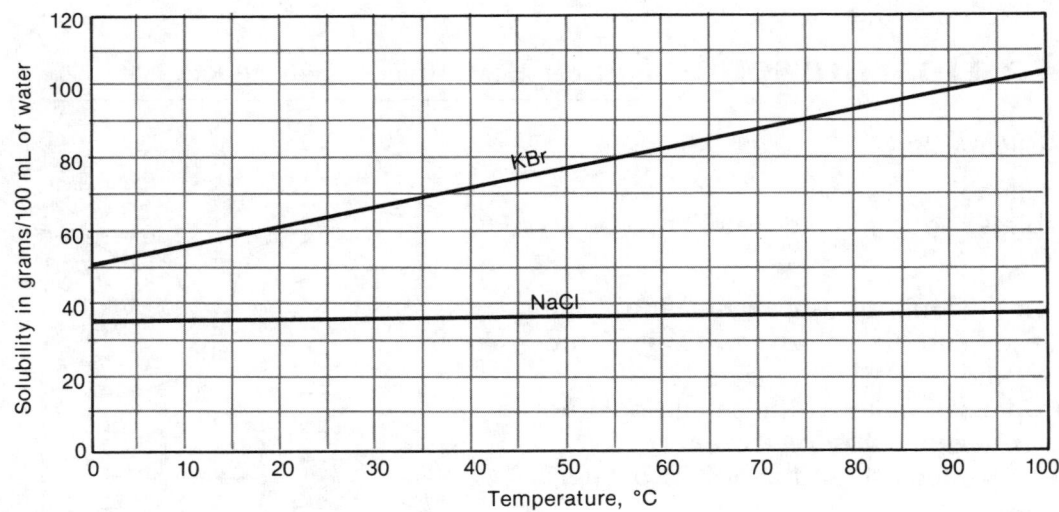

IV. Answer in sentences.

A scientist has large salt crystals which he wants to dissolve quickly in water. Suggest at least two ways to solve this problem.

_____

_____

_____

What Are Suspensions?

Exploring Science

"Shake Well Before Using!" This was the warning that appeared on many old medicine bottles. Today, collectors are paying very high prices for these containers. Many of the mixtures in these bottles had fascinating names like "Daffy's Elixir," "Dr. Porter's Healing Oil for Man and Beast," and "Cuforhedake Brane-Fude." Obviously, the manufacturer wanted this last item to translate into "cure for headache: brain food." Another substance called "Chinese Stones," claimed to cure toothache, cancer, and the bites of mad dogs and rattlesnakes.

Some of these products contained only harmless ingredients. Often, they were only slightly soluble in water. Pills were hard to manufacture. Capsules were unknown. So, the solid was ground up, by hand, and added to alcohol and water.

Sometimes as much as 30% alcohol was used. Even then, the solid ingredients still did not dissolve. Thus, the label had a warning. The patient had to shake the bottle before each dose.

Twenty-two deaths were blamed on an ingredient called acetanilid (us-ee-TAN-ihl-id). It affected patients' red blood cells. In small amounts, it was supposed to be harmless. Perhaps these were cases where a good shake could have saved lives.

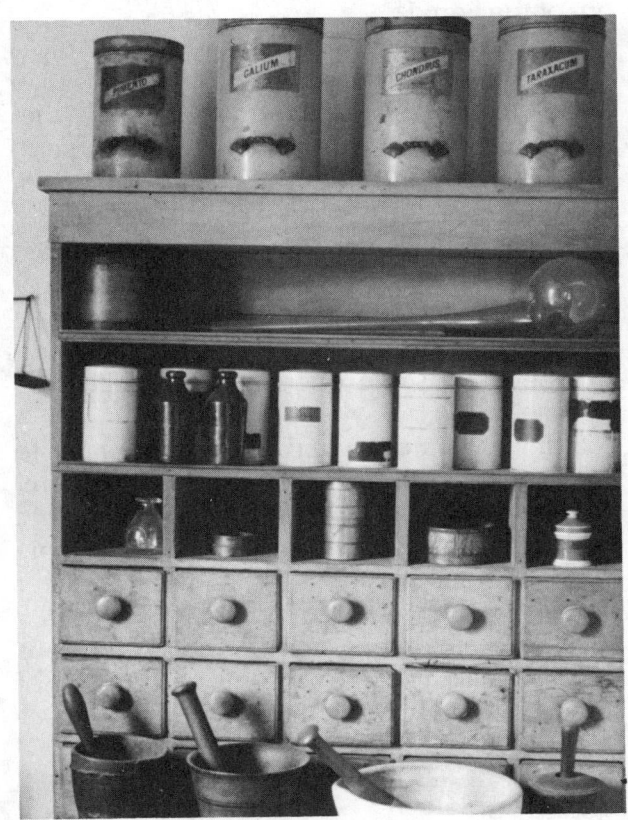

Pharmacists used mortars and pestles like those shown in the photo to grind up substances before adding them to medicines.

● Why was alcohol used in medicines?

Suspension

Imagine what happened when an ingredient was not harmless. Arsenic was frequently used in medicines. Arsenic is very dense. If a bottle was not shaken well enough, the first few doses might have little arsenic. However, by the time the patient got to the bottom of the bottle, he might drink a poisonous dose.

If all materials were soluble, there would be no problem. Remember, solutions are uniform. The top teaspoonful has the same amount of solute as the bottom teaspoonful. There is no settling out of the solute.

A **suspension** (suhs-PEN-shun) is a mixture in which the particles in a liquid are larger than molecules or ions. The particles are large enough to block light. So, suspensions are cloudy. The particles in a suspension can settle out of the liquid.

Consider a stream. As the water flows, it picks up many materials. Some compounds like $CaCO_3$, can dissolve and form a solution. Other substances, like clay, are insoluble. Clay particles are pulled into the water as a suspension. The stream water becomes cloudy because of the clay. The clay particles remain suspended because of the force of the stream. But, what happens if the stream empties into a pool?

Then, the water becomes still. The clay particles begin to settle out. The water becomes clear again.

In the laboratory, we can use filter paper to remove the particles in a suspension. Remember, the particles of solute in a solution were not filterable. But a suspension is filterable. The big pieces in a suspension are caught on the filter paper.

So far, we have only considered suspensions of solids in liquids. Let's look at suspensions of liquids in liquids. Remember, some liquids are immiscible. They will not dissolve in one another. They will not form a solution. Oil and water are immiscible. So, if you pour oil into water, it layers. The oil is less dense than the water. The top layer, therefore, is oil. Shake it! For a few moments, the oil forms a suspension in the water. Then the oil and water separate, again forming two distinct layers.

Milk, before processing, is also a suspension. The fat particles of raw milk come to the surface. This layer would be cream. During processing, however, the fat particles are made smaller. We say that the milk has been "homogenized" (hum-ah-jen-EYEZD). It has been made more uniform.

Homogenized milk is an example of an emulsion. An **emulsion** (ee-MULL-shun) is a mixture which can stay uniformly mixed for long periods of time. The particles are still larger than those in a solution. But they do not settle out of the liquid as quickly as they do in an untreated suspension.

Sometimes, a special substance is added to two immiscible liquids. The mixture of the three substances can form an emulsion. For example, detergent can be added to oil and water. The detergent causes the oil to break up into tiny droplets. The result is an emulsion. Similar processes have been developed for making medicines.

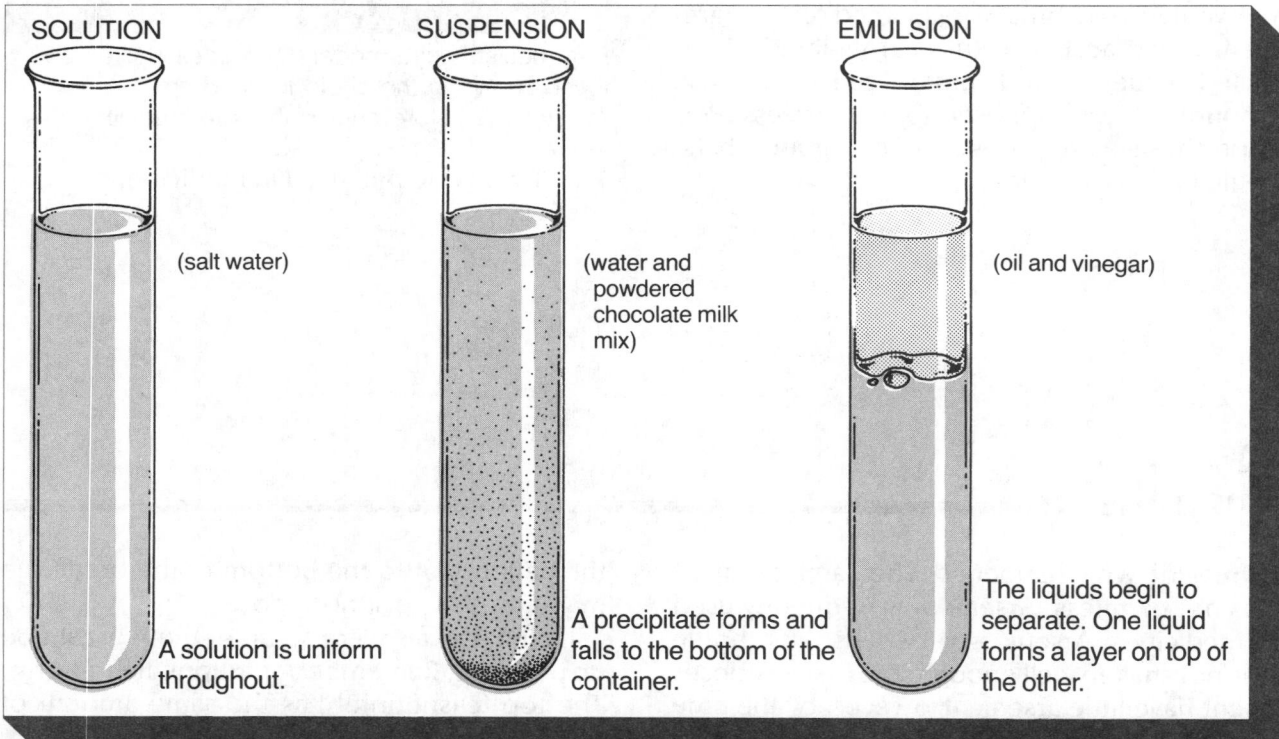

SOLUTION

(salt water)

A solution is uniform throughout.

SUSPENSION

(water and powdered chocolate milk mix)

A precipitate forms and falls to the bottom of the container.

EMULSION

(oil and vinegar)

The liquids begin to separate. One liquid forms a layer on top of the other.

To Do Yourself   How Do You Mix Two Substances That Don't Mix?

You will need:

Vinegar, salad oil, liquid detergent, three test tubes with stoppers.

1. Put small, but equal, amounts of vinegar and oil into a test tube. Shake vigorously.
2. Put a small amount of vinegar in a second test tube. Add a drop of liquid detergent and shake.
3. Put a small amount of oil in a third test tube. Add a drop of liquid detergent and shake.
4. Add a drop of liquid detergent to the first test tube and shake.

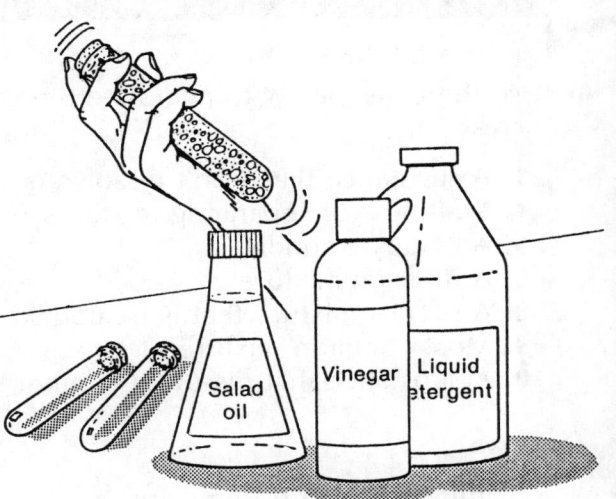

Questions

1. Did the vinegar and oil mix in step 1? _____

2. Does vinegar mix with detergent? _____Oil? _____
3. What happened when you mixed the vinegar and oil together with the liquid detergent? _____

Review

I. Fill in each blank with the word that fits best. Choose from the words below.

suspension   separate   oil   emulsion   immiscible   permanent

Vinegar and _____ are _____ liquids.

With egg yolk, they form an _____ . The result,

mayonnaise, will not _____ . It is a fairly _____

mixture.

II. Which statement seems more likely to be true?

A. _____ All suspensions are mixtures.

B. _____ All mixtures are suspensions.

III. Answer in sentences.

What might you find suspended in river water?

_____

_____

_____

Review What You Know

A. Use the clues below to complete the crossword.

Across

1. A substance that won't dissolve is _____
4. Methods for separating mixtures
5. A cloudy solution
6. A uniform mixture
8. A special mixture that is treated so it will stay mixed
9. Most common mixture of gases
10. A substance that dissolves another substance

Down

1. Liquids that do not mix are _____
2. A mixture of metals
3. Method for separating salt from salt water
4. Gases dissolve when this is increased
7. A substance that is dissolved

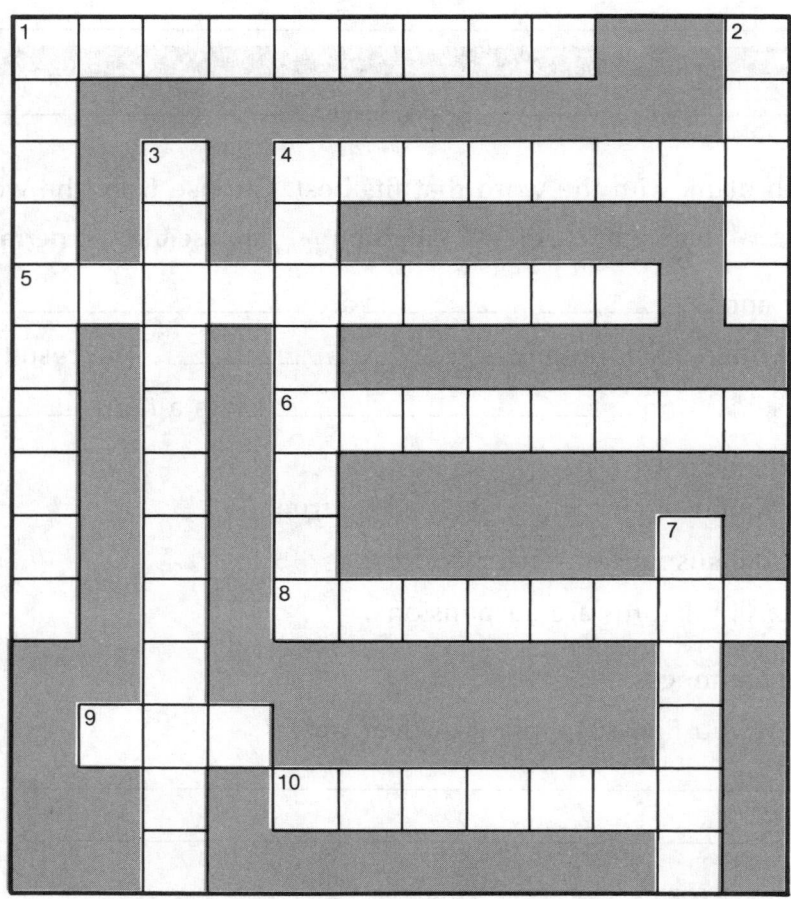

B. Write the word (or words) that best complete each statement.

1. _____ The substances in mixtures can be separated by **a.** shaking **b.** burning **c.** mechanical methods

2. _____ Liquids that mix are **a.** miscible **b.** immiscible **c.** emulsions

3. _____ The best solvent for most inorganic substances is **a.** salt **b.** water **c.** alcohol

4. _____ All solutions are **a.** liquids **b.** gases **c.** mixtures

5. _____ A substance that dissolves another substance is called a **a.** solute **b.** solution **c.** solvent

6. _____ A substance that is dissolved is called a **a.** solute **b.** solution **c.** solvent

7. _____ A solution that cannot dissolve any more solvent is **a.** saturated **b.** unsaturated **c.** supersaturated

8. _____ A substance that cannot be dissolved is **a.** soluble **b.** insoluble **c.** immiscible

9. _____ Most solutions can accept more solute if **a.** heat is added **b.** heat is removed **c.** the solution is stirred

10. _____ All mixtures are **a.** elements **b.** compounds **c.** matter

11. _____ A solution that has been forced to hold more solute than it could normally hold is **a.** unsaturated **b.** saturated **c.** supersaturated

12. _____ The gold in an alloy of gold and silver **a.** is still gold **b.** becomes gold silvide **c.** becomes a different element

13. _____ Mixtures in which the particles in a liquid are larger than molecules or ions are called **a.** suspensions **b.** miscible **c.** compounds

14. _____ A mixture that is treated to keep particles from coming out of solution is a(n) **a.** suspension **b.** emulsion **c.** alloy

C. Apply What You Know

Look at the data table below. Then write the word (or words) that best completes each statement on page 110.

Data Table   Solubility in Grams per 100 Grams of Water

	Temperature				
	0°C	20°C	50°C	80°C	100°C
Cane sugar	179.2 g	203.9 g	260.4 g	362.1 g	487.2 g
Copper sulfate	14.3 g	20.7 g	33.3 g	55.0 g	75.4 g
Sodium nitrate	73.0 g	88.0 g	114.0 g	148.0 g	180.0 g

a. The solvent in each solution is _____ .

b. 100 g of water at 20°C can hold _____ of sodium nitrate.

c. 200 g of water at 20°C is saturated with _____ of cane sugar.

d. 100 g of water at 20°C is saturated with _____ of cane sugar.

e. A sodium nitrate solution at 100°C is unsaturated if less than

_____ of solute is dissolved.

f. If 220 g of sodium nitrate are dissolved in 100 g of water at 100°C the

solution is _____ .

g. Of the three substances tested, the most soluble is

_____ .

h. Of the three substances tested, the least soluble is

_____ .

D. Find Out More

1. Make a solubility curve for copper sulfate. Draw a vertical (up-and-down) line on a piece of graph paper. Then draw a horizontal (across) line at the bottom of the graph paper. Each of these lines is called an axis.

   On the horizontal line, draw a temperature scale that ranges from 0°C to 100°C. Make a scale from 0 grams to 100 grams on the vertical axis. Then plot the points that show the solubility of copper sulfate. Use the data given in the data table in part C of this review. After all of the points are plotted, draw a curve by connecting the points with a line.

2. Make a chart of solutions, suspensions, and emulsions. Make four columns on a piece of paper. Label the columns: Solutions, Suspensions, Emulsions, and "?". Review the lessons in this unit to find examples of solutions, suspensions, and emulsions. Write each example in the correct column of your chart.

   Now look at the jars in the refrigerator and cabinets of your home. Add new examples of solutions, suspensions, and emulsions to your chart. If you do not know which category to place an item in, write it in the column marked "?". Compare your chart to the chart of a classmate. Discuss the items in the column marked "?". Decide which column of the chart these items should be placed in. When you are sure about an item, move it to the proper column.

Careers in Physical Science

Prescriptions and Doses. Does your neighborhood have a drug store? It is more proper to call this store a pharmacy (FAR-muh-see). In many towns, pharmacies are now a part of supermarkets and other large stores. The shelves of these stores are filled with many medicines and first-aid supplies.

Records about the use of medicines date back as far as 3600 B.C. Physicians (fih-ZIHSH-unz) made their own prescriptions (prih-SKRIP-shunz) back then. They invented their own medicines. Most of these medicines were made from plants. In the 16th century, physicians turned to chemical remedies. The preparation of these remedies became a full-time occupation.

Pharmacists. A pharmacist is a health care professional. Pharmacists have a great knowledge about medications and their proper usage. Some pharmacists make up prescriptions in local stores. Some work in hospitals. Others choose careers in pharmaceutical (far-muh-SOOT-ih-kul) research. These pharmacists work to develop new kinds of medicines.

All pharmacists must complete a college program. But the education of a pharmacist does not stop there. These specialists must constantly read and study information about new medications. It is very important for a pharmacist to keep up-to-date.

Medical Service Representatives. The men and women in this field are involved in sales. They help to keep doctors and pharmacists informed about new medications. Medical sales representatives visit the offices of doctors and pharmacists several times a year. These people are health information specialists.

Many medical sales representatives do not have college degrees in the sciences. They are often trained by their companies. Their training must be very thorough. The health of many people depends on their knowledge.

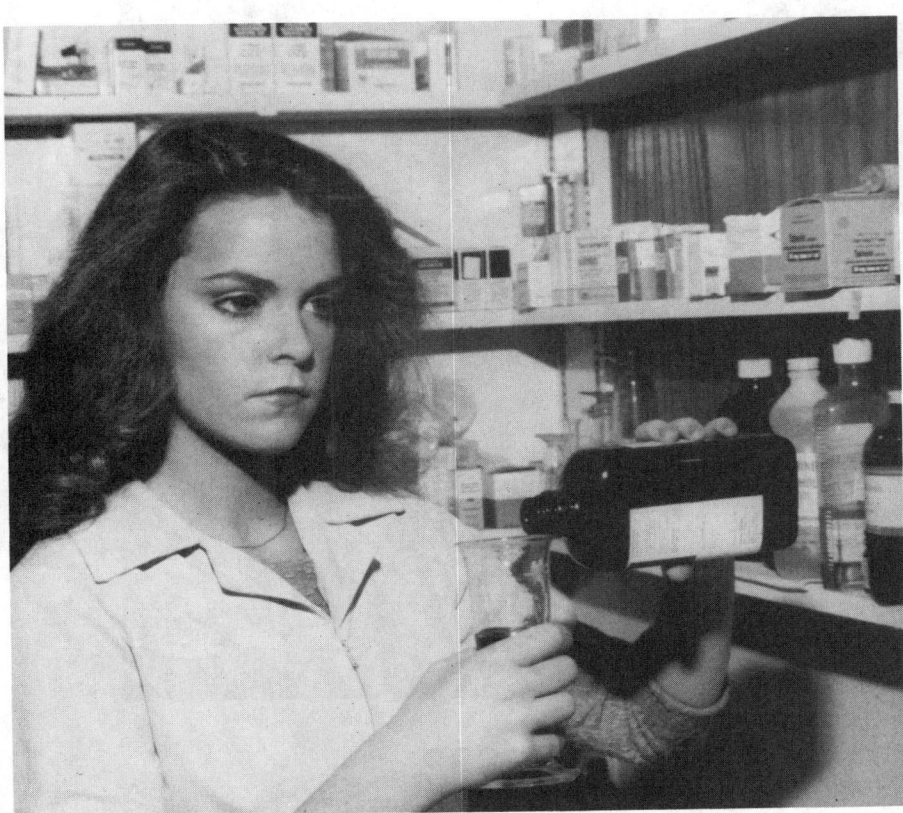

This pharmacist is preparing a prescription.

UNIT 5

CHEMICAL REACTIONS

How Can Matter Change?

Exploring Science

Science and Criminology. Suppose a hand gun is found at the scene of a crime. Or perhaps a stolen car is pulled out of layers of mud. How might police be able to find the owners of these or other objects? Police often use serial numbers to find the owners of lost or stolen property.

Very often criminals try to make such identification difficult. They try to change or destroy any serial numbers on guns or other things that they steal. Criminals can even remove molded numbers that are part of the metal.

Today, scientists can find these numbers again. Let's look at one example. A mold for a sports car engine is made in a factory. The serial number is part of the mold. Let's suppose, the mold for this engine is made with the numbers 2671. The molten metal is poured into the mold.

After cooling, the serial number is part of the engine.

Six months later, a thief steals the car. He uses a file to destroy the numbers. He works on it until no sign of the numbers remain.

There is one thing the thief does not know. When the engine cooled down in the factory, the metal under each number changed just a little.

Scientists from the F.B.I. place acid on the metal and the places that were changed by the thief disappear. Chemical activity reveals the original serial number. The number 2671 is 'visible' once more.

● Do you think a thief could change the number 2671 to the number 2611 without scientists being able to find out?

SERIAL NO. 78572389

How could this serial number be changed or destroyed?

Physical Change and Chemical Change

Much of the study of science is based on the changes we observe. Since chemistry is the study of matter, we need to study substances and observe how they change. Chemists know that there are different kinds of changes.

Some changes in matter occur quickly. These changes can be exciting to watch. Other changes occur very slowly. Some changes are so common that we hardly notice them. Others are so rare, that few people ever observe them.

Years may pass before a change is even noticed.

To a scientist, the amount of time or frequency of a change is not the most important factor. A scientist asks, "Have the molecules of matter changed?" If the molecules of matter have changed, then one or more new substances have formed. When a new substance forms, a **chemical change** has taken place.

When hydrochloric acid is dropped on zinc, hydrogen gas is released. Some of the zinc combines with chlorine. A new substance has formed. Mixing zinc and hydrochloric acid produces zinc chloride, a chemical change.

In a chemical change, the bonds between the atoms in the substance break. New bonds form. To reverse this change, another chemical change must occur. The atoms must be bonded back into their original positions.

Let's look at a molecule of water, H_2O. The two hydrogen atoms are bonded to a single oxygen atom. We can use electrical energy to break the bonds between these atoms. This causes a chemical change to occur. Two new substances, hydrogen gas and oxygen gas are produced.

To reverse this change, we have to break the new bonds. Then, we must correctly link the atoms together again. This can only be done with another chemical change.

Chemical changes always involve energy. Sometimes energy is given off, or released, during a chemical change. For example, when hydrochloric acid is dropped on zinc, the dish holding the zinc gets very warm. This chemical change releases heat. Heat is a form of energy.

In order to change a molecule, electrical energy must be added. This energy is taken in, or absorbed, by the atoms in a water molecule. They use this energy to break the bonds that hold them together.

Not all changes in matter are chemical changes. Sometimes the physical properties of

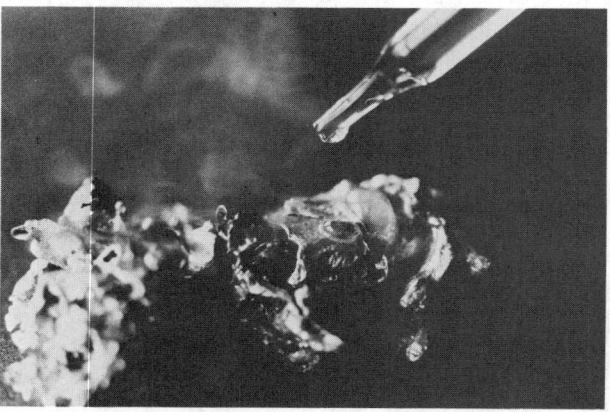
When hydrochloric acid is dropped on zinc, a rapid change occurs.

The formation of rust is a slow change.

Water boiling is a common change.

A volcanic eruption is a rare change.

a substance change. For example, solid ice can melt to form liquid water. During such a change, no molecules change. No new substances form. All the molecules remain as H_2O molecules. The substance is always water. Changes in which no new substances form are **physical changes.**

Recall, that ice melting is a change in phase. All phase changes are physical changes. For example, when water evaporates, the colorless liquid becomes a colorless vapor. There is a change, but the molecules of H_2O are still the same. No new substance has formed. So evaporation is a physical change.

When sugar dissolves in water, the sugar dis-appears. But if the liquid is allowed to evaporate sugar crystals are left behind. The sugar molecules have not changed. No new substances have formed. Dissolving, then, is another example of a physical change.

Recall the case of the stolen car. Filing off the serial number was a physical change. The reaction of the metal to the acid was a chemical change.

Generally, chemical changes are more drastic than physical changes. Therefore, chemical changes are more difficult to reverse than physical changes. For example, it is easier to refreeze a melted ice cube than it is to separate the zinc and chlorine of zinc chloride.

To Do Yourself   How Do Physical and Chemical Changes Differ?

You will need:

Two M&M halves, wooden splint, wooden-handled probe, candle, matches, safety glasses, heat-resistant mitt, heat-resistant pad

1. Place one half of an M&M between your thumb and index finger. Apply pressure for two minutes. Observe what happens.
2. Place the other half of the M&M on the end of the probe. **CAUTION: Use great care when working with sharp objects.**
3. Put on your safety glasses.
4. Hold the M&M in the candle flame until it ignites. **CAUTION: Use great care when heating substances. Always wear a heat-resistant mitt when handling hot objects. Never set hot objects directly on a desk or table top. Place them on a heat-resistant pad.**
5. Observe what happens to the M&M while it is being heated.

Questions

1. What effect did pressure have on the M&M? _____
2. What changes occured in the M&M when it burned? _____
3. Did burning the M&M cause a chemical change or a physical change? Explain your answer. _____

Review

I. Fill in each blank with the word that fits best. Choose from the words below.

physical   chemical   energy   molecules   bonds   reverse

A _____ change produces a new substance. New links or

_____ between atoms are formed. Such a change is

difficult to _____ . The melting of an ice cube is an

example of a _____ change.

II. Label the following examples as physical or chemical changes. Use the letters
P (for a physical change) and **C** (for a chemical change).

A. _____ Salt is dissolved in water.

B. _____ Serial numbers are filed from a surface.

C. _____ Sulfuric acid releases a gas when it touches magnesium.

D. _____ A piece of metal rusts.

E. _____ A marshmallow is toasted.

F. _____ An icicle melts.

III. Answer in sentences.

A new type of first aid pack is kept at room temperature. In an emergency it
can be squeezed and turns very cold. It can be used only once. Does the first
aid pack work because of a physical or a chemical change? Explain your
answer.

_____

_____

_____

How Can Elements Combine?

Exploring Science

Lead Hazard. The toddler played in his play-pen. He was bored. He amused himself with his favorite game. He used his tiny fingers to pick peeling paint from the apartment wall. He ate the sweet white flakes. Then he reached for more.

Eventually, this child may not be able to play this game. He may lose control of his hands and show other serious symptoms of illness. He might get lead poisoning.

If the toddler has lead poisoning, the ions from the paint will be in his bloodstream. These lead ions resemble ions that the body needs, such as calcium and magnesium ions. The lead ions trick the body. They take the place of these helpful calcium and magnesium ions. They become part of the body's molecules and make them useless.

Until recently, this child would have had only a 5% chance of surviving. Fortunately, doctors have a new drug for treating lead poisoning. They can give the victim an injection of EDTA (calcium disodium edetate). This drug has improved the chances of survival to 65%.

EDTA travels through the bloodstream until it finds a lead ion. It "catches" it, forms a bond with the ion, and then travels on to find another. Then the EDTA with the bonded lead ions leaves the body.

There is still no instant cure for lead poisoning. EDTA works slowly. But for the toddler a slow cure is better than a painful death.

● How do you think the body is tricked by the lead ion? (Use Periodic Table)

Bright Life
Consumer Benefits

- Many beautiful colors.
- Beautiful satin gloss finish.
- Fade, spot and stain resistant.
- Durable — Scrubbable.
- One coat covers similar colors.
- Free of lead hazards, safe for children's rooms.
- Easy to apply.

Why is it important to know if a can of paint contains lead?

Synthesis Reactions

Lead, magnesium, and calcium are all metals. Remember, metal ions have a positive charge. Lead, magnesium, and calcium all have the same valence number—two. This means that each of these atoms can give up two electrons when combining with a nonmetal.

When elements join to form a compound, a chemical change occurs. This change is called a **direct union** or **synthesis** (SIN-thih-sis) **reaction.** For example, the direct union of calcium with sulfur results in the putting together, or synthesis, of calcium sulfide.

This reaction can be written as:

calcium + sulfur → calcium sulfide

This method for writing a chemical reaction is called a **word equation.** An equation shows the **reactants** (ree-AK-tunts), or raw materials, on the left side. The **products,** or results, are shown on the right side. The arrow which separates them is read as "yields." So, calcium plus sulfur yields calcium sulfide.

Once the chemical symbols and formulas of compounds are known, a **symbolic equation** can be written. A chemist would write a symbolic equation for the formation of calcium sulfide as:

$$Ca + S \rightarrow CaS$$

Ca is the symbol for calcium. S is the symbol for sulfur. CaS is the formula for calcium (+2) sulfide (−2).

If the synthesis reaction involved lead instead of calcium the equation would be:

lead + sulfur → lead sulfide
$$Pb + S \rightarrow PbS$$

In this reaction, lead and sulfur are the reactants and lead sulfide is the product.

Suppose we wanted to make magnesium chloride. This would involve the direct union of the metal magnesium with the nonmetal, chlorine. The word equation for this reaction is:

magnesium + chlorine → magnesium chloride

We have to be careful when writing the symbolic equation for this reaction. The symbol for magnesium is Mg. But, remember that chlorine is one of the diatomic elements. In nature, chlorine is found as Cl_2. So, the left side of the equation is $Mg + Cl_2$.

Since magnesium is Mg^{+2} and chlorine is Cl^{-1}, the formula for magnesium chloride is $MgCl_2$. The equation is written:

$$Mg + Cl_2 \rightarrow MgCl_2$$

One atom of magnesium combines with one molecule (2 atoms) of chlorine. These elements are the reactants. The product is one molecule of magnesium chloride, $MgCl_2$.

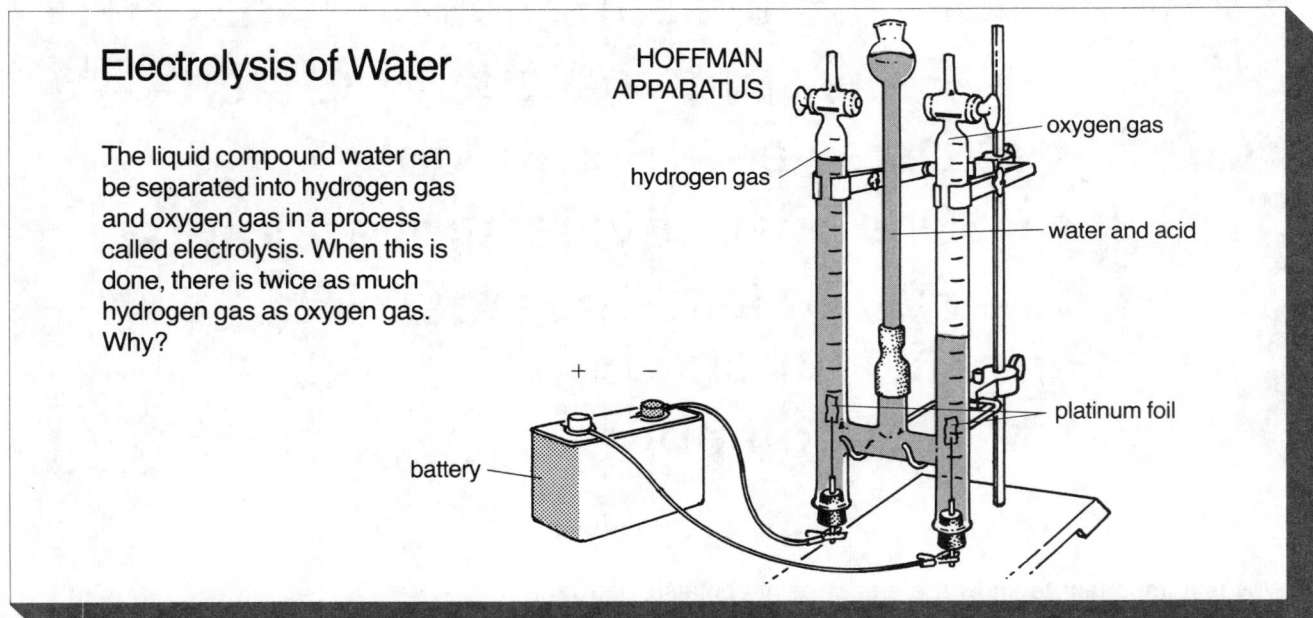

Electrolysis of Water

HOFFMAN APPARATUS

The liquid compound water can be separated into hydrogen gas and oxygen gas in a process called electrolysis. When this is done, there is twice as much hydrogen gas as oxygen gas. Why?

hydrogen gas

oxygen gas

water and acid

platinum foil

battery

+ −

In a chemical reaction, the bonds between atoms change, but atoms are *never lost*. Atoms are also *never created*. This illustrates a basic law of chemistry—the **Law of Conservation** (kon-sur-VAY-shun) **of Matter.** This law states that matter can neither be created nor destroyed, only changed.

In a synthesis reaction, atoms are bonded together to become molecules. For example, when a shiny iron fence rusts, the iron atoms are not destroyed. They bond with oxygen atoms. This direct union of iron atoms and oxygen atoms produces iron oxide. The production of iron oxide is a synthesis reaction. Rust would not form if air did not supply the oxygen.

$$Iron + Oxygen \rightarrow Iron\ Oxide$$

When invisible hydrogen gas combines with invisible oxygen gas, water forms. It may appear that matter was created, but it was only changed.

$$hydrogen + oxygen \rightarrow water$$

Writing the symbolic equation for this synthesis reaction is a little tricky. We know that hydrogen and oxygen are both diatomic molecules. The reactants, then, are H_2 and O_2. The product is water, H_2O. We *know* these formulas are correct.

Using these formulas, we get the following equation:

$$H_2 + O_2 \rightarrow H_2O$$

Now let's check to see if all the atoms that reacted are present in the product. On the left we have 2 hydrogen atoms and 2 oxygen atoms. On the right, we have 2 hydrogen atoms but only 1 oxygen atom. One oxygen atom seems to have disappeared. But, the law of conservation of matter tells us this cannot happen. What we have here is an **unbalanced equation.** The number of atoms of each element on the left do not equal the numbers of atoms of each element on the right. We have to balance the equation.

The first thing that comes to mind is to drop the subscript 2 from the oxygen reactant. But we can't do that. The formula for oxygen gas is O_2. *You can never change the formula of a substance to get a symbolic equation to balance.*

You cannot change the number of atoms in a molecule. But, you CAN change the number of molecules. Let's try using 2 molecules of hydrogen gas.

$$2H_2 + O_2 \rightarrow H_2O$$

Now we have 4 atoms of hydrogen and 2 atoms of oxygen on the left. We have 2 atoms of hydrogen and 1 atom of oxygen on the right. The equation is still not balanced. Let's increase the number of water molecules formed.

$$2H_2 + O_2 \rightarrow 2H_2O$$

Now we have 4 hydrogen atoms and 2 oxygen atoms on the left. We have 4 hydrogen atoms and 2 oxygen atoms on the right. The equation is balanced!

What this equation tells us is that in a synthesis reaction of hydrogen gas and oxygen gas, TWO molecules of hydrogen gas combine with ONE molecule of oxygen gas to produce TWO molecules of water.

The numbers written before the formulas in an equation are called coefficients. **Coefficients** (koh-uh-FISH-unts) show the number of particles (molecules or atoms) that take part in a reaction. The coefficient 1 is always understood. It is never written.

Look at this equation for a synthesis reaction:

$$H_2 + Cl_2 \rightarrow HCl$$

First check to see if the equation is balanced. There are 2 hydrogen atoms and 2 chlorine atoms on the left. There is 1 hydrogen atom and 1 chlorine atom on the right. The equation is not balanced. Let's try using 2 molecules of HCl:

$$H_2 + Cl_2 \rightarrow 2HCl$$

Now the number of atoms of each element is the same on both sides of the equation. The equation is balanced.

Now read the equation. One molecule of diatomic hydrogen plus one molecule of diatomic chlorine yields two molecules of hydrogen chloride.

I. Fill in each blank with the word that fits best. Choose from the words below.

**bond   yields   unbalanced   synthesis   metal   nonmetal
equation   balanced**

$Zn + Cl_2 \rightarrow ZnCl_2$ is an _____ . It shows the

_____ of a compound. A _____ and

a nonmetal have formed a _____ . The equation is

_____ .

II. Which statement seems more likely to be true?

A. _____ Atoms can never be destroyed.

B. _____ Molecules can be changed.

III. Place an **X** next to the equations which are balanced.

A. _____ $2Na + S \rightarrow Na_2S$

B. _____ $K + S \rightarrow K_2S$

C. _____ $Ag + Cl_2 \rightarrow AgCl$

D. _____ $Al + Cl_2 \rightarrow AlCl_3$

E. _____ $H_2 + S \rightarrow H_2S$

F. _____ $2Al + 3S \rightarrow Al_2S_3$

IV. Answer in sentences.

Why do you need 40 grams of calcium and 32 grams of sulfur to make 72 grams of calcium sulfide?

_____

_____

_____

Lesson Three

How Can Compounds Change Chemically?

Exploring Science

From Fat to Thin. Fad diets promise quick weight loss. Many teens who dream of a new figure have a new hope. They buy new fad diet "drinks." They imagine that their fat molecules will disappear.

Instead, they begin to feel depressed. They don't even want to see their friends. They are dizzy and tired most of the time. Finally, they seek help.

The people who sell the diet drinks gain from such misery. They earn more than 10 billion dollars a year. They promise shortcuts to a thin body. But what can actually happen?

Diet "drinks" often contain complex compounds. The body can turn these compounds into ketones (KEY-tones). Ketones are **toxic,** or poisonous. Ketones can make a person feel tired and dizzy.

How can a person become thin safely? Research shows that exercise and three small, nutritionally balanced meals a day are still the best answer. With good eating habits and proper exercise, chubby teens shape up and stay healthy!

Too much dieting can be very dangerous. Singer, Karen Carpenter died at age 32 from complications caused by over-dieting.

● What errors did the dieter make?

Decomposition Reactions

If dieters only knew the Law of Conservation of Matter, they would know that molecules cannot disappear. But, how can molecules chemically change?

A compound can be forced to separate into its elements. In order to do this, the bonds between atoms have to be broken. When this happens a **decomposition** (dee-KOM-puh-ZISH-un) **reaction** takes place. Decomposition, or taking apart, is the opposite of synthesis, or putting together.

Suppose a scientist needed the element mercury, Hg. He might heat mercuric (mer-KYOOR-ik) oxide, HgO. Heating mercuric oxide breaks the bonds that hold the compound together.

Once these bonds are broken, the elements mercury and oxygen are separated. The scientist can then collect the mercury.

The symbolic equation for this decomposition reaction would look like this:

$$\text{mercuric oxide} \xrightarrow{\triangle} \text{mercury} + \text{oxygen}$$
$$\text{HgO} \xrightarrow{\triangle} \text{Hg} + \text{O}_2 \text{ (unbalanced)}$$
$$2\text{HgO} \xrightarrow{\triangle} 2\text{Hg} + \text{O}_2 \text{ (balanced)}$$

The triangle written above the arrow is a symbol for heat. Heat energy was used to break the bonds. The mercury ions, Hg^{+2} were forced to separate from the oxygen ions, O^{-2}.

Sometimes heat will not cause decomposition. Suppose we are trying to decompose water. Heating would only make it change its phase. The water would be changed to steam. The liquid H_2O would become the gas H_2O. This is a physical change.

But, if we used electricity, the bonds between the hydrogen and oxygen atoms would break. Let's look at the decomposition equation for this process.

water $\xrightarrow{\text{electricity}}$ hydrogen + oxygen

H_2O $\xrightarrow{\text{electricity}}$ $H_2 + O_2$ (unbalanced)

$2H_2O$ $\xrightarrow{\text{electricity}}$ $2H_2 + O_2$ (balanced)

Electricity is needed for this reaction. So, the word "electricity" is written above the arrow.

Sometimes, compounds do not break apart completely. This happens with molecules involving a radical. Remember, these compounds contain more than a metal and a nonmetal. Let's look at the decomposition of calcium carbonate.

calcium carbonate $\xrightarrow{\triangle}$
calcium oxide + carbon dioxide

$Ca\,CO_3 \xrightarrow{\triangle} CaO + CO_2$ (balanced)

In this decomposition, one compound, $CaCO_3$, broke down to form two simpler compounds, CaO and CO_2.

Review

I. Fill in each blank with the word that fits best. Choose from the words below.

synthesis   **chlorine**   **unbalanced**   **decomposition**   **reaction**
chloride   **balanced**

An equation always describes a _____ . Hydrogen

chloride → hydrogen + _____ . This is an example of a

_____ reaction. $HCl \rightarrow H_2 + Cl_2$ is

a(n) _____ equation. The opposite of decomposition

is _____ .

II. Which statement seems more likely to be true?

   A. _____ The reactant in a decomposition reaction is always a compound.

   B. _____ The products in a decomposition reaction are always elements.

III. Place an **S** before synthesis reactions. Place a **D** before decomposition reactions.

_____ $H_2S \rightarrow H_2 + S$

_____ $H_2 + Cl_2 \rightarrow 2HCl$

_____ $2KClO_3 \rightarrow 2KCl + 3O_2$

_____ $C_6H_{12}O_6 \rightarrow 6CO_2 + 6H_2O$

_____ $2Na + Cl_2 \rightarrow 2NaCl$

IV. Answer in sentences.

Why is decomposition different from destruction?

How Can Elements React With Compounds?

Exploring Science

She's a Pretty Lady Again! At the beginning of the twentieth century, many people came to the United States from Europe. They traveled across the vast ocean by ship. The trip was long, and not very pleasant.

Then, the Statue of Liberty came into view. She stood tall and strong. Her torch was held high. The Statue of Liberty welcomed the weary travelers. She became a symbol of freedom and the beginning of a new life, bringing to an end their long, hard journey.

By 1980, this view had changed. As "The Lady" neared her 100th birthday, she was no longer perfect. She became weaker and weaker. Her copper skin was affected by acid rain. Her iron strappings had reacted with the copper skin they supported. In some places, only half the thickness of the iron remained.

Millions of Americans gave money to restore the Statue of Liberty to her original beauty. Damaged parts, like her torch, were rebuilt. The iron supports were replaced. A new metal replaced the iron in the strappings. This new metal would not react with "The Lady's" copper skin.

The Statue of Liberty is strong and sturdy again. The American spirit has made "The Lady" a perfect symbol once more.

● What was one problem which faced chemists on this project?

In 1985 and 1986, the Statue of Liberty was restored to her original beauty.

Single Replacement Reactions

Early scientists observed the physical properties of metals. They found that lead is soft and malleable. Silver conducts heat well. Gold has a high luster.

Later, scientists studied the chemical properties of metals. They needed to find out how metals will react with different compounds. The problem with the Statue of Liberty is just one example of this need.

Iron reacts with copper compounds. It can push copper out of a molecule and replace it. This is an example of a single replacement reaction. In a **single replacement reaction,** an element takes the place of another element in a compound.

Consider a reaction between the element calcium, Ca, and the compound zinc oxide, ZnO. The equation for this reaction would look like this.

$$\text{calcium} + \text{zinc oxide} \rightarrow \text{calcium oxide} + \text{zinc}$$
$$Ca + ZnO \rightarrow CaO + Zn \text{ (balanced)}$$

In this reaction, the calcium forces the zinc out of the compound and combines with the oxygen.

If we try to reverse the reaction, nothing happens.

$$\text{zinc} + \text{calcium oxide} \rightarrow \text{no reaction!}$$
$$Zn + CaO \rightarrow NR$$

In this case, there is no chemical change. Zinc is not active enough to push the calcium out of the compound.

Chemists can predict what will happen in single replacement reactions. They have made a list of metals and placed them in order of their chemical activity. The list is called the **Electromotive** (ih-lek-truh-MOH-tiv) **Series of Metals.**

Look at the Electromotive Series of Metals. Find zinc on the list. Now find calcium. Calcium is higher on the list. It is a more active metal. Calcium can replace zinc, but zinc cannot re-

To Do Yourself — Can One Metal Replace Another in a Compound?

You will need:

Iron nail, mossy zinc, copper strip, copper sulfate solution, silver nitrate solution, three test tubes

1. Half-fill two of the test tubes with copper sulfate solution. Half-fill the other test tube with silver nitrate solution. **CAUTION: Use great care when working with chemicals. Pour these solutions very slowly so that they do not splash onto your skin or your clothing.**
2. Place an iron nail in copper sulfate.
3. Place a piece of zinc in copper sulfate.
4. Place a copper strip in silver nitrate.
5. Observe each test tube for a few minutes.

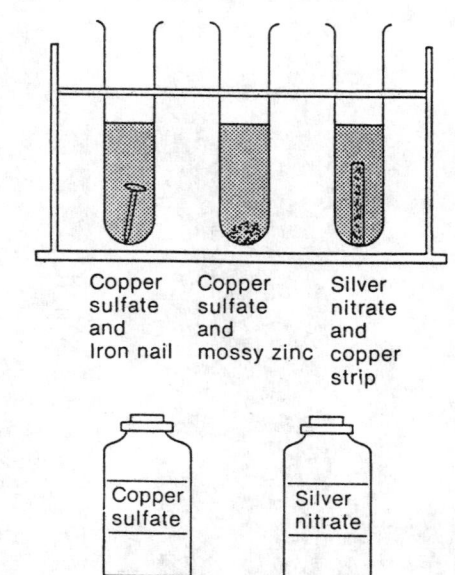

Copper sulfate and Iron nail   Copper sulfate and mossy zinc   Silver nitrate and copper strip

Copper sulfate

Silver nitrate

Questions

1. What metal formed on the iron and the zinc? _____

2. What metal formed on the copper strip? _____

3. Where did these new metals come from? _____

124

place calcium. The Electromotive Series makes predicting single replacement reactions easier.

Many single replacement reactions involve hydrogen. Find hydrogen on the Electromotive Series. It is low on the list. This means that many metals can push hydrogen out of a compound.

ELECTROMOTIVE SERIES OF METALS

Element		Symbol
lithium	**most active**	Li
potassium		K
barium		Ba
calcium		Ca
sodium		Na
magnesium		Mg
aluminum		Al
zinc		Zn
iron		Fe
tin		Sn
lead		Pb
hydrogen		H
copper		Cu
silver		Ag
platinum		Pt
gold	**least active**	Au

Remember, the compounds in which hydrogen acts as a metal (H^{+1}) are acids. Find zinc and hydrogen on the list. Let's look at a reaction involving zinc and hydrogen.

$$Zn + H_2SO_4 \rightarrow ZnSO_4 + H_2 \text{ (balanced)}$$

Zinc was active enough to push hydrogen out of the compound.

Let's place sodium, Na, in hydrochloric acid, HCl.

sodium + hydrochloric acid →
        sodium chloride + hydrogen

$$Na + HCl \rightarrow NaCl + H_2 \text{ (unbalanced)}$$

$$2Na + 2HCl \rightarrow 2NaCl + H_2 \text{ (balanced)}$$

This reaction is very violent. Sodium is so active, it replaces hydrogen easily.

If we put copper, Cu, in hydrochloric acid, HCl, we observe no chemical change.

copper + hydrochloric acid → No Reaction!
$$Cu + HCl \rightarrow \text{No Reaction}$$

This is no surprise. Copper is lower than hydrogen in the Electromotive Series.

Review

I. Fill in each blank with the word that fits best. Choose from the words below.

replacement   decomposition   active   element   compound   nonmetal
        zinc   lithium

Calcium can replace _____ in a _____.

Calcium is more _____ . This type of reaction is called a

_____ reaction. Calcium bonds with the

_____ . This reaction leaves zinc as a(n) _____ .

II. Which statement seems more likely to be true?

    A. _____ Copper can replace silver.   **B.** _____ Silver can replace copper.

III. Label each reaction type. Use the letters **S** (synthesis), **D** (decomposition), and **SR** (single replacement).

    _____ $Cu + S \rightarrow CuS$ _____ $2HCl \rightarrow H_2 + Cl_2$

    _____ $Cu + AgNO_3 \rightarrow CuNO_3 + Ag$ _____ $CaCO_3 \rightarrow CO_2 + CaO$

    _____ $Fe + CuSO_4 \rightarrow FeSO_4 + Cu$

How Can Compounds React?

Exploring Science

Compound Magic. The clown approached the group of birthday party guests. This was his first job. He had tried his best tricks already, but the kids had not been thrilled. The containers he had hidden in his costume were his last hope.

He filled a clear plastic cup with water and stirred in some salt from a shaker. The kids watched closely. He turned his back and added a few drops of liquid. He faced the kids again. "It's milk," they squealed.

Next, he took a bar of soap from his pocket. He dunked it into another glass of water ten times. He then added a few drops of a second liquid. The "water" became bright pink!

The clown was a big hit. He left the party with cheers and two new jobs.

● What new properties might have made the kids think that the water had turned to milk?

After the clown added the salt to the water, the liquid in the cup turned white. What may have caused this change?

Double Replacement Reactions

The "tricks" performed by the clown involved chemical compounds. The "Pink" trick involved an indicator. Soap made the solution basic. The pH was probably about 10. The indicator used by the clown turned bright pink at such a high pH.

You may have performed the "Milk" trick in Lesson Two. This chemical change uses two compounds as reactants. It is a double replacement reaction.

In **double replacement reactions,** the elements from two compounds change place with one another. The products are two new compounds. Let's look more closely at the "Milk" trick. The reactants were silver nitrate and sodium chloride. In this reaction, the silver and the sodium switched positions. Let's look at the equation for this reaction.

sodium chloride + silver nitrate →
    silver chloride + sodium nitrate

$$NaCl + AgNO_3 \rightarrow AgCl + NaNO_3$$

As the silver and the sodium switch positions, two new products are formed. The two new products are silver chloride and sodium nitrate. Silver chloride is insoluble. Insoluble substances, remember, do not dissolve in a solvent. So as it forms, it comes out of solution. Since it is white, it looks like milk. An insoluble solid which forms in a solution is called a **precipitate** (prih-SIP-ih-tayt). A precipitate is shown by an arrow pointing down ($\downarrow$). Let's show the precipitate, silver chloride, in the equation.

$$NaCl + AgNO_3 \rightarrow AgCl \downarrow + NaNO_3$$

Let's try another case. This time, we'll use the reactants potassium carbonate and calcium chloride.

potassium carbonate + calcium chloride →
    calcium carbonate + potassium chloride

$$K_2CO_3 + CaCl_2 \rightarrow CaCO_3 \downarrow + KCl$$
                    (unbalanced)

$$K_2CO_3 + CaCl_2 \rightarrow CaCO_3 \downarrow + 2KCl \text{ (balanced)}$$

Now, calcium carbonate is the precipitate. Since this is a double replacement reaction, we have two new compounds.

Let's look again at the compounds called acids and bases. Remember, acids always involve H^+ ions and bases involve OH^- ions. Let's try to react an acid with a base.

hydrochloric acid + sodium hydroxide →
sodium chloride + water

$$HCl + NaOH \rightarrow NaCl + H_2O \text{ (balanced)}$$

The H^+ of the acid bonded with the OH^- of the base. The result is water and the salt, NaCl.

A double replacement reaction involving an acid and a base is called a **neutralization reaction.** Remember, when neutralization occurs, the pH is changed. In fact, the result of such a reaction should be a pH = 7. Remember, a pH of 7 is neutral.

Let's look at another example of a neutralization reaction. This time we'll use sulfuric acid and calcium hydroxide as our reactants.

sulfuric acid + calcium hydroxide →
calcium sulfate + water

$$H_2SO_4 + Ca(OH)_2 \rightarrow CaSO_4 \downarrow + H_2O$$
(unbalanced)

To Do Yourself   How Can a Scientist Tell That a Reaction Has Occurred?

You will need:

Four test tubes numbered 1, 2, 3, and 4, solutions of sodium sulfate, barium nitrate, lead nitrate, calcium nitrate, and copper nitrate, dropper.

1. Half-fill each of the four test tubes with each of the nitrate solutions (one solution per test tube). **CAUTION: Use great care when working with chemicals. Pour these solutions very slowly so that they do not splash onto your skin or your clothing.**
2. Add 10 drops of sodium sulfate solution to the first test tube.
3. Observe the test tube for a few minutes.
4. Repeat steps **2** and **3** for the second test tube.
5. Repeat steps **2** and **3** for the third test tube.
6. Repeat steps **2** and **3** for the fourth test tube.

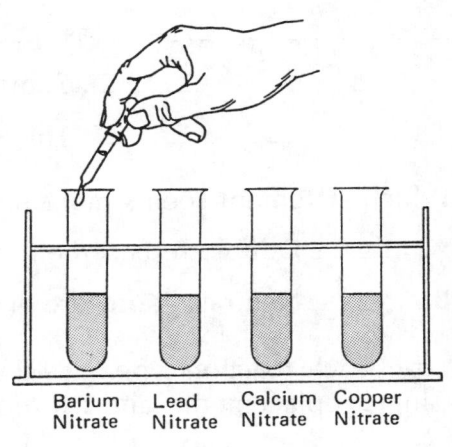

Barium Nitrate   Lead Nitrate   Calcium Nitrate   Copper Nitrate

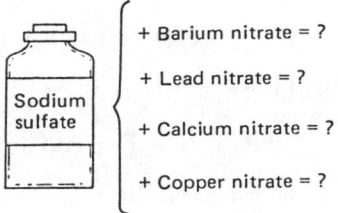

Sodium sulfate
+ Barium nitrate = ?
+ Lead nitrate = ?
+ Calcium nitrate = ?
+ Copper nitrate = ?

Questions

1. In which test tubes did a reaction take place? _____
2. What is one way of telling when a reaction has taken place between two compounds? _____
3. Are precipitates soluble or insoluble in water? _____

This equation is not balanced. There are more hydrogens and oxygens on the left side than there are on the right side. How can the equation be balanced? A second water molecule makes the equation balanced.

$$H_2SO_4 + Ca(OH)_2 \rightarrow$$
$$CaSO_4 \downarrow + 2H_2O \text{ (balanced)}$$

In this reaction, the calcium and the hydrogen switched positions. The products of the reaction are calcium sulfate (a salt) and water.

Neutralization reactions *always* have an acid and a base as reactants. The products are *always* a salt and water.

Suppose the clown added some vinegar to his pink soap solution. Vinegar, remember, is a weak acid. Neutralization would take place. The bright pink color would fade. Perhaps he could use this trick at his next party!

Review

I. Fill in each blank with the word that fits best. Choose from the words below.

compounds   **reactions**   **products**   **precipitate**   **soluble**   **insoluble**

Double replacement _____ always have two

_____ as reactants. Two new compounds are formed as

_____ . Sometimes one of the products is a

_____ . This solid is always _____ .

II. Which statement seems more likely to be true?

A. _____ Double replacement reactions are always neutralizations.

B. _____ Neutralizations are always double replacement reactions.

III. Label each reaction type. Use the letter **S** (synthesis), **D** (decomposition), **SR** (single replacement), and **DR** (double replacement).

_____ $2H_2O \rightarrow 2H_2 + O_2$

_____ $HCl + KOH \rightarrow H_2O + KCl$

_____ $2Li + CaSO_4 \rightarrow Li_2SO_4 + Ca$

_____ $AgNO_3 + KCl \rightarrow AgCl + KNO_3$

_____ $H_2 + I_2 \rightarrow 2HI$

IV. Answer in sentences.

How does adding lime (a base) to acidic soil help plants grow?

_____

_____

_____

What Are Chemical Wastes?

Exploring Science

No More Dead Zones. Look down your street on garbage pick-up day. Now imagine how many homes there are in the U.S.A. Then add all the wastes from all our mines, farms, and factories. Now you may be able to picture a huge number. America produces about 4.5 billion tons of waste each year!

The amount of waste produced each year may be the simple part of the problem. Some industrial wastes are very dangerous. Areas where these wastes were dumped in the past can be a serious problem. Many are now badly polluted. Some of these dumping sites are so dangerous that the federal government calls them "Dead Zones." Humans can no longer live in these areas.

Manufacturers are no longer allowed to dump their wastes. They now have to find some way to safely dispose of toxic products.

One solution is the "waste exchange." Lists of companies and their chemical wastes are published. For example, a company which makes an artificial sweetener lists an unwanted product which is toxic. A plastic company a thousand miles away can use this product as a raw material. The two companies arrange a waste exchange.

Fortunately, companies like these are now thinking of America's future as well as their own profits.

● What problems do you see in the "waste exchange" solution?

Can you think of an alternative method of disposal for these toxic wastes?

Chemical Wastes

No company wants to produce chemical wastes. If chemical processes were 100% efficient, there would be no wastes. But, most reactions involve more than one product.

Let's suppose that a plant wanted to make compound QZ. The easiest way to make QZ might be a synthesis reaction.

$$Q + Z \rightarrow QZ$$

But Q and Z are not usually available inexpensively. For example, Q might be in QJ. Z might only be available in KZ. Now the company must develop a process to separate Q and Z from their compounds. A double replacement reaction might provide the answer.

A double replacement reaction for these compounds might look like this:

$$QJ + KZ \rightarrow QZ\downarrow + KJ$$

In this case QZ is a precipitate. Great! But KJ is also a product.

Perhaps this company can find a use for KJ. If not, maybe "waste exchange" can find a company that can use the KJ. Meanwhile with every batch of QZ, an unwanted volume of KJ solution is produced.

Sometimes, decomposition is possible.

$$JK \rightarrow K + J.$$

Perhaps other companies can use K or J. Every possibility must be tried, especially if KJ or its decomposed products are toxic.

If all else fails, disposal must be considered. KJ could be burned in an incinerator. It could be buried in special dumpsites. Disposal of toxic substances must be done with great care.

To Do Yourself   Can Products Become New Reactants?

You will need:

Marble chips, dilute hydrochloric acid, test tube, one-hole stopper with bent glass tubing, two beakers, calcium hydroxide solution, water, safety glasses

1. Put on your safety glasses.
2. Place a few marble chips in the test tube. Slowly add some hydrochloric acid to the test tube until it is half-full. **CAUTION: Use great care when working with chemicals. Pour solutions very slowly so that they do not splash onto your skin or your clothing.**
3. Insert the stopper with the glass tubing into the test tube.
4. Place the free end of the glass tubing in a beaker about half-filled with water. The end of the tube should be under water. Observe what happens.
5. Now bubble the gas into a beaker of calcium hydroxide solution. Observe.

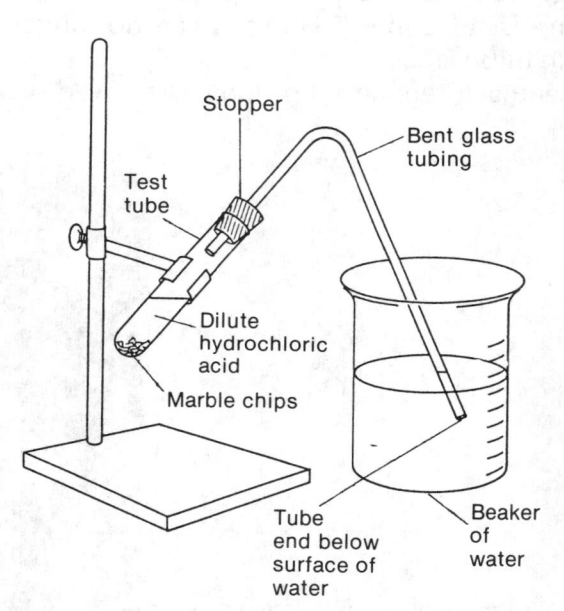

Questions

1. What happened when you placed the end of the tube in the beaker of water?

_____

2. Does the gas react with water? _____

3. Does the gas react with the calcium hydroxide solution? _____

4. What kind of substance formed in the calcium hydroxide solution? _____

_____

Review

I. Fill in each blank with the word that fits best. Choose from the words below.

disposal　　**reactant**　　**toxic**　　**reaction**　　**product**　　**waste**

Every _____ yields at least one _____ .
Sometimes it is _____ , or poisonous. If such a compound
is a _____ product, then proper _____
may be a problem.

II. Which statement seems more likely to be true?

A. _____ Toxic wastes should be physically changed.

B. _____ Toxic wastes should be chemically changed.

III. Place a check (✔) next to the two solutions which best solve chemical waste disposal problems.

A. _____ Bury the waste in magnesium containers.

B. _____ Exchange the waste with another company for use as a raw material.

C. _____ Use the waste product as a reactant in another process.

D. _____ Pour the waste far out into the ocean.

IV. Answer in sentences.

Why is the transport of toxic wastes a government concern?

_____

_____

_____

A. Hidden in the puzzle below are nine words related to chemical reactions. Use the clues to help you find these words. Circle each word you find in the puzzle. Then write each word on the line next to its clue.

```
P C O M P O U N D S H
R S D W A S T E R T E
E L L Q U I Z O R E A
C O E F F I C I E N T
I O I W A S T R A P L
P H Y S I C A L C F R
I Y P R O D U C T W E
T S E C O N Y S I Z I
A I E R H T S D O N G
T L P S I D E T N I O
E S Y N T H E S I S P
```

Clues:

1. In a reaction, it is shown as △. _____

2. The large number before a formula. _____

3. A chemical change. _____

4. The reactant in a decomposition reaction. _____

5. A reaction that changes elements to a compound. _____

6. It is shown as ⎯⎯→. _____

7. An unwanted product. _____

8. An insoluble solid that forms in a solution. _____

9. The end result of a chemical reaction. _____

B. Write the word (or words) that best completes each statement.

1. _____ A change in matter in which no new substance is formed is a   **a.** chemical change   **b.** nuclear change   **c.** physical change

2. _____ The direct union of elements to form a compound is a   **a.** replacement reaction   **b.** synthesis reaction   **c.** decomposition reaction

3. _____ The Electromotive Series lists the activeness of   **a.** metals   **b.** nonmetals   **c.** noble gases

4. _____ In a symbolic equation, a precipitate is shown as   **a.** →   **b.** △   **c.** ↓

5. _____ The 2 in H_2O is called the   **a.** coefficient   **b.** subscript   **c.** balance

6. _____ The 3 in $3H_2SO_4$ is called the   **a.** coefficient   **b.** subscript   **c.** balance

7. _____ Substances that result from chemical changes are called   **a.** products   **b.** reactants   **c.** reactions

8. _____ After a chemical reaction has occurred, the mass of the product(s) must equal the mass of   **a.** one reactant   **b.** all reactants   **c.** hydrogen

9. _____ Matter can be   **a.** created   **b.** destroyed   **c.** changed

10. _____ In a symbolic equation, heat is shown as   **a.** $\rightarrow$ **b.** $\triangle$ **c.** $\downarrow$

11. _____ The products of a neutralization reaction are always   **a.** an acid and a salt   **b.** an acid and water   **c.** water and a salt

12. _____ Neutralization reactions are   **a.** synthesis reactions   **b.** decomposition reactions   **c.** double replacement reactions

13. _____ The reactant in a decomposition reaction is always a(n)   **a.** compound   **b.** element   **c.** salt

14. _____ $HCl \rightarrow H_2 + Cl_2$ is a(n)   **a.** balanced equation   **b.** unbalanced equation   **c.** double replacement reaction

15. _____ $Pb + S \rightarrow PbS$ is a   **a.** synthesis reaction   **b.** decomposition reaction   **c.** single replacement reaction

C. Apply What You Know

1. Study the drawings below. Then answer the questions on page 134.

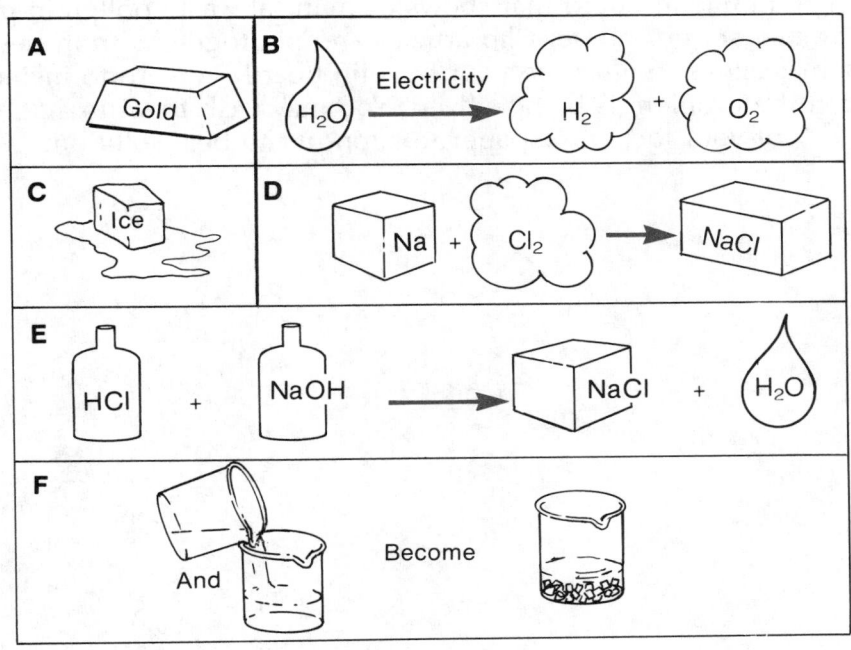

133

a. A physical change is shown in drawing _____.

b. A precipitate forms in drawing _____.

c. A synthesis reaction is shown in drawing _____.

d. A decomposition reaction is shown in drawing _____.

e. A balanced equation is shown in drawing _____.

f. No change is occurring in drawing _____.

2. Study the equation below. Then complete each statement.

$$Na + AgCl \rightarrow NaCl + Ag$$

a. _____ This is a (single replacement/decomposition) reaction.

b. _____ The mass of $Na + AgCl$ must equal the mass of (Ag/$NaCl + Ag$).

c. _____ One of the reactants in this equation is (Na/$NaCl$).

d. _____ One product of this equation is ($AgCl$/$NaCl$).

e. _____ This equation is (balanced/unbalanced).

f. _____ The nonmetal ion in this equation is (Ag^{+1}/Cl^{-1}).

D. Find Out More

1. Collect labels from various chemical products that you have in your home. Read the "Danger" or "Cautions" labels on these products. These statements sometimes warn consumers about unwanted chemical reactions. Make a poster out of the warning labels. Share the information you collect with your parents or classmates.

2. Design a bulletin board that shows a chemical waste pollution problem in an area near your home. Clip articles and photographs from newspapers and magazines to illustrate your bulletin board. Be sure to include ways in which the problem is being solved. You may wish to write a "Letter to the Editor" of your local newspaper to support the best solution.

Summing Up
Review What You Have Learned So Far

A. Study the chemical equations. On line **a**, indicate whether the equation is balanced or unbalanced. On line **b**, label the type of chemical reaction shown. Choose from these labels.

single replacement reaction   **neutralization reaction**   **synthesis reaction**
decomposition reaction

1. $2HgO \rightarrow Hg + O_2$

   a. _____

   b. _____

2. $H_2 + O_2 \rightarrow H_2O$

   a. _____

   b. _____

3. $HCl + NaOH \rightarrow NaCl + H_2O$

   a. _____

   b. _____

4. $Zn + CuSO_4 \rightarrow ZnSO_4 + Cu$

   a. _____

   b. _____

B. Each statement below refers to the solubility graph. Circle the underlined word or phrase that makes each statement true.

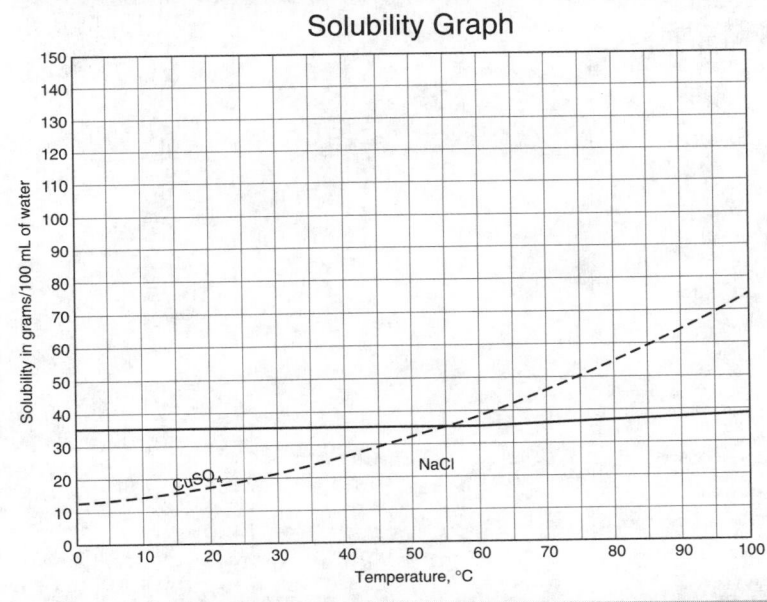

Solubility Graph

1. At 20°C, (21 g/38 g) of $CuSO_4$ will dissolve in 100 mL of water.
2. The solvent in this solubility graph is ($CuSO_4$/water).
3. At 20°C, NaCl is (more soluble/less soluble) in water than $CuSO_4$.
4. At 0°C, $CuSO_4$ is (more soluble/less soluble) in water than NaCl.
5. If a solution of $CuSO_4$ and 100 mL of water contains 20 grams of solute at 40°C, the solution is (unsaturated/supersaturated).
6. $CuSO_4$ and NaCl are (solutes/solvents).

135

UNIT

6

NUCLEAR CHANGE

Lesson One

What Is Radioactivity?

Exploring Science

An Invisible Curse. Jobs were scarce on the Navajo reservation (nav-uh-HO rez-ur-VAY-shun). Seventeen year old Tony Light was happy to get a job. He worked in the uranium mines of Red Valley, Arizona. He earned $1.75 an hour.

Tony's crew hauled uranium ore from the mine shaft, 200 feet below, to the surface. Then they set off explosives in the green and yellow rock to free more ore. Each blast filled the tunnel with dust. There was little ventilation in the tunnel. The men had to wait for the dust to settle before they could drag out the ore. They often ate their lunches in the four-foot high shaft. It saved them a long trip to the surface.

Tony only worked in the mines for a few weeks. Other members of his tribe could not give up their salaries. They had wives and children. They continued to work in the mines. Year after year they blasted the rock and inhaled the dust.

Tony saw many of the miners die. Many of their children died at birth. Many of the babies that did not die, had serious birth defects.

The problems of the miners and their families have not gone unnoticed. Several legislators have tried to get the government to take action. Scientists are trying to find a link between these deaths and illnesses and the work in the mines. The mines are now closed. But the miners and their families still need help. So far, they have not received the help they need.

Agencies like the March of Dimes have gathered information. Someday the data might prove that there is a link between the work in the uranium mine and the deaths and illness in this Navajo community.

● What questions might researchers ask in Arizona?

This Navajo mine worker is tending to the grave of his son.

137

Radioactivity

There are several theories that might explain the deaths of the Navajo miners. The most accepted theory involves the mine dust. The inhaled particles of dust and gases became trapped in the miners' lungs. These particles affected the Navajos for the rest of their lives.

Let's look at an uranium atom.

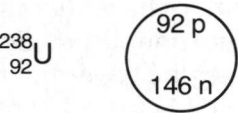

$$^{238}_{92}U \qquad \boxed{\begin{matrix} 92\ p \\ 146\ n \end{matrix}}$$

The nucleus of a uranium atom is filled with particles. Remember, the protons are positively charged. Each of these + charges pushes away every other + charge. Remember, like charges repel each other. **Nuclear** (noo-CLEE-ar) **force** holds these particles together.

Sometimes, the nuclear force isn't strong enough. Some particles escape from the nucleus. Particles can leave the nucleus in different ways. Sometimes, 2 protons and 2 neutrons are released. The combination of 2 protons and 2 neutrons is called an **alpha** (AL-fuh) **particle.** The diagram shows an alpha particle being released from a uranium atom.

$$\underset{\substack{\text{uranium}\\\text{atom}}}{\left(\begin{matrix}92\ p\\146\ n\end{matrix}\right)} - \underset{\substack{\text{alpha}\\\text{particle}}}{\left(\begin{matrix}2\ p\\2\ n\end{matrix}\right)} \rightarrow \underset{\substack{\text{"new"}\\\text{element}}}{\left(\begin{matrix}90\ p\\144\ n\end{matrix}\right)} + \text{Energy}$$

This spontaneous (spon-TAY-nee-us) release of particles and energy is called **radioactivity** (RAY-dee-oh-AK-tiv-ih-tee). Spontaneous means that this nuclear change happens naturally. Uranium atoms were shooting off particles long before people were on Earth.

Look at the nucleus which is left after the uranium atom gives off, or **emits** (EE-mits) an alpha particle. It is no longer uranium. The element has changed. This nucleus has only 90 protons. Look at the Periodic Table. What element has 90 protons? The uranium has become thorium, $^{234}_{90}$Th.

The spontaneous change of an atom of one element into an atom of another element is called **radioactive decay.** Uranium is only one of the radioactive elements. In fact, our new thorium atom is very radioactive.

Within days, the thorium atom will probably change. This time a beta (BAY-tuh) particle will be emitted from the nucleus. A **beta particle** is an electron. This seems strange. There are no electrons in the nucleus, only protons and neutrons. Electrons, remember, are found orbiting around the nucleus.

How did an electron get into the nucleus of the thorium atom? The electron is in the nucleus because a neutron has broken up into a proton and an electron. When this happens, a beta particle (electron) is emitted from the nucleus. The diagram below shows what happens when a thorium atom emits a beta particle.

$$\underset{\substack{\text{thorium}\\\text{atom}}}{\left(\begin{matrix}90\ p\\144\ n\end{matrix}\right)} - \underset{\substack{\text{beta}\\\text{particle}}}{e} \rightarrow \underset{\substack{\text{"new"}\\\text{element}}}{\left(\begin{matrix}91\ p\\143\ n\end{matrix}\right)} + \text{Energy}$$

Remember, the neutron broke up into an electron and a proton. What happens to the new proton that was formed? The new proton is added to the 90 other protons of the thorium atom. The atomic number is now 91. Another new element has formed. This new element is proactinium (PRO-ak-tin-ee-um), $^{234}_{91}$Pa.

Proactinium is also an unstable element. The nuclear force of an unstable atom cannot keep the particles together. It will probably emit a beta particle within minutes.

In fact, our original $^{238}_{92}$U atom can change fourteen times. This means that 14 new elements will be formed. These 14 elements make up a group of elements known as the **radioactive series of elements.**

The radioactive series begins with the uranium isotope U-238. Sometimes the unstable U-238 atom shoots off alpha particles. Sometimes it emits beta particles. Eventually, it reaches a stable form. This stable form is the lead isotope, $^{206}_{82}$Pb. The nuclear force of the stable nucleus is able to hold the protons and neutrons together. When this happens, the radioactive decay comes to a stop.

Radioactivity can take another form—**gamma** (GA-muh) **radiation.** This occurs when gamma rays are emitted by an atom. A **gamma ray** is energy that is often released during a nuclear change. Gamma rays *do not* consist of nuclear particles.

Gamma radiation travels at high speeds because it has no mass. It is difficult to stop. Thick lead shields are needed to stop gamma rays. Gamma rays can do a great deal of damage.

Alpha particles travel much more slowly. Alpha particles remember, are 2p + 2n. They have a weight of 4 atomic mass units. They are fairly heavy so they are easier to stop. A thick piece of paper can stop an alpha emission. Alpha particles, therefore, do not cause much damage.

Beta particles travel much faster than alpha particles. However, they do not travel as fast as gamma rays. A beta emission can go through paper. But it is stopped by a thick piece of wood.

Think back to the Navajo miners. Many of the miners inhaled uranium atoms. Other unstable atoms were also trapped in their lungs. Remember, radioactivity is spontaneous. Nuclear changes can happen any time. An alpha particle can be emitted. A beta particle can be released. Gamma rays can pass right through the body.

This type of radioactivity is called **ionizing radiation.** This type of radiation changes the atoms it hits into ions. The damage it causes can be fatal.

Ionizing radiation is very harmful to an individual. It can also hurt future children. Changes in the body of the parent can result in birth defects in their children.

Review

I. Fill in each blank with the word that fits best. Choose from the words below.

stable   nucleus   beta   unstable   decay   gamma   number

emit

Some elements can change by radioactive _____ . A(n) _____ atom will _____ radiation. The _____ can give off an alpha or _____ particle. This changes the atomic _____ of the atom.

II. Which statement seems more likely to be true?

A. _____ Radioactivity is an invention of man.

B. _____ Radioactivity is a natural event.

III. Label each change as **P** (physical), **C** (chemical), or **N** (nuclear).

A. _____ Ice becomes water.

B. _____ Uranium changes to thorium.

C. _____ Calcium and chlorine become calcium chloride.

D. _____ Radon emits an alpha particle.

E. _____ Hydrogen and oxygen are formed from water.

F. _____ A beta particle is emitted by lead -210.

IV. Answer in sentences.

Why are some atoms radioactive?

What Are Radioactive Isotopes?

Exploring Science

Isotopes to Order. A doctor in Chicago needs radioactive iodine. He will use it to help diagnose a patient's thyroid problem.

A therapist in Alaska needs radioactive cobalt. She will use it to treat a cancer victim.

A biologist in New York needs radioactive oxygen. She will use it to trace the chemical changes in a green plant.

A researcher in Hawaii needs radioactive magnesium. He will use it to trace the metal's involvement in pineapple growth.

Today's nuclear laboratories can meet these needs. They fill orders for radioactive sources every day.

This was not the case in 1913. In that year two chemists, Hevesy and Paneth, used the first radioactive tracer. They put some radioactive lead into soil. Later, they tested plants growing in the soil to find out how much lead the plants had absorbed.

Lead can harm a plant. Today, scientists can

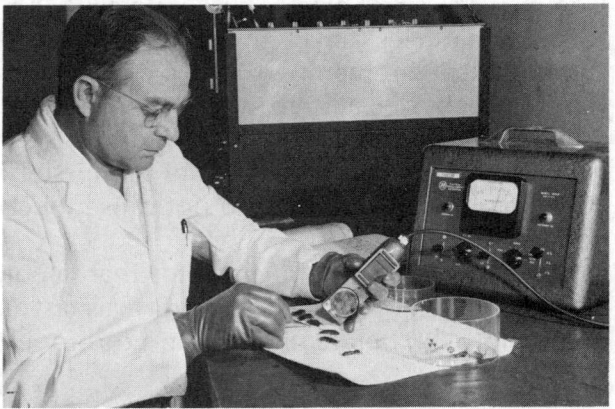

This scientist is using a Geiger counter to identify radioactive cockroaches.

choose tracers that are safer. These tracers are made from elements normally used by the living thing.

● What problems might scientists have to face when filling orders for radioactive isotopes?

Radioactive Isotopes

At first, scientists studied only natural radioactive elements. Many of these were heavy atoms. They had high atomic numbers and large atomic masses.

Scientists found an easy way to test for radioactivity. A sample of an element was placed on an unexposed photographic plate. The film was then developed. If the developed film showed a dark spot, the element was radioactive. Scientists soon learned, however, that for some elements, only some isotopes of the element are radioactive.

Today, scientists use a **Geiger** (GY-gur) **counter** to detect radioactivity. A Geiger counter is much more accurate than a photographic plate. It is also much quicker and easier to use.

Imagine that a scientist has samples of two isotopes of lead—Pb-214 and Pb-206. A Geiger counter would show that the Pb-214 is very radi-

oactive. It would also show that Pb-206 is not radioactive at all.

Pb-214   Pb-206

The isotope of lead which has 132 neutrons is very unstable. It emits beta particles. An unstable form of an element is called a **radioactive isotope.**

Hundreds of radioactive isotopes occur naturally. Many occur in very small amounts. Some of them exist for only a short period of time.

Ernest Rutherford was the first scientist to change the nucleus of an atom. He aimed an alpha particle (2p + 2n) at a nitrogen nucleus.

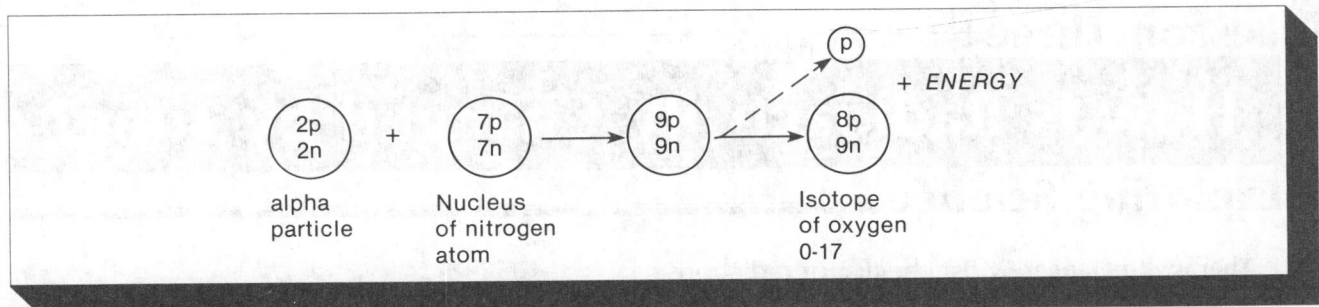

alpha particle + Nucleus of nitrogen atom → Isotope of oxygen O-17 + ENERGY

As soon as $\frac{9p}{9n}$ forms, it loses a proton and becomes $\frac{8p}{9n}$ This new atom has an atomic number of 8. It is a rare form of oxygen, $_{8}^{17}O$.

Before Rutherford's experiment, only natural radioactive elements existed. Rutherford's experiments led to an entirely new field of science. Now scientists can change the nucleus of an atom to create new radioactive elements.

Today, nuclear laboratories shoot many different particles at atoms. These particles can be aimed with great accuracy. They can also travel at very high speeds.

Review

I. Fill in each blank with the word that fits best. Choose from the words below.

laboratory   isotopes   heavy   nucleus   naturally   light

Many elements are _____

radioactive. Most of these atoms are _____ . Today,

_____ can be made in a _____ .

The _____ of the atom is actually changed.

II. Which statement seems more likely to be true?

A. _____ Carbon-14 is radioactive; all isotopes of carbon are radioactive.

B. _____ Carbon-14 is radioactive; other isotopes of carbon may not be radioactive.

III. Mark a **T** next to the statements that are true. Use the Periodic Table of Elements.

A. _____ All oxygen atoms have an atomic number of 8.

B. _____ All oxygen atoms have 8 protons.

C. _____ All oxygen atoms have an atomic mass of 16.

D. _____ Electrons must be taken from orbits to change nitrogen to oxygen.

E. _____ The nucleus must be changed to change nitrogen to oxygen.

F. _____ A radioactive isotope of oxygen has an unstable nucleus.

IV. Answer in sentences.

Why is equipment needed to change the nucleus of an atom?

Lesson Three

How Is Radioactivity Changed by Time?

Exploring Science

Therapy to Danger. Most checks for radiation levels show "normal." But a test in Los Alamos, Mexico, in 1984, showed "high radiation." This result started an urgent investigation.

A total of 109 houses were found to contain radioactive material. The radioactivity was traced to steel rods in the walls. The steel contained cobalt-60, a highly radioactive metal.

A Mexican steel factory made the steel rods. This company purchased raw materials from many places. They had to find out where the materials used in the Los Alamos batch came from.

The contaminated metal came from a junkyard in northern Mexico. It sold scrap metal to the steel factory. One of the scrapped items came from a medical facility months earlier. The item was a cancer therapy machine. It contained cobalt-60 which is used for radiation treatments in cancer patients. This machine was in the scrap metal that the steel company bought.

The steel company melted down the machine with tons of other metal. All of the steel was contaminated with radiation.

All of the contaminated houses in Los Alamos, Mexico were torn down. But the problem does not end here. The steel has shown up in forty of the United States. It has also been found in Canada.

● Why is the contaminated steel a threat?

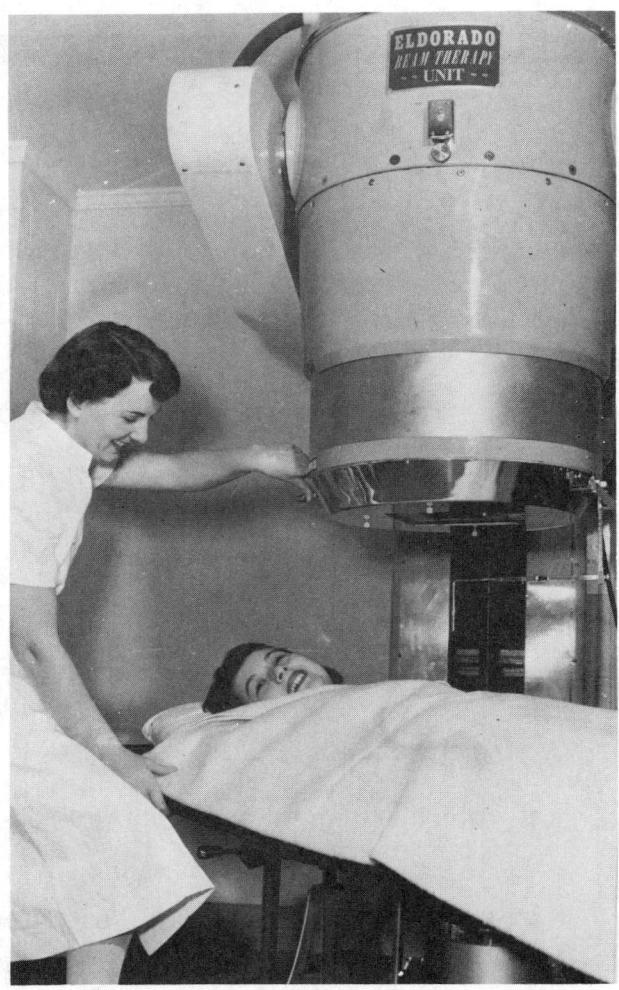

A cobalt machine similar to this one was responsible for the tearing down of 109 homes in Mexico, Canada, and the United States.

Half-Life

The radioactive material might have been in the machine for a year. It was in the junkyard for months. More time passed while it was made into steel and sent to Los Alamos. How long does an element stay radioactive?

The answer is different for every radioactive

isotope. Remember, radioactive substances are unstable, and are constantly making spontaneous changes into other substances. The lifetime of a radioactive substance is not complete until all of it has decayed into a stable element.

Scientists use the term half-life to describe the amount of time it takes a radioactive isotope to decay. A **half-life** is a measure of the length of time it takes for one half of the sample to undergo radioactive decay.

Let's look more closely at U-238. The half-life of U-238 is 4.5 billion years. If you had 1000 atoms of U-238, in 4.5 billion years only 500 atoms of the U-238 would remain.

What happened to the other 500 atoms of U-238? They have changed into atoms of other elements. The original sample is no longer pure U-238.

In 9 billion years (4.5 + 4.5 billion), only 250 atoms of the original U-238 would remain in the

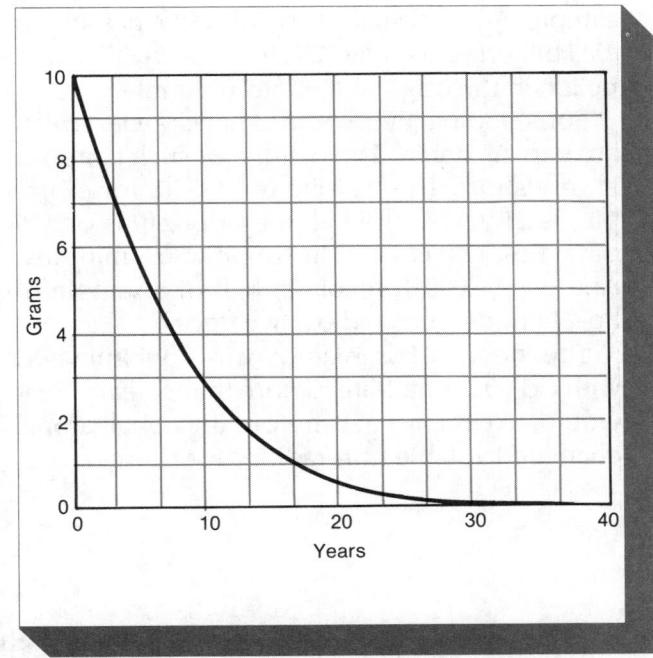

To Do Yourself   How Do Two Models of Half-Life Compare?

You will need:

Shoe box with lid, 100 paper fasteners, graph paper, marker

1. Label the sides of the shoebox **1, 2, 3,** and **4.**
2. Place the fasteners inside the box.
3. Shake the box 1 time in a vertical direction.
4. Remove all fasteners pointing to side **1.** Record how many fasteners remain in the box.
5. Shake the shoebox a second time. Remove all fasteners pointing to side **1.** Record how many fasteners remain. Do this until no fasteners remain.
6. Repeat the procedure. This time remove fasteners pointing either to side **1** or to side **2** after each shake.

Questions

1. What is the half-life of the first model? _____

2. What is the half-life of the second model? _____

3. Which model represents a faster decay rate? _____

sample. Another half-life would have passed. In 13.5 billion years only 125 atoms of U-238 would be left in the original 1000 atom sample.

Some radioactive isotopes have nuclei which are very unstable. The half-life of such isotopes is very short. The half-life of Pb-210, for example, is 21 years. The half-life of Bi-210 is only 5 days. Po-218 has a half-life of only 30.5 minutes. And every .00016 seconds, half of a sample of Po-214 undergoes radioactive decay.

The steel in Los Alamos was contaminated with Co-60. Its half-life is more than 5 years. So it would take many years before the cobalt atoms changed to stable non-radioactive forms.

Half-Life for Some Isotopes

Isotope	Half-Life
C-14	5730 y
Co-60	5.3 y
Cs-137	30.23 y
H-3	12.26 y
I-131	8.07 d
K-40	1.28 billion y
K-42	12.4 h
P-32	14.3 d
U-235	7.1 billion y
U-238	4.51 billion y

y = years; d = days; h = hours

Review

I. Fill in each blank with the word that fits best. Choose from the words below.

radioactive   half-life   42   one-half   50

The length of time an isotope stays _____ is different for each isotope. The length of time it takes for _____ of a sample to undergo a nuclear change is called the _____ of the isotope. If a pure sample of Pb-210 weighs 100 grams, in 21 years only _____ grams will still be radioactive. After _____ years 75 grams will be changed to another element.

II. Which statement seems more likely to be true?

a. _____ After two half-lives, ¼ of radioactive atoms remain.

b. _____ After two half-lives, no radioactive atoms remain.

III. Complete the table with data from the text.

Radioactive Isotope	Half-life
Co-60	
Po-218	
Po-214	
U-238	
Bi-210	

IV. Answer in sentences.

Why are radioactive wastes a serious problem?

Lesson Four

How Is Radioactivity Changed by Distance?

Exploring Science

Doomsday Test. From a plane, the large brown circle looks strange. The rest of the Long Island landscape is green. Over 20 years ago, Dr. George Woodwell started an experiment. He put a tall pole in an oak and pine forest near the Brookhaven National Laboratory. He put a source of radiation, cesium-137, at the top of the pole.

For fifteen years, Woodwell and his staff observed the effects of the radioactive cesium on the forest. They measured the amount of radioactivity that reached different parts of the forest area.

Their instruments measured 19,000 units of radiation near the cesium. All life near the pole was killed. The radiation ionized necessary atoms in the living cells. Nothing survived.

Some distance away from the pole, 1000 units of radiation were measured. Some plants, like mosses and lichens, were living there. Further away, 150 units of radiation were measured. Shrubs grew in this area. Only where the instruments measured 2 units of radiation or less was the forest undamaged.

After a while, some new plants began to grow closer to the pole. But the graceful pines and oaks could not survive in the high radiation area.

● What do you notice as you compare the distance from the source and the units of radiation?

The radiation from the cesium that Dr. George Woodwell placed near the Broookhaven National Laboratory caused the death of the surrounding plants.

Inverse Square Law

Let's suppose a scientist is at the edge of the forest. His Geiger counter reads 1 radiation unit per day. He is quite far from the source of radiation.

The scientist knows that the radiation will increase closer to the source. In fact, if the distance is cut in half (½), the amount of radiation will not just double. The radiation will be four times (2 × 2) as much.

Suppose the new distance is only one quarter (¼) of the original distance. The new radiation readings would be sixteen (4 × 4) times the original level.

This data shows **The Inverse Square Law.** To find the change in levels of radiation with distance:

(1) Invert the change in distance (when a fraction is inverted, it is turned upside-down). For example, ¼ becomes ⁴⁄₁ or 4.

(2) Then square the number you get in step **1** (when a number is squared, it is multiplied by itself). For example, 4^2 is equal to 4 × 4 which is 16.

Let's try another example. Suppose the scientist were ¹⁄₁₀ the original distance from the source.

(1) Invert the change in distance. ¹⁄₁₀ becomes 10
(2) Then square the new value. $10^2 = 10 × 10 = 100$

The new readings would be 100 times higher than the original readings. Even a small change in distance can result in a great change in radiation.

To Do Yourself   How Does Distance Affect a Light Beam?

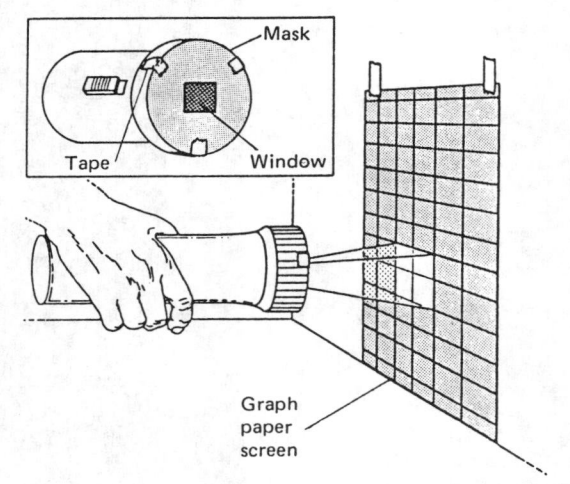

You will need:

Flashlight with a 1 cm² window, metric ruler, graph paper screen

1. Turn on the light. Position the square beam so that it covers 16 boxes on the screen (4 × 4).
2. Record the distance (in centimeters) from the screen to the flashlight.
3. Double the distance of the flashlight from the screen. Count the number of lighted boxes.
4. Double the distance from step **3**. Count the number of lighted boxes again.

Questions

1. How many boxes were lighted in step **3**? _____
2. When the distance is doubled does the number of lighted boxes double?

_____

3. Predict the number of boxes that would be lighted if the distance in step **4** is doubled. _____

4. What happened to the brightness of the light on the paper as the distance was increased? _____

I. Fill in each blank with the word that fits best. Choose from the words below.

change   invert   small   5   distance   square   great   25

The _____ from a source of radiation to you is cut to ⅕.

First you _____ the number to make it 5. Then you

_____ the value. A small _____ in

distance can result in a _____ change in radiation. In this

case, the radiation is _____ times stronger.

II. Which statement seems more likely to be true?

a. _____ Amounts of radiation can be measured.

b. _____ Radiation dose is not affected by distance.

III. Label the following statements as **T** (true) or **F** (false).

	3,000 meters	6,000 meters	12,000 meters
Source	↓	↓	↓
	?	?	1 radiation unit

a. _____ At 6,000 meters, the radiation is 2 units.

b. _____ At 6,000 meters, the radiation is 4 units.

c. _____ At 3,000 meters, the radiation is 1 unit.

d. _____ At 3,000 meters, the radiation is 16 units.

e. _____ At 15,000 meters, the radiation is greater than 1 unit.

IV. Answer in sentences.

Why must the distance from the patient to the source of radiation be carefully measured during radiation treatments?

_____

_____

_____

How Is Energy Released from an Atom?

Exploring Science

A Lady of Imagination. Lise Meitner was a Jewish scientist from Austria, who from 1907 until 1938 worked often with the German scientist Otto Hahn in Berlin. Together, they specialized in radioactivity, and as early as 1917, they discovered a previously unknown element, protactinium.

Early in the 1930s, the Italian physicist Enrico Fermi learned how to bombard uranium atoms with slow neutrons. He, too, thought he had found a new element, which he called Uranium X. Fermi hoped that Uranium X would turn out to be Element 93—but he could not prove it. Meitner and Hahn decided to redo Fermi's work more carefully. They soon found that they had unexpected results.

The year 1938 was very important to all three scientists. Fermi won the Nobel Prize for Physics for his work. Hahn published a description of the strange results he and Meitner were getting. Meitner, because she was Jewish, had to flee Germany. She went to Stockholm, Sweden. She was no longer able to perform experiments because she had lost her laboratory in Berlin.

Meitner considered the strange results she and Hahn had seen and of Fermi's earlier experiments. Then she wrote to a British science journal. She suggested that the slow neutron "bullet" did not stay in the uranium nucleus. Instead, it caused the nucleus to split in two. She used the term *fission* to describe this splitting.

Meitner's idea was right. Lise Meitner did not win a Nobel Prize. Her idea, however, helped other scientists to achieve success.

● How is *fission* different from *atom smashing*?

Dr. Lise Meitner was a pioneer in nuclear physics.

Fission

The world was at war. The work of scientists like Fermi and Meitner was the beginning of a new branch of science. Soon, the products of fission reactions became less important. The energy released when the nucleus split became the concern. The fission reaction was seen as a new source of energy.

Remember, fission is not "atom smashing.

Fission (FIZ-shun) is the splitting of a large nucleus into two smaller nuclei. Let's look at a drawing of the fission of one atom of U-235.

A neutron is shot at the nucleus of a U-235 atom. The neutron hits the nucleus. The nucleus immediately splits into two smaller nuclei. It emits 3 neutrons at a very high speed. There is also a release of great energy.

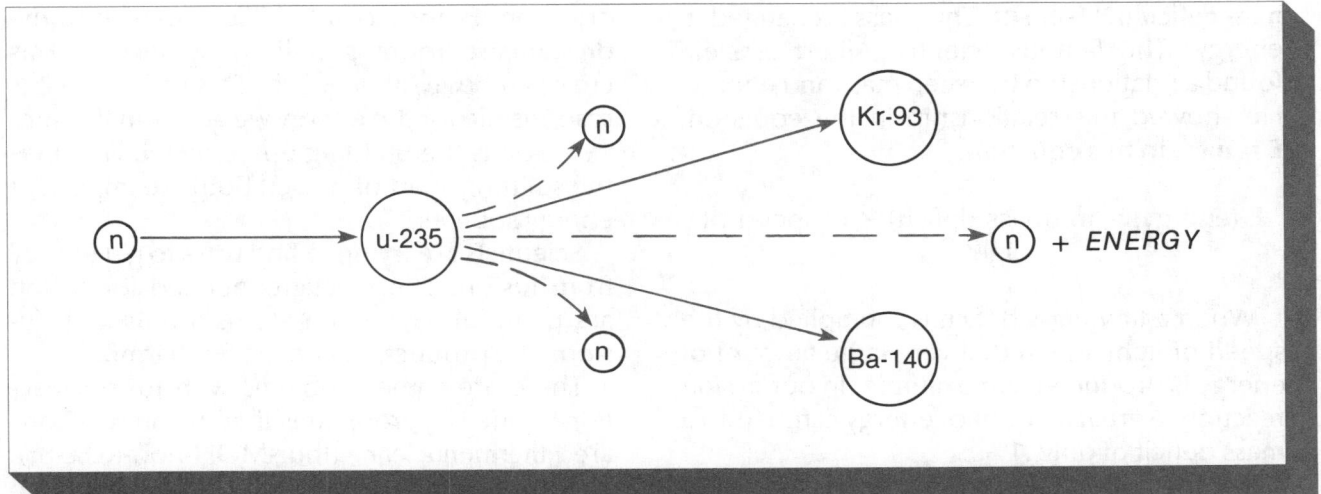

What happens to the 3 neutrons that are released? If other U-235 atoms are nearby, the neutrons will cause them to split. This can keep happening as long as there are U-235 atoms present. This type of reaction is called a **chain reaction.** Once started, a chain reaction can continue by itself.

Imagine putting a burning piece of charcoal on a layer of other coals. The heat from the first will ignite several others. These will light others. Eventually, all the coals will be on fire. This is a chain reaction, also.

Imagine the energy that is released. Atom after atom undergoes fission. Each gives off energy. But how much energy is there?

Let's consider how much matter was there before the nuclear change. One neutron (1 a.m.u.) plus one U-235 atom (235 a.m.u.) gives us a total mass of 236 a.m.u.

After the reaction, there is a Kr-93 atom and a Ba-140 atom. These atoms have a total mass of 233 a.m.u. There are also 3 neutrons with a total mass of 3 a.m.u. The products of this reaction therefore have a total mass of 236 a.m.u. (233 + 3). It seems like there was no change in mass.

However, very careful measurement shows a loss of 0.1% of the mass. This loss is called a

The cyclotron (also called a nuclear accelerator) fires particles into the nuclei of atoms.

mass deficit (DEF-ih-sit). This mass is changed to energy. The famous scientist Albert Einstein found a relationship between mass and energy. He showed this relationship in the equation, $E = mc^2$. In this equation:

$$E \text{ (energy)} = m \text{ (mass deficit)} \times c \text{ (speed of light)}^2$$

When a tiny mass deficit is multiplied by the speed of light squared, a very large amount of energy is produced. For example, in our fission reaction, a great amount of energy came from a mass deficit of only .1%.

Another type of reaction has a mass deficit of .5%. This mass deficit occurs in fusion. **Fusion** (FYOO-shun) is the joining together of light atoms to form one heavier atom. Fusion, like fission, involves a nuclear change.

The energy we get from the sun is the result of fusion. Every second, millions of tons of hydrogen are forming helium by fusion. This creates a huge mass deficit. This mass deficit is responsible for the energy we get from the sun.

Fusion is the building up of nuclei. Fission is the splitting apart of nuclei. Both are important energy sources.

Scientists are trying to find ways to get energy from fusion. The reactants needed for fusion are plentiful. Hydrogen atoms are used in fusion. The product, helium, is not harmful.

There are some problems with fusion reactions. One big problem is that fusion reactions are **thermonuclear** (thur-MOH-noo-KLEE-ur). This means that they require heat. Hydrogen atoms must be heated to several billion degrees Celsius to produce helium. Another problem is finding a container that can hold something this hot without melting. The technology needed for practical use of fusion is just not available yet.

To Do Yourself   What Happens When an Atom is Split?

You will need:

Walnut, 8 oz. fishing sinker, cardboard tubing, tape, balance, paper

1. Construct a 2m long piece of tubing. Stand the tubing up on the paper.
2. Find the mass of the walnut. Record the mass.
3. Place the walnut inside the bottom of the tube with the pointed end down.
4. Position the sinker (atomic bullet) at the top of the tube. Drop the sinker through the tube.
5. Carefully collect and take the mass of the resulting material. Record the mass.

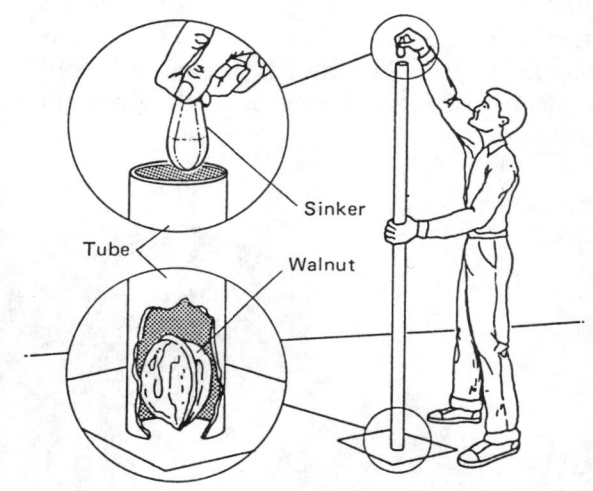

Sinker

Tube

Walnut

Questions

1. What causes the bullet (sinker) to speed up? _____

_____

2. How does the mass compare before and after? _____

_____

3. How is splitting a walnut similar to splitting an atom? _____

_____

I. Fill in each blank with the word that fits best. Choose from the words below.

heat   energy   light   chain   nucleus   nuclei   fission   fusion
reaction

_____ is a nuclear reaction in which one nucleus splits

to form two lighter _____ . This reaction releases a great

deal of _____ . If the neutrons released in this reaction

continue to strike other atoms, a _____ reaction will take
place. The joining together of light atoms to produce heavier atoms is called

_____ . This type of reaction is described as thermo-

nuclear because it requires _____ .

II. Which statement seems more likely to be true?

    A. _____ Fusion starts with light nuclei.

    B. _____ Fusion starts with heavy nuclei.

III. Label the following statements as **T** (true) or **F** (false).

    A. _____ Fission and fusion release energy.

    B. _____ Fission is the building of a heavier nucleus.

    C. _____ Uranium is one fission fuel.

    D. _____ A chain reaction goes on forever.

    E. _____ In $E = mc^2$, m refers to mass that is changed.

IV. Answer in sentences.

Why might fusion be a good energy source for the future?

_____

_____

_____

What Is a Nuclear Reactor?

Exploring Science

The Public Speaks. In the 1950s, American scientists became very excited about nuclear energy. A nuclear power plant could produce electricity without some traditional fuel problems. Nuclear plants produce little or no air pollution. Nuclear plants do not depend on oil, so, the U.S. would not need to depend on other countries for fuel. The chairperson of the U.S. Atomic Energy Commission predicted that electricity from nuclear power would be "too cheap to meter."

By 1957, the first nuclear power plants in the U.S. were producing electricity. They were extremely expensive, but the government felt that the benefits were worth the high cost.

Slowly, the public became aware of the dangers. Radioactive fuel and wastes had to be transported through communities. Even more frightening was the danger of a nuclear explosion. People grew to fear the results of a human error in operating a nuclear plant or in designing the safety system. Any accident would be a great danger to workers. Escaping radioactive chemicals would be a danger to the public.

Citizens' groups demonstrated against nuclear plants, and the politicians and power companies listened. Some plants were closed while others did not open. No nuclear power plant has been ordered in the U.S. since 1978.

The next year, 1979, saw a serious accident at the Three Mile Island nuclear plant in Pennsylvania. Although no one was injured and only a small amount of radiation escaped, the nuclear reactor was destroyed by immense heat—a meltdown. Power-plant operators pointed to how well the public had been protected by safety features at the plant. Demonstrators against nuclear power emphasized that a more serious accident could occur.

The fears of the demonstrators were shown to be real in 1986. A nuclear plant in Chernobyl, a city in Ukraine, was the scene of an explosion. About 30 people died from radiation sickness. Perhaps as many as 20,000 may be killed by cancer.

● List the advantages and disadvantages of nuclear power.

Are these people *for* or *against* nuclear energy? How do you know?

Nuclear Reactors

Nuclear energy has many uses. It can supply electricity or run a submarine. The energy potential of fission is very great.

A **nuclear reactor** (REE-ak-tohr) is a plant in which a controlled fission chain reaction takes place. It is the heart of any nuclear energy system. The designs of reactors vary. However, all reactors must have fuel, a coolant, controls, and safety features.

FUEL. A fission reaction must start with heavy atoms that can be split. These atoms are called **fissionable** (FISH-shun-uh-bull) material. Uranium and plutonium are two such fuels.

COOLANT. The energy released during fission must be carried from the fuel cells. This energy is in the form of heat. Many materials have been used, but water is still the most common coolant.

The coolant absorbs heat at the fission site. Then it is pumped to other heavy equipment.

CONTROLS. Controlling a chain reaction depends upon the number of free neutrons present. After reactor start-up, the number of neutrons must be limited. Remember, each fission reaction may release several neutrons. Control rods in the reactor absorb neutrons. Without controls the reaction would 'go wild'. The placement of control rods depends upon the number of neutrons. If many neutrons are recorded, maximum control is used. If few neutrons are available, the control rods are pulled back.

SAFETY. The control system is the most important safety feature of a nuclear reactor. Accurate instruments give information to the plant operators. Safety back-up systems also help to protect the environment. All control features can be operated manually by a trained person, or automatically, by computers.

All reactors are capable of a slow shut-down. They also have a "scram" system. A scram system is used for a rapid shut-down. This system activates in any emergency.

To Do Yourself   How is Nuclear Fission Controlled?

You will need:

15 wooden blocks (2 cm × 4 cm × 1 cm),
2 wooden blocks (2 cm × 10 cm × 1 cm)

1. Stand the blocks in rows like bowling pins. The last row should have 5 blocks.
2. Push row 1 into row 2 and observe.
3. Line up the blocks again. Slide one larger block (control rod) halfway behind row 2. Push the first block again and observe.
4. Repeat step 3. This time, slide one control rod behind row 4. Push the first block. Observe.
5. Line up the blocks again. Now use 2 control rods, one halfway behind row 2 from the right, and the other halfway behind row 4 from the left. Push the first block. Observe.

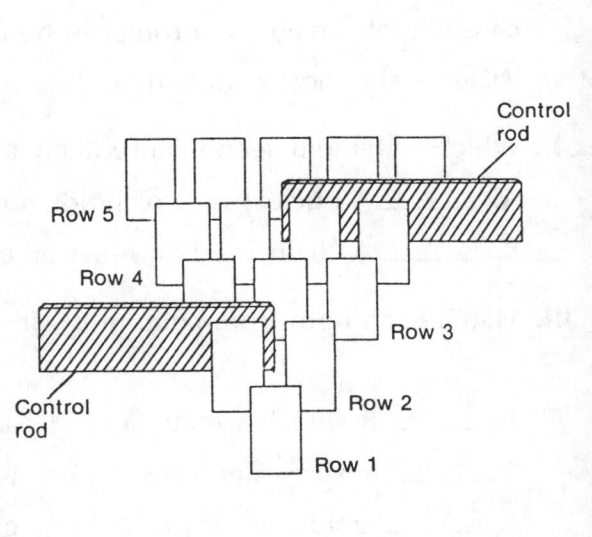

Questions

1. When was the chain reaction uncontrolled? _____

2. What is the function of control rods? _____

3. How can a chain reaction be stopped? _____

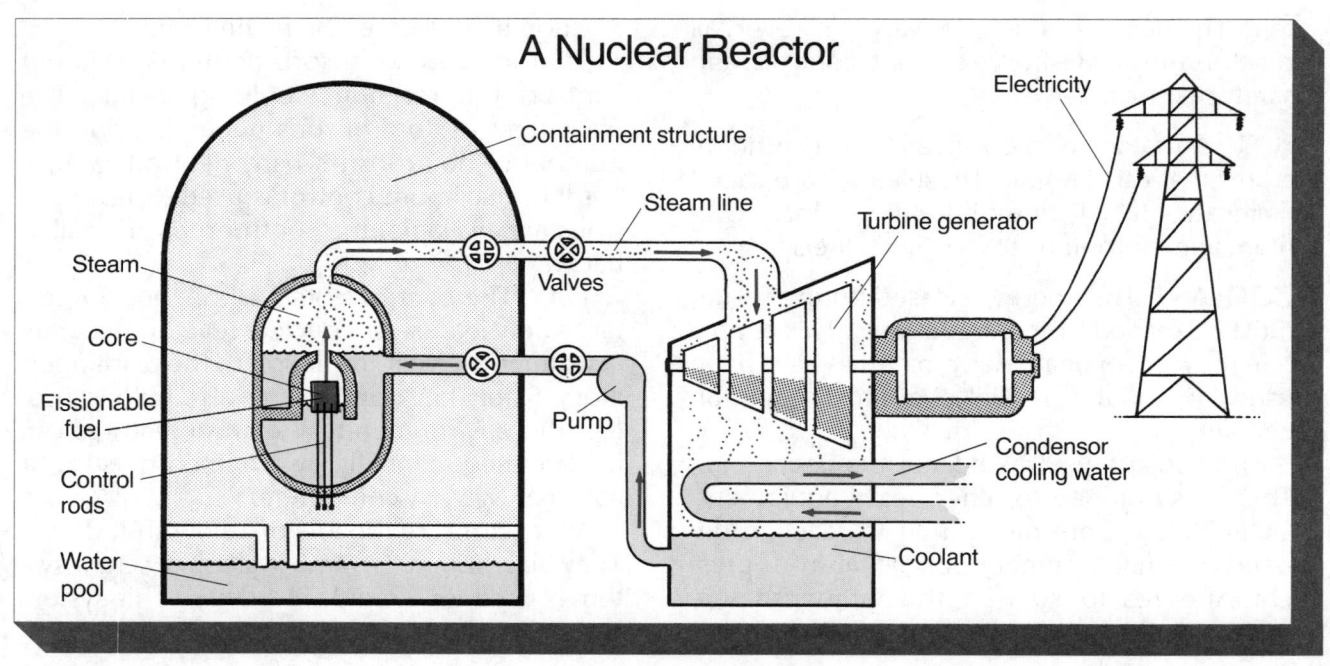

A Nuclear Reactor

Containment structure

Steam line

Turbine generator

Electricity

Steam

Core

Fissionable fuel

Control rods

Water pool

Valves

Pump

Condensor cooling water

Coolant

Review

I. Fill in each blank with the word that fits best. Choose from the words below.

fissionable   **fusion**   **chain**   **neutrons**   **control**   **coolant**

A reactor controls the _____ reaction of

_____ fuel. The number of free _____ is

carefully observed. Neutrons can be absorbed by _____ rods.

Water is the most common _____ .

II. Which statement seems more likely to be true?

A. _____ A nuclear reactor holds no dangers.

B. _____ A chain reaction must be controlled.

III. Match each term from column **B** with a term from column **A**.

	A		B
1.	_____ fissionable material	**a.**	scram
2.	_____ absorbs neutrons	**b.**	control rods
3.	_____ rapid shut-down	**c.**	uranium-235
4.	_____ careful observation	**d.**	water
5.	_____ coolant	**e.**	accurate instruments

IV. Answer in sentences.

Why are some citizens against nuclear power plants?

Review What You Know

A. Use the clues below to complete the crossword.

Across

1. Plant for controlled fission reaction
4. Joining together of two light nuclei
6. Prefix meaning "against"
8. In $E = mc^2$, c is the speed of _____ squared.
10. Detects radioactivity
14. Particle of 2 protons and 2 neutrons
15. Atoms with the same atomic number, but different atomic masses
16. The E in $E = mc^2$
17. The m in amu

Down

1. Gamma radiation travels as these
2. Reaction that continues by itself
3. Spontaneous release of particles and energy
4. Splitting a large nucleus into two smaller nuclei
5. Dense center of an atom
7. Most common coolant
9. Thermonuclear reactions require this
10. Radiation that is not made of particles
11. Smallest piece of an element
12. Ionizing radiation changes atoms into these
13. Particle equated with an electron

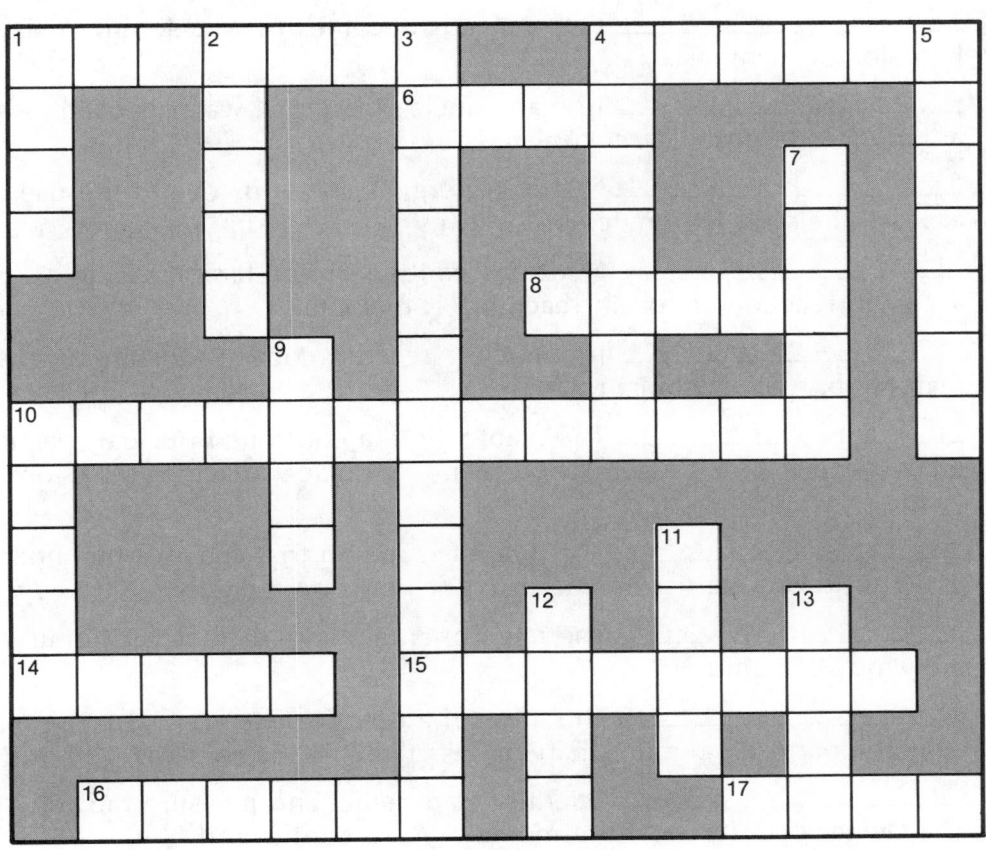

B. Write the word (or words) that best completes each statement.

1. _____ Radioactive atoms are **a.** unstable **b.** stable **c.** ions

2. _____ The particle with a +2 charge is **a.** an alpha particle **b.** a gamma ray **c.** a beta particle

3. _____ Radioactive atoms emit particles from the **a.** valence shell **b.** inner shell **c.** nucleus

4. _____ In a nuclear reaction, "lost" mass is changed to **a.** matter **b.** energy **c.** a gas

5. _____ The particle that is emitted as an electron is **a.** an alpha particle **b.** a gamma ray **c.** a beta particle

6. _____ The type of radiation that travels in wave patterns is **a.** alpha **b.** ionizing **c.** gamma

7. _____ Thermonuclear reactions require **a.** water **b.** heat **c.** light

8. _____ Two nuclei become one nucleus in **a.** fission **b.** fusion **c.** synthesis

9. _____ Atoms that have the same atomic number, but different atomic masses are **a.** ions **b.** isotopes **c.** gamma rays

10. _____ The amount of time it takes for one-half of a radioactive element to decay is its **a.** heat loss **b.** mass deficit **c.** half-life

11. _____ One nucleus becomes two nuclei in **a.** fission **b.** fusion **c.** synthesis

12. _____ In many nuclear reactors, water is used as a **a.** reactant **b.** product **c.** coolant

13. _____ Nuclear energy is an alternative to the use of **a.** fossil fuels **b.** fission **c.** radioactivity

14. _____ Neutrons emitted during fission set up a **a.** fusion reaction **b.** chain reaction **c.** fuel cell

15. _____ In a nuclear reactor, control rods are a **a.** safety feature **b.** fuel **c.** coolant

16. _____ The plant where a controlled nuclear reaction takes place is a **a.** desalinization plant **b.** nuclear reactor **c.** fission chamber

17. _____ The type of radiation that changes the atoms it hits into ions is **a.** fission **b.** fusion **c.** ionizing radiation

18. _____ The mass of an alpha particle is **a.** 1 amu **b.** 2 amu **c.** 4 amu

19. _____ The mass of a beta particle is _____ the mass of an alpha particle **a.** more than **b.** less than **c.** the same as

20. _____ Scientists can detect and measure radiation with **a.** a Geiger counter **b.** a thermometer **c.** a control rod

C. Apply What You Know

1. Study the diagram. Then answer the questions.

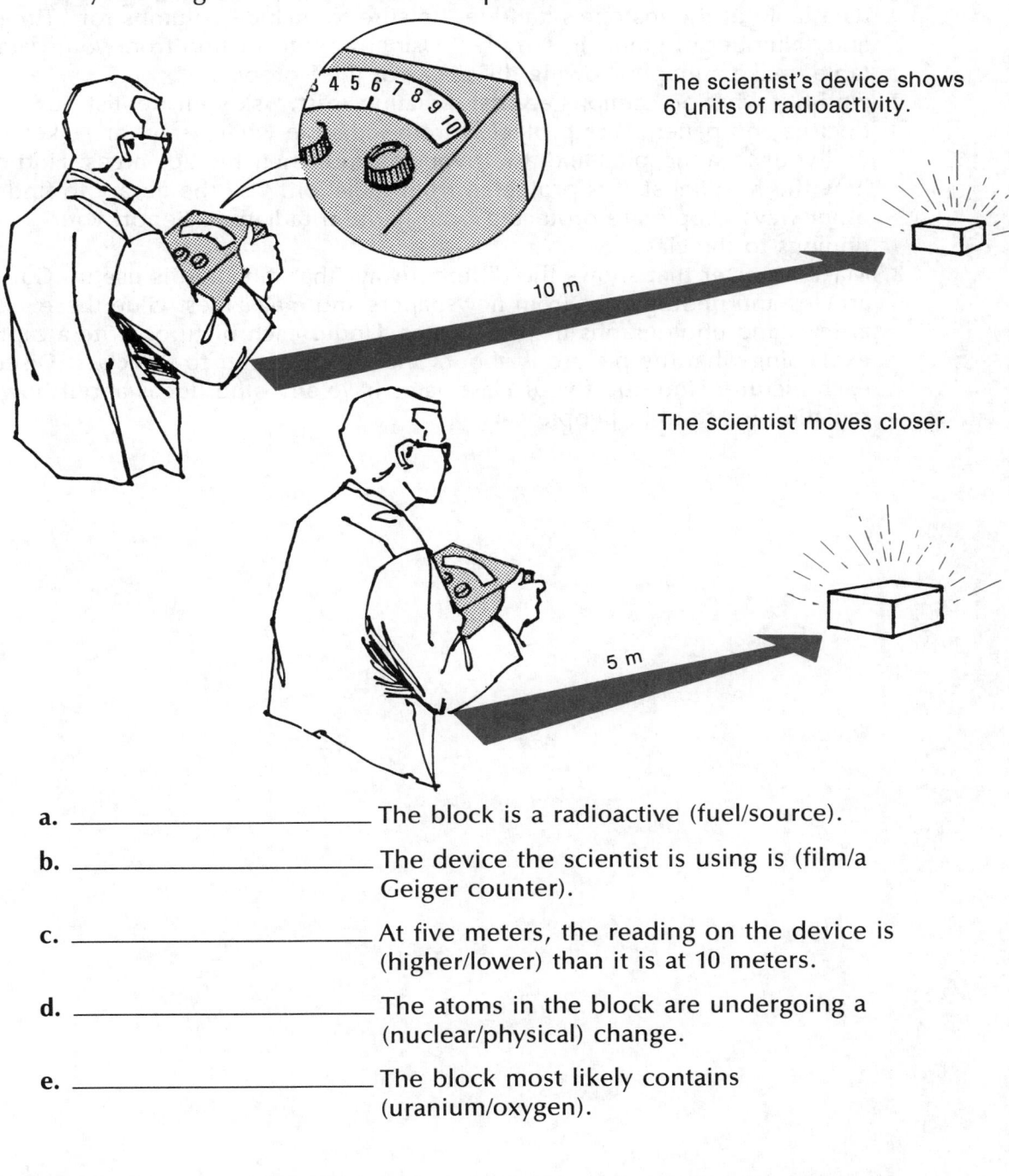

The scientist's device shows 6 units of radioactivity.

10 m

The scientist moves closer.

5 m

a. _____ The block is a radioactive (fuel/source).

b. _____ The device the scientist is using is (film/a Geiger counter).

c. _____ At five meters, the reading on the device is (higher/lower) than it is at 10 meters.

d. _____ The atoms in the block are undergoing a (nuclear/physical) change.

e. _____ The block most likely contains (uranium/oxygen).

2. Cesium-137 has a half-life of 30 years. A 100-gram sample of Cs-137 was collected in 1960.

a. _____ Only 50 grams of the sample will be radioactive in (1990/2010).

b. _____ Only 25 grams of the sample will be radioactive in (2010/2020).

c. _____ Cs-137 is (a stable/an unstable) element.

D. Find Out More

1. Choose a radioactive isotope. Find out the half-life of the isotope. Make a data table of the iostope's half-life. Be sure to include columns for "Time" and "Number of grams in sample." Using the information from your data table, make a graph showing the decay of the isotope.

2. Find out what precautions against radiation exist. Ask your dentist how doctors and patients are protected from radiation when x-rays are taken. Call your local hospital and ask if you can tour their therapy areas. Find out how the hospital staff is protected from radiation. Use the library to find other ways people are protected from harmful radiation. Report your findings to the class.

3. Make a poster that shows the different ways that radiation is useful. Cut articles and photographs from newspapers and magazines. Glue these articles and photographs to your poster. Under each picture, write a caption explaining what the picture is about. Show your poster to the class. Discuss each picture. Find out if your classmates have any other ideas about how radiation is useful to people.

Careers in Physical Science

"We Need To Take Some Tests." This statement is often heard in hospitals. One of the most common medical tests in the x-ray. X-rays are pictures taken with radiation. They look like the negative of a photograph.

X-rays have been part of medicine for many years. Many lives have been saved because of the information provided by x-rays.

Almost all hospitals, and many doctors offices, have some type of x-ray equipment. In most hospitals, x-rays are taken in the Radiology (ray-dee-OL-oh-gee) Department. X-rays are just one of the tests taken by this department. Other tests involve the use of radioactive substances, magnetic fields, and ultrasound.

Radiologists. The doctors who work in a radiology department are radiologists (ray-dee-OL-oh-jists). Radiologists work very closely with doctors in almost every other branch of medicine. Some radiologists examine patients. But most radiologists interpret test results. Doctors in all branches of medicine rely heavily on the diagnoses (DY-ag-NOH-sees) made by radiologists.

All Radiologists have college degrees. They must also attend medical school. After medical school, a radiologist must complete four years of training in radiology.

X-ray Technicians. There are many technicians working in a radiology department. If you have ever had an x-ray taken, it was probably taken by an x-ray technician. In addition to taking x-rays, x-ray technicians care for equipment, film, and patient files. Technicians also make certain that patients are properly protected from harmful radiation.

X-ray technicians must complete special training programs. Most technicians have completed at least two years of college. In many hospitals, x-ray technician are also called **radiologic technologists.**

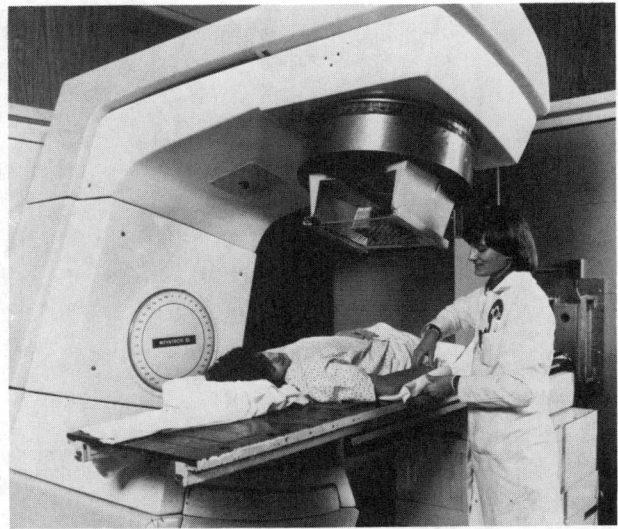

This x-ray technician is making sure the patient is properly positioned for her x-ray.

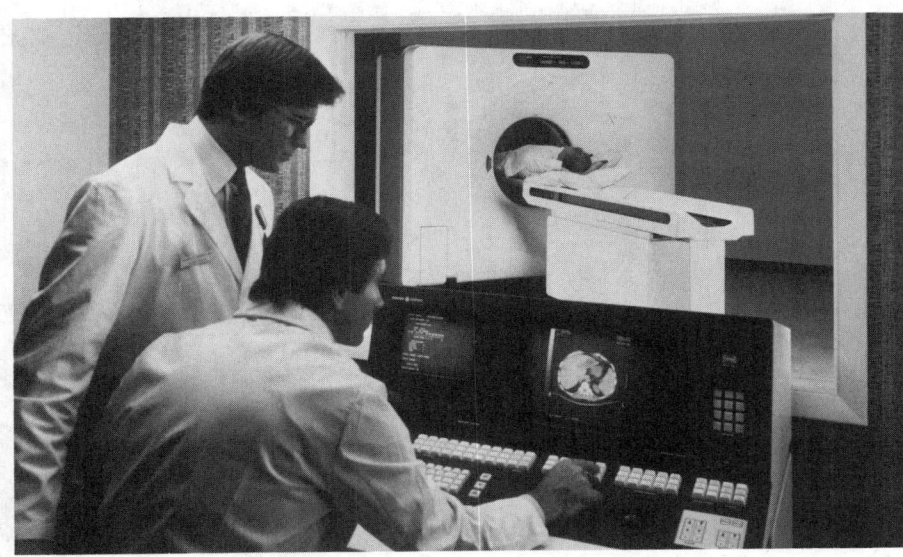

These radiologists are evaluating a special kind of x-ray called a CAT scan.

UNIT 7

THE NATURE OF ENERGY

Lesson One

What Is Energy?

Exploring Science

A Windmill Alternative. Oil became a source of world control in the twentieth century. Countries rich in oil became very powerful. They had something other countries needed.

Governments asked everyone to conserve, or save, energy. Scientists were asked to find other sources of energy. Some scientists worked with new technology. Others tried improving older energy sources. The windmill farms of Altamont Pass, California, are one result of these efforts.

Most American windmills are found in Altamont Pass and similar places in California. A hundred years ago, wooden windmills in Altamont Pass and on farms in the Midwest were used to pump water. Farm windmills are mostly gone. Today thousands of metal windmills in California "wind farms" use wind power for another purpose. Each windmill provides the electric power for about 20 homes.

Will our nation ever be covered by windmills? Probably not. Windmills don't turn unless wind speeds are over 16 kilometers per hour. Also, for a windmill to be of practical use, the winds must blow steadily. Most places cannot depend on winds as a source of power. However, the winds do blow almost all the time in Altamont Pass. Windspeeds can sometimes reach up to 130 kilometers per hour.

Scientists are continuing to do research to improve windmills. They have developed windmills that can operate at lower speeds and have even built huge superwindmills. A superwindmill can provide electricity for as many as 1500 homes, but there are only a few places on Earth with the right winds for such a giant windmill.

People from all walks of life are concerned about future energy needs. Students from one high school in New England have put their concern to work. They have made a windmill to furnish the energy needs of their school.

● Could your town or city use windmills for its energy needs?

What is one advantage of using windmills as a source of power? What is one disadvantage?

Forms of Energy

Think about how much energy your school uses in one day. How many uses come to mind? Overhead lighting, ringing bells, and central heating are just a few uses. But what is energy?

In some ways, it is easier to describe what energy is not. Energy is not matter. It does not have mass or volume. Energy is defined by what it can do. **Energy** is the ability to do work.

Gasoline must contain energy. It can move a car. Wind can turn the blades of a windmill. It must have energy. Steam must have energy. It can force the lid off a pot. Electricity can light a lamp or heat an oven, so it must also have energy.

As you can see, energy can take many forms. Some of these forms are described below.

CHEMICAL ENERGY. **Chemical energy** is the energy stored in chemical bonds. This energy can be released when the chemical bonds are broken. **Fuels** (FYOOLS) are sources of chemical energy. They contain many chemical bonds. When fuels are chemically changed, these bonds are broken and a great deal of energy is released. The burning of oil or gasoline releases chemical energy.

ELECTRICAL ENERGY. **Electrical** (ee-LEK-trih-kul) **energy,** or electricity, is created by electrons moving through a conductor. We depend heavily on electricity. Refrigerators, televisions, radios, and many other appliances use electrical energy.

MECHANICAL ENERGY. **Mechanical energy** is the energy of moving things. Much of the work of our society is done using mechanical energy. Any time an object moves, mechanical energy is used. For example, a wrecking ball striking a brick building is using mechanical energy. A waterfall turning electrical turbines uses mechanical energy. Wind moving a sailboat also uses mechanical energy.

HEAT ENERGY. **Heat energy** is created by molecules in motion. Recall the kinetic-molecular theory. It stated that molecules are *always* in motion and constantly bumping into one another. The result of this motion is heat. The faster the motion, the greater the heat.

Examples of heat energy include the moving of pistons in an engine and erupting volcanoes. Gases are heated in an engine. These heated gases move the pistons. Volcanoes erupt because of heat rising from within the Earth.

NUCLEAR ENERGY. **Nuclear energy** is the energy stored in the nucleus of an atom. It is often *incorrectly* called atomic energy. Recall that energy is released by the sun in fusion reactions. Fusion, remember, is the joining together of light atoms to form heavier atoms. Fission also involves nuclear energy. Fission is the splitting of a large nucleus into two smaller nuclei. Fission reactions take place in a nuclear reactor.

LIGHT ENERGY. **Light energy** travels in a special pattern. Much of the sun's energy, or **solar energy,** reaches us as light energy. Light passing through a magnifying lens can burn a piece of paper. Lasers are special light forms. They can

How would your life be different if you did not have electricity?

This Iranian farmer is using mechanical energy to plow his land.

turn a rock into vapor. Many of our high-tech advances depend on light energy. Light energy will be discussed more in a later unit.

SOUND ENERGY. **Sound energy** is produced by vibrating objects. Sound energy also travels in a special pattern. All moving objects produce sounds. Sound can cause rocks to tumble in an avalanche. Sound can also cause a glass to crack. Sound energy will be discussed more in a later unit.

People have always had energy needs. Wood was an early source of energy. Wood is a fuel, so it is a source of chemical energy. Later, coal and oil were found to be more convenient fuels than wood. But, supplies of these fuels are limited. We need to find other sources of energy.

To Do Yourself   What Causes a Pinwheel to Move?

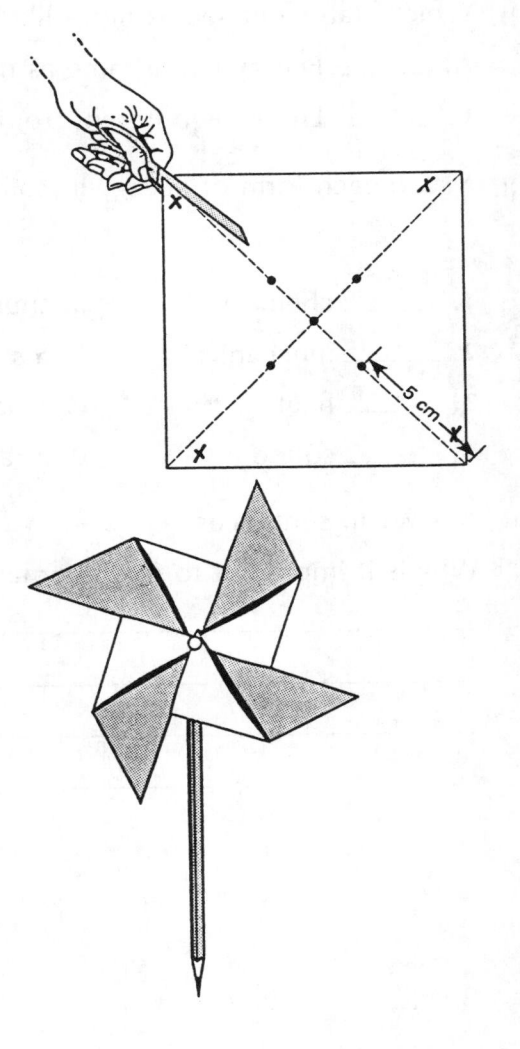

You will need:

Construction paper, metric ruler, scissors, pencil with eraser, straight pins, two plastic drinking straws

1. Using the scissors, cut a piece of construction paper into a square that measures 10 cm by 10 cm. **Caution: Use great care when working with sharp objects.**
2. Draw two diagonal lines across the construction paper. Mark an X on each corner of the paper as shown in the sketch.
3. Make a dot on each diagonal line 5 cm from the outside of the square. Cut along each of these lines until you reach the dot.
4. After cutting each line, bring each of the marked corners to the center of the square. Push a pin through all four corners and the center of the paper.
5. Fasten the pinwheel to a pencil eraser with the straight pin.
6. Move the pinwheel through the air at different speeds. Observe what happens.
7. Blow air through one straw toward the pinwheel. Observe what happens.
8. Place both straws in your mouth and blow air toward the pinwheel. Observe what happens.

Questions

1. What causes the pinwheel to move? _____
2. When did the pinwheel move the fastest? _____

   Why? _____
3. What kind of energy caused the pinwheel to move? _____

_____

Review

I. Fill in each blank with the word that fits best. Choose from the words below.

mechanical   energy   forms   light   fuels   work

A moving object has _____ energy. Energy is the ability

to do _____ . Heat and _____ are other

_____ of energy. Chemical energy is supplied by

_____ .

II. Which statement seems more likely to be true?

A. _____ Energy is the same as matter.

B. _____ Energy is different from matter.

III. Match each form of energy in column **A** with an example from column **B**.

	A		**B**
1. _____	chemical	a.	thunder shakes rocks
2. _____	mechanical	b.	a stick of dynamite is used
3. _____	heat	c.	a bat strikes a ball
4. _____	sound	d.	a radiator cap explodes

IV. Answer in sentences.

Why is it important to find alternative sources of energy?

_____

_____

_____

_____

How Is Energy Changed?

Exploring Science

Solar-Powered Flight. Much recent research has involved solar energy. Recall that solar energy is the heat and light energy provided by the sun.

One person working with solar power is Paul MacReady. His unusual plane, the *Gossamer Penguin* (GOSS-uh-mur PEN-gwin) has made a two-mile, solar-powered flight. MacReady's airplane used solar cells for energy instead of an oil-powered engine.

MacReady later built the *Solar Challenger*. This was also a solar-powered plane. The wings of this plane were covered with 16,000 solar cells.

MacReady and his team are considered amateurs. They are model plane builders. However, scientists at NASA are also using solar cells to power their aircraft.

The space shuttle *Discovery* raised the first solar sail in September 1984. This giant was ten stories tall. Scientists at NASA hope that such sails can be placed in orbit in the future to provide power for space stations.

This solar panel from the space shuttle *Discovery* transforms light energy from the sun into electricity.

● What is similar between the experiments performed by MacReady and those done by NASA?

During its test flight, the *Solar Challenger* flew 36 meters above the ground at speeds of 30-40 kilometers per hour.

The Law of Conservation of Energy

Recall the Law of Conservation of Matter. A similar law applies to energy. It is called the **Law of Conservation of Energy.** It states that energy cannot be created or destroyed. It can only be changed, or **transformed.**

Most of our technology depends on transforming energy from one form to another. For example, MacReady's airplanes change solar energy, in the form of heat and light, to mechanical energy. In the space station, solar energy is transformed to electricity.

The change of energy from one form to another is never perfect. Think about hitting a golf ball. The golfer wants *all* of the mechanical energy of the club to go to the ball. But, we hear the "crack" when the club hits the ball. Some of the mechanical energy has been changed to sound energy. Also, when the club hits the ball, some heat energy is produced. So, *most* of the mechanical energy of the club does go to the ball. However, some of this energy is changed to sound and heat. As far as the golfer is concerned, this is "lost" energy.

Let's look at another example of "lost" energy. Suppose you heat your house with a wood fire. You want *all* of the chemical energy in the wood to change to heat energy. But as the wood burns, some of the energy is changed to light energy. And the "crackling" of the fire tells you that some of the chemical energy is being changed to sound energy. For heating your house, the light and sound are useless. These forms of energy are being "lost."

According to the law of conservation of energy, energy cannot be created or destroyed. When we talk about energy loss, or "lost" energy, we do not mean that energy is destroyed. We mean that some form of energy that we cannot use is produced. This "unwanted" form of energy is often heat. That is why so many machines have cooling systems or vents. Think of all the items you use that convert energy to different forms. Electric motors, televisions, cars, and lawn mowers are just a few. How many of these items have cooling systems or vents?

Look at the illustration on the right. It shows examples of one kind of energy being transformed into another kind of energy. Can you think of any other kinds of energy that are produced by each of the items shown?

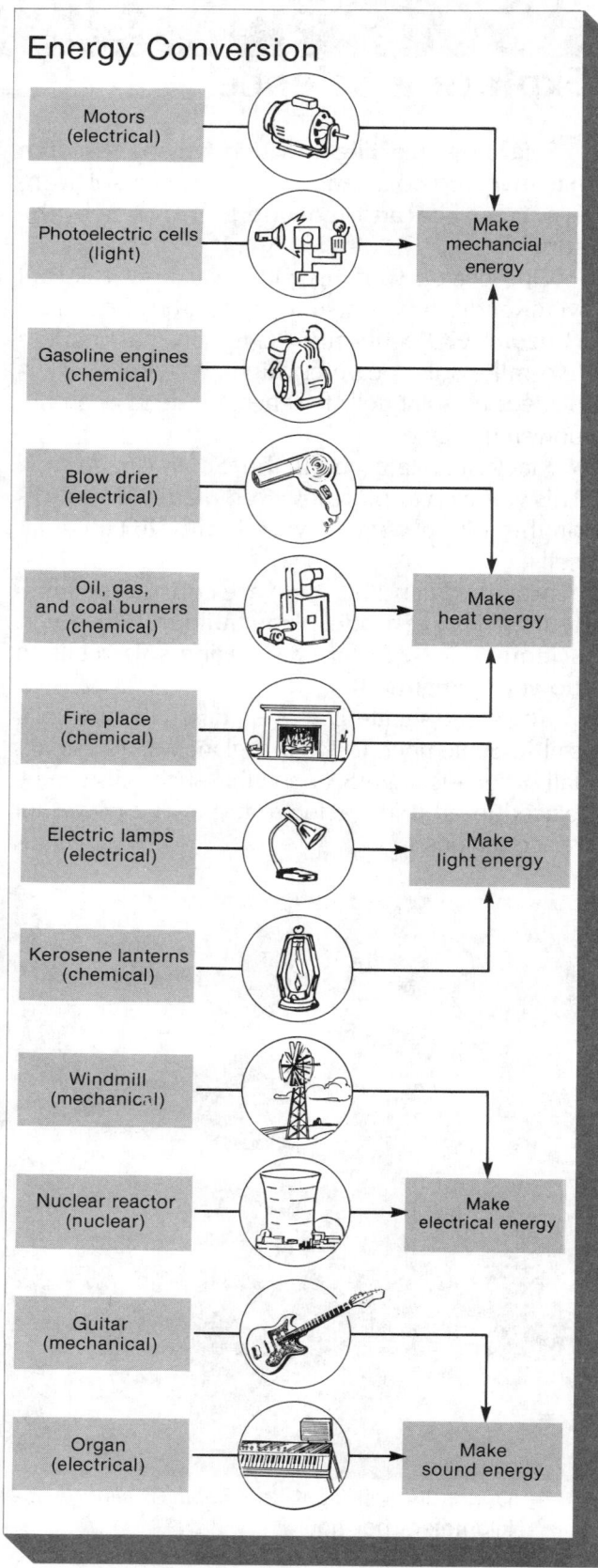

Energy Conversion

Motors (electrical) → Make mechanical energy

Photoelectric cells (light) → Make mechanical energy

Gasoline engines (chemical) → Make mechanical energy

Blow drier (electrical) → Make heat energy

Oil, gas, and coal burners (chemical) → Make heat energy

Fire place (chemical) → Make heat energy

Electric lamps (electrical) → Make light energy

Kerosene lanterns (chemical) → Make light energy

Windmill (mechanical) → Make electrical energy

Nuclear reactor (nuclear) → Make electrical energy

Guitar (mechanical) → Make sound energy

Organ (electrical) → Make sound energy

To Do Yourself   How Does Energy Change from One Form to Another?

You will need:

Aluminum foil, scissors, metric ruler, straight pin, pencil with eraser, matches, candle, candle holder, water, seltzer tablet, test tube, heat-resistant mitt, heat-resisant pad

1. Make a pinwheel out of aluminum foil. Follow the directions in the last activity.
2. Place the candle in the holder. Light the candle. **Caution: Use great care when working with fire. Keep hair and clothing away from the flame. Always wear a heat-resistant mitt when handling hot objects. Never set hot objects directly on a desk or table top. Place them on a heat-resistant pad.**
3. Hold the pinwheel about 6 cm above the flame. Observe what happens.
4. Break a seltzer tablet in half. Drop one-half of the seltzer tablet into a test tube.
5. Fill the test tube one-third full with water. Quickly place your thumb over the opening in the test tube.
6. After a few seconds, remove your thumb from the test tube. Hold the pinwheel over the flame again. Quickly aim the test tube over the flame. Observe what happens.
7. Repeat steps 4–6 using the other half of the seltzer tablet. Observe what happens.

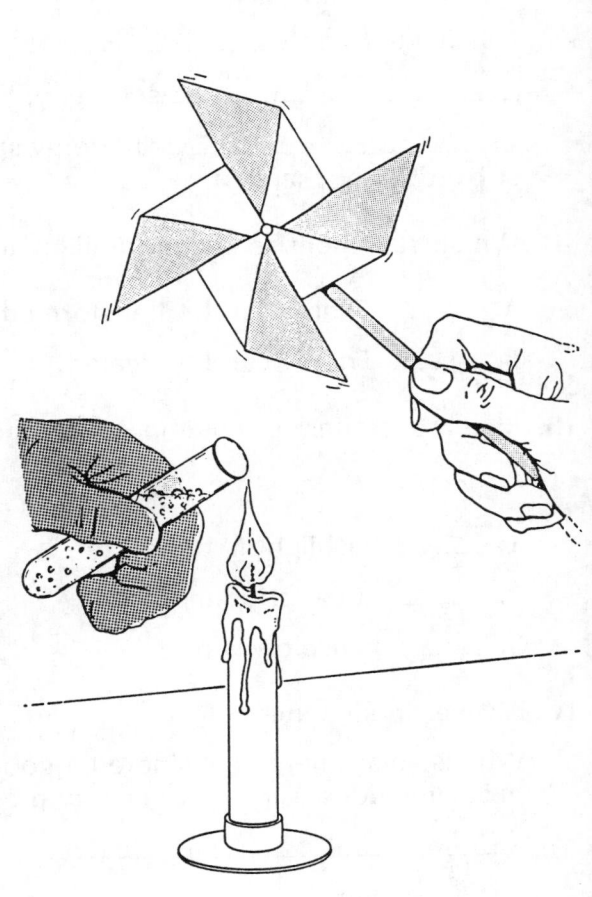

Questions

1. What happened when you held the pinwheel over the flame? _____

    _____

2. What happened when you held the test tube with the seltzer tablet and the pinwheel over the flame? _____

    _____

3. In step **3**, _____ energy was transformed to _____ energy.

4. In step **6**, _____ energy was transformed to _____ energy.

Review

I. Fill in each blank with the word that fits best. Choose from the words below.

sun   nuclear   light   sound   forms   transform   Conservation

Solar energy is produced in the forms of heat and _____.
It is made by the _____ . Special cells allow us to
_____ solar energy to other forms. The Law of
_____ of Energy states that energy is not created or destroyed, only changed.

II. Which statement seems more likely to be true?

A. _____ Energy can be transformed.

B. _____ Energy can be created.

III. Match each item in column **A** with the desired energy change in column **B**.

A	B
1. _____ flashlight is turned on	**a.** electrical to mechanical plus heat
2. _____ guitar strummed	**b.** chemical to light
3. _____ hair blower on	**c.** mechanical to sound

IV. Answer in sentences.

Why is solar energy considered a good alternative to burning fuels? What problems does using solar energy present?

_____

_____

_____

_____

How Is Energy Classified?

Exploring Science

Danger and Rescue. Building construction is common in most large cities. It is not unusual to see large cranes at construction sites. The cranes are used to lift large loads of building materials to construction workers.

One warm May day, Brigete Gerney was passing a construction site in New York City. Suddenly, the crane working at the site toppled over. Ms. Gerney was trapped under the 35-ton piece of equipment.

The crane had been lifting a two-ton load of steel rods. The load was too large for the crane to handle. The crane toppled over at the edge of the foundation pit. When it fell, Mrs. Gerney was pinned at the edge of the pit.

Rescue workers were afraid to make a mistake. If they moved anything the wrong way, the crane might fall into the pit. Mrs. Gerney would be killed. For six hours, millions of Americans listened and watched TV for news about the rescue attempt. Hundreds watched as Mrs. Gerney put her trust in police and rescue workers.

After several hours, rescue workers gave up trying to move the crane. They decided to dig Mrs. Gerney out instead. The digging was done very slowly. The crane could not be allowed to slip.

Finally, in the late afternoon, Mrs. Gerney was freed. Her legs were badly injured, but she was alive! Americans praised the courage of a brave lady and a great rescue team.

● Why was the rescue so slow?

What might happen if this crane fell into the pit?

Potential and Kinetic Energy

All forms of energy can be classified into two types—potential energy and kinetic energy. **Potential** (POH-ten-chul) **energy** is energy that is stored in some way. The two most common ways that energy is stored is by *position* and by *composition* (kom-POH-zih-shun). Composition means the atoms that make up the substance and the bonds that hold them together.

Kinetic (kih-NET-ik) **energy** is energy due to *motion*. *All* moving things have kinetic energy.

Let's look back at the crane accident. The crane had already fallen. Why were rescuers so careful while trying to lift the crane? Because of the crane's *position* at the edge of the pit, it was still very dangerous. The crane had a great deal of potential energy.

Suppose that the digging had caused the crane to drop into the pit. The potential energy of the crane would have been released. It would have changed to kinetic energy, the energy of motion.

Think about a wind-up toy. When the spring is unwound, the toy has no energy. Kinetic energy is used to turn the key that winds the spring. When the spring is tightly wound, the energy used to wind it is stored in the spring. It is potential energy.

When the toy is turned on, the spring starts to unwind. The toy moves. The potential energy in the spring changes to kinetic energy.

Chemical energy is a common form of potential energy. It is stored because of the *composition* of a substance. If you used a battery-powered race car instead of a wind-up toy, the potential chemical energy in the batteries would change to kinetic energy when the car was turned on. Any time something moves, it has kinetic energy.

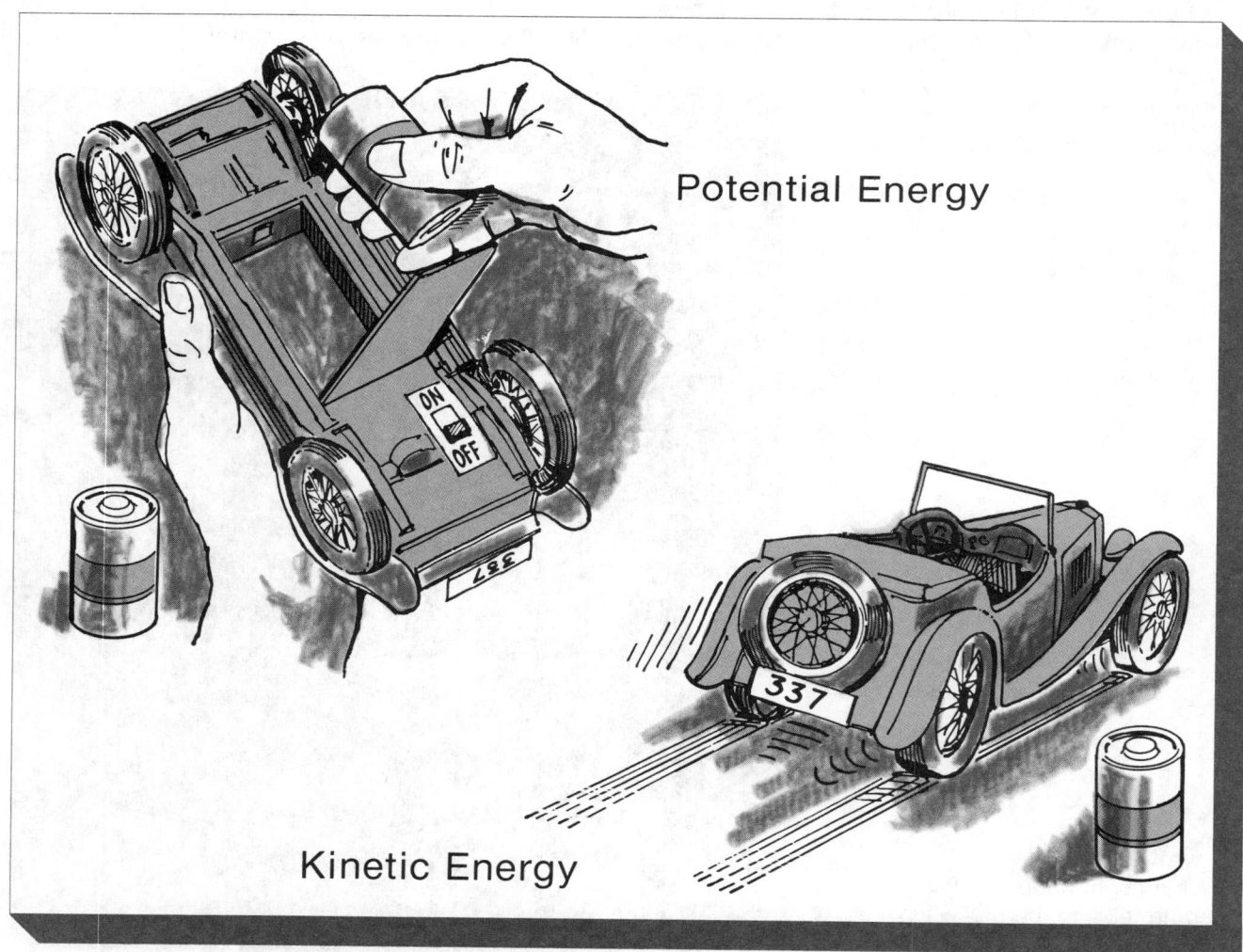

Potential Energy

Kinetic Energy

Review

I. Fill in each blank with the word that fits best. Choose from the words below.

potential   **kinetic**   **stored**   **motion**   **chemical**   **mechanical**

Energy has two forms. It is either kinetic or _____ .

Potential energy is _____ energy. One form of potential

energy is _____ energy. _____ energy is

a form of kinetic energy.

II. Which statement seems more likely to be true?

A. _____ Potential energy must be a weak energy source.

B. _____ Potential energy can be a powerful energy source.

III. Match each item in column **A** with its description in column **B**.

A	**B**
1. _____ a barrel of oil	**a.** high in potential energy
2. _____ a boulder in a pit	**b.** high in kinetic energy
3. _____ a 20 ton truck moving at 55 miles per hour	**c.** low in potential energy

IV. Answer in sentences.

Jenny is batting .200 this season. Her coach says, "You have the potential to be a .350 batter." Explain what the coach means by this statement.

_____

_____

_____

_____

How Can We Represent Forces?

Exploring Science

Thoroughbred Miracles. Horse shows and race tracks can be dangerous places. Riders can be thrown and injured. Horses can tumble to the ground.

Until recently, if a horse broke its leg, it had to be killed. There was no other solution. The injured leg could not be set and put in a cast like the leg of a human. Horses cannot lie down for weeks while a bone heals. A horse must stand to breathe and digest its food properly.

A splint on the leg does not work either. A horse needs all four legs to carry its weight. If one leg cannot function, the others collapse.

Dr. Lawrence Bramlage has found a new solu-

tion to this problem. He performs surgery on the injured leg. He inserts metal plates, screws, and wires into the injured bones. These metal parts reduce the amount of stress on the bone. They provide some extra support.

Of the first 40 adult horses that have had this surgery, 22 have survived. Dr. Bramlage has had even greater success with younger horses. In fact, some of his patients have actually competed in shows and races again.

● Why might success be more likely in a younger horse?

Jockey, Carlos Astorga was not injured when his horse, Sudden Danger, fell to the ground. But the horse suffered a fractured right foreleg.

Forces and Vectors

Dr. Bramlage approached the problem of supporting a broken leg the way an engineer designs a bridge. He identified the forces acting on the broken bone. Then, he tried to change

the amount of stress, or force, on the bone.

A **force** is a push or a pull. You cannot see a force. You can only see its effect. You know that hurricane winds have great force. They can blow down fences and trees. Winds have a force that pushes. You know that a tow truck has great force. It can move a car or a bus. A tow truck has a force that pulls.

Force is measured in metric units called **newtons** (NOO-tunz). The symbol for a newton is **N**. Sir Isaac Newton was a scientist who studied forces. The newton is named after him.

Many times it is not enough to know only the size, or strength, of a force. You also need the *direction* of the force. A combination of force and direction is called a **vector** (VEK-tur). An arrow can be used to show a vector. When an arrow is used, it is called a **vector diagram**. The arrowhead shows the direction of the force. The length of the arrow shows the size. In order to draw the arrow to the correct length, a scale must be given with the diagram.

Let's look at a vector diagram.

W ←————————————————

1 cm = 10 newtons

The arrowhead shows that the force is to the west. The length of the arrow represents the size of the force. We must use the scale to find out how strong the force is.

The scale tells us that each centimeter is equal to 10 newtons of force. Now measure the length of the arrow. It is 6 cm long. If 1 cm is equal to 10 newtons, then 6 cm must be equal to 6×10, or 60 newtons. The vector diagram shows a force of 60 newtons to the west.

Let's try another vector diagram. Suppose a wind is blowing toward the east with a force of 500 newtons. How would you go about drawing a vector diagram for that wind?

First, you know that the arrowhead will point toward the east (to the right). The length of the arrow must show the size of the force. To figure out how long to make the arrow, you need to choose a scale. Let's have 1 cm = 100 newtons.

How do we use this scale to figure out how long to make the arrow? The force of the wind is 500 newtons. If 1 cm on our scale is equal to 100 newtons, how many centimeters will equal 500 newtons? To find out, divide 500 by 100: $500 \div 100 = 5$. The arrow must be 5 cm long. The vector diagram for this wind is

————————————————→ E

In the everyday world, we rarely deal with a single force. Think about a tug-of-war. Ruth's team is pulling to the left with a force of 100 N.

Ron's team is pulling to the right with a force of 80 N. First, we must choose a scale for the vectors. Let 1 cm = 20 N.

The scale must be the same for both teams. Since both forces are acting on the same object (the rope), we can put the two diagrams together.

Ruth's team   Ron's team
←————————————————•————————————→

Obviously, Ruth's team will win the tug-of-war. But much of their force is cancelled by Ron's team. The resulting force is called the resultant (REE-zul-tant).

The resultant can be shown in a vector diagram also. The length of the resultant can be found by subtracting the length of the shorter vector from the length of the longer vector. *This can only be done if the same scale is used!*

The vector diagram for the resultant in our tug-of-war example would look like this.

←————

The resultant shows a vector that looks like Ruth's team alone applied a force of 20 N to the left.

Suppose the two teams applied the same amount of effort to the west to move a stalled car. The vector diagrams would look like this.

Ruth's team ←————————————
Ron's team ←————————————
Let 1 cm = 20 N

In this case, the two forces are moving in the same direction. Thus, the resultant force is equal to the two forces added together. That sum is equal to 100 N + 80 N = 180 N. Our diagram of the resultant would look like this.

W ←————————————————————

The arrow is 9 cm long. Remember, our scale was 1 cm = 20 N. The resultant force is equal to 180 N. Thus, the arrow is $180 \div 20 = 9$ cm long.

So, if the *forces are opposite,* the resultant is found by *subtracting the forces.* If the *forces act in the same direction,* the resultant is found by *adding the forces.*

Now let's look at a more difficult case. Suppose you are pushing a table east with a force of 100 N. At the same time, your friend is pushing the same table north with a force of 60 N. How would the resultant for these two forces be shown?

First, we must choose a scale.

Let 1 cm = 20 N

Next, we can draw each of the vectors, starting from the same point. Each vector must show the direction of the force.

The first drawing on the next page shows how such a vector diagram would look, with the two forces acting on the table. The next two vector diagrams show how the forces combine.

To Do Yourself   How Are Forces Represented With Vectors?

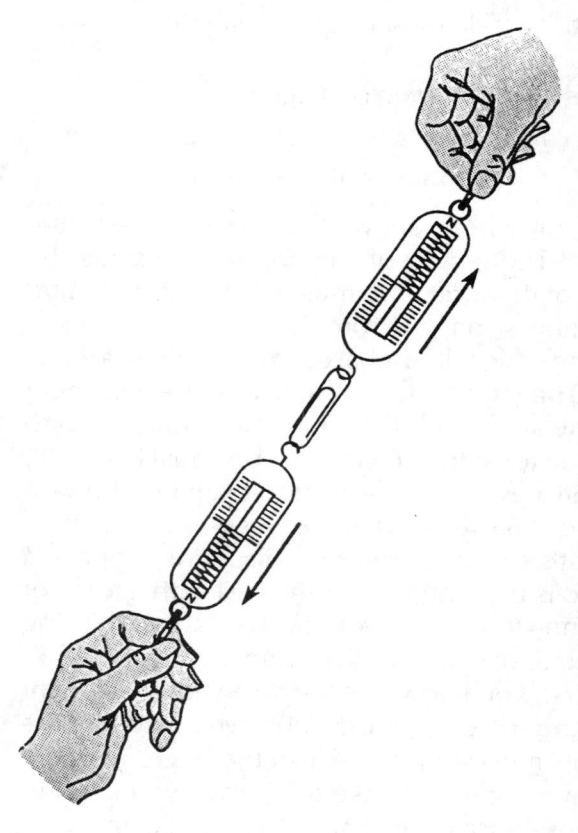

You will need:

Two spring scales, 500 gram weight, paper clip, graph paper, metric ruler

1. Attach the weight to the spring scale. Pull the weight across the graph paper from left to right. Record the force (in newtons) needed to pull the weight across the paper.
2. Draw a vector diagram to show the direction and the force of the moving weight. Let one space on the graph paper equal .5 N.
3. Attach a paper clip to a spring scale. Attach another spring scale to the other side of the paper clip.
4. Pull the spring scales in opposite directions with a force of 3 newtons on each side. Draw a vector diagram to represent each of the spring scales.
5. Repeat step **4** exerting a force of 2 N to the right, and record the force needed to move the paper clip to the left. Draw a vector diagram showing the forces exerted on the paper clip.

Questions

1. What two things must a force have to be drawn as a vector diagram? _____

_____

2. What happened to the paper clip in step **4?** _____

_____

3. What would the resultant in step **5** show? _____

_____

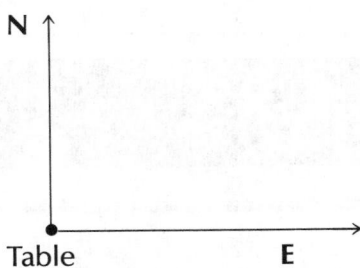

Table   E

Next, we draw a rectangle, connecting the two vectors with dashed lines.

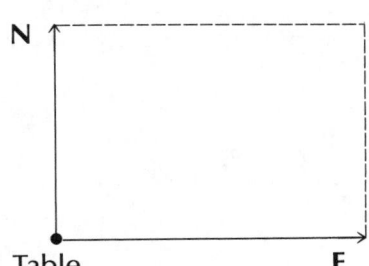

Table   E

To find the resultant force, the diagonal of the rectangle is drawn from the starting point to the northeast corner of the rectangle. Thus, our resultant looks like this:

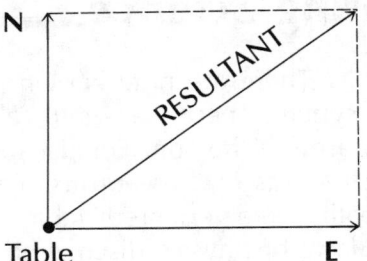

Table   E

The resultant is 5 cm long. The amount of force used to move the table is therefore 100 N (5 cm × 20 N). Its direction is northeast. So, the table would move to the northeast as if it were moved by a single force of 100 N.

Review

I. Fill in each blank with the word that fits best. Choose from the words below.

direction   arrow   resultant   force   vector   size   object

A _____ is a push or a pull. Forces can be drawn in

_____ diagrams. The length of the _____

shows the _____ of the force. If more than one force

acts on an object, the total force is called a _____

II. Which statement seems more likely to be true?

A. _____ If two vectors are in opposite directions, the resultant is found by subtraction.

B. _____ If two vectors are in opposite directions, the resultant is found by addition.

III. Match each term in column **A** with its description in column **B**.

	A		**B**
1.	_____ a scale	**a.**	———————→
2.	_____ a unit of force	**b.**	Let 2 cm = 5 N
3.	_____ a vector	**c.**	newton
4.	_____ opposing forces	**d.**	←———•———→

Lesson Five

What Is Gravity?

Exploring Science

Planet X? There are nine known planets in our solar system. These planets travel in paths, or orbits, around the sun. For decades, scientists kept records of these orbits. They found that the paths were very predictable. The orbits followed laws of physics discovered centuries ago. However, the data on Neptune and Uranus showed small differences between the calculated path and the actual orbits.

Some scientists predicted that a tenth planet exists. They called this planet Planet X. They thought that Planet X would be 100 times farther from the sun than the Earth is from the sun. They thought that its great mass could be pulling on Neptune and Uranus. This pull could explain the strange data about their orbits.

After a hundred years of arguments, Planet X is now declared "dead." Actually, scientist Myles Standish says it never existed. Using new facts from NASA, he explains the orbit data in a different way. NASA found that Neptune has less mass than we thought. Since mass affects the orbit, Dr. Standish made new calculations.

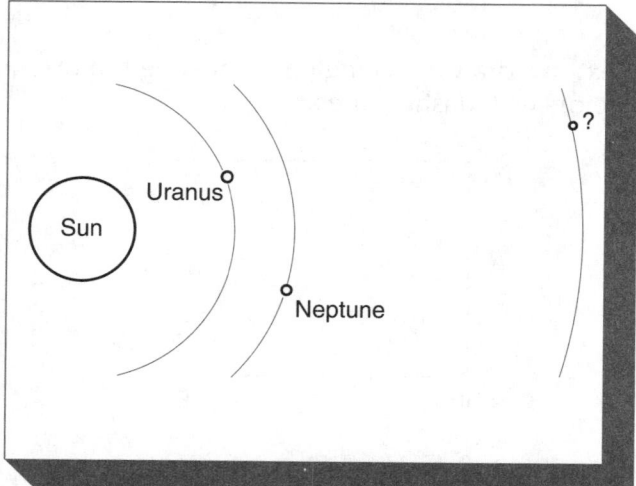

The strange orbits now match the predicted path.

Although there have been mini-planets found, no Planet X is likely to exist.

● Why was Planet X called a theory?

Gravity

Why do planets stay in orbit about the sun? How could Planet X have affected the orbits of the other planets? The explanation is based on a force called *gravity* (GRAV-uh-tee).

Gravity is the force of attraction that exists between any two objects. The size of the force depends on the masses of the objects. If the mass of an object is very great, its attractive force will be very great.

In our solar system, the sun is the most massive body. So, every other object in the solar system is attracted to the sun. That is why the planets remain in orbit around the sun.

Another factor that affects the force of gravity is the distance between the objects. The greater the distance, the smaller the force. In fact, Newton found that the force of an object

decreases in an amount equal to the square of the increase in the distance between the two objects. Recall the discussion of Inverse Square Law and radioactivity. Newton's findings work in the same way.

Objects on Earth are attracted more strongly to Earth than they are to the sun. On Earth, gravity usually means the pull of an object to the Earth. The strength of this force changes with the mass of the object being pulled.

Weight is the measure of the Earth's gravitational pull on the mass of an object. An object on a less massive body, such as the moon, would not be attracted as strongly as it would on Earth. Therefore, the weight of the object would be less. However, its mass would remain the same.

Weight is a measure of Earth's gravitational pull on the mass of an object. On Earth, this man weighs 40 newtons. On the moon, he only weighs 6 newtons. Can you explain this difference?

To Do Yourself   How Can Gravitational Force Be Measured and Drawn?

You will need:

Spring scale, various weights, graph paper, metric ruler, pencil

1. Hang each weight from the spring scale. Record the gravitational force on each weight in newtons.
2. Using the graph paper, draw a vector diagram for each weight. Let one space on the graph paper equal .5 N.

Earth

Questions

1. What two things determine an object's weight? _____

_____

2. If the mass of an object is increased, what happens to its weight? _____

_____

3. Why do objects on the moon weigh less than they do on Earth? _____

_____

4. What do the different lengths of the vectors in your diagrams tell you about the

masses of the objects? _____

_____

_____

_____

_____

Review

I. Fill in each blank with the word that fits best. Choose from the words below.

attraction   **weight**   **mass**   **force**   **gravity**   **distance**

There is a force of _____ between any two objects. This

_____ is called _____ . The strength of

the force depends on the _____ of the objects. On Earth,

the effect of gravity is called the _____ of an object.

II. Which statement seems more likely to be true?

A. _____ The moon is less massive than the Earth, so an object's mass is less.

B. _____ The moon is less massive than the Earth, so an object's weight is less.

III. The weight of object A is 200 newtons. Draw a vector diagram to show its force. Let 1 cm = 20 N.

IV. Answer in sentences. (Include a vector diagram).

A boy holds a 10-newton notebook in his hand. The book does not move. What forces are acting on the book?

_____

_____

_____

_____

Review What You Know

A. Unscramble the groups of letters to make science words. Write the words in the blanks.

1. TIVARYG (force of attraction between two objects) _____

2. HMECCAIL (form of energy in fuels) _____

3. CKITENI (energy in motion) _____

4. ROSAL (energy from the sun) _____

5. VERTNANOICSO (basic law of energy) _____

6. TRVCOE (combination of a force and a direction) _____

7. OLPAITTEN (stored energy) _____

8. CEORF (a push or pull) _____

B. Write the word (or words) that best completes each statement.

1. _____ The ability to do work is **a.** force **b.** energy **c.** conservation.

2. _____ Potential energy is energy that is **a.** in motion **b.** from the sun **c.** stored

3. _____ The metric unit of force is the **a.** watt **b.** pound **c.** newton

4. _____ A vector diagram shows a force's **a.** size **b.** direction **c.** size and direction

5. _____ The metric unit of weight is the **a.** watt **b.** pound **c.** newton

6. _____ The resultant of forces moving in the same direction are **a.** added together **b.** subtracted **c.** multiplied

7. _____ The resultant of forces moving in opposite directions are **a.** added **b.** subtracted **c.** multiplied

8. _____ Gravity on the moon is less than gravity on Earth because the moon has **a.** less volume **b.** less mass **c.** more mass

9. _____ Energy can be **a.** created **b.** transformed **c.** destroyed

10. _____ The form of energy that results from electrons moving through a conductor is **a.** mechanical energy **b.** chemical energy **c.** electrical energy

11. _____ The form of energy that results from vibrating objects is **a.** sound energy **b.** light energy **c.** electrical energy

12. _____ The energy stored in the nucleus of an atom is **a.** chemical energy **b.** nuclear energy **c.** electrical energy

13. _____ The force of attraction that exists between any two objects is **a.** weight **b.** gravity **c.** a vector

14. _____ A measure of Earth's gravitational pull on the mass of an object is **a.** weight **b.** gravity **c.** mass

15. _____ The mass of an object on Earth is _____ the mass of the same object on the moon **a.** more than **b.** less than **c.** the same as

C. Apply What You Know

1. Study the drawings below. Then answer the questions.

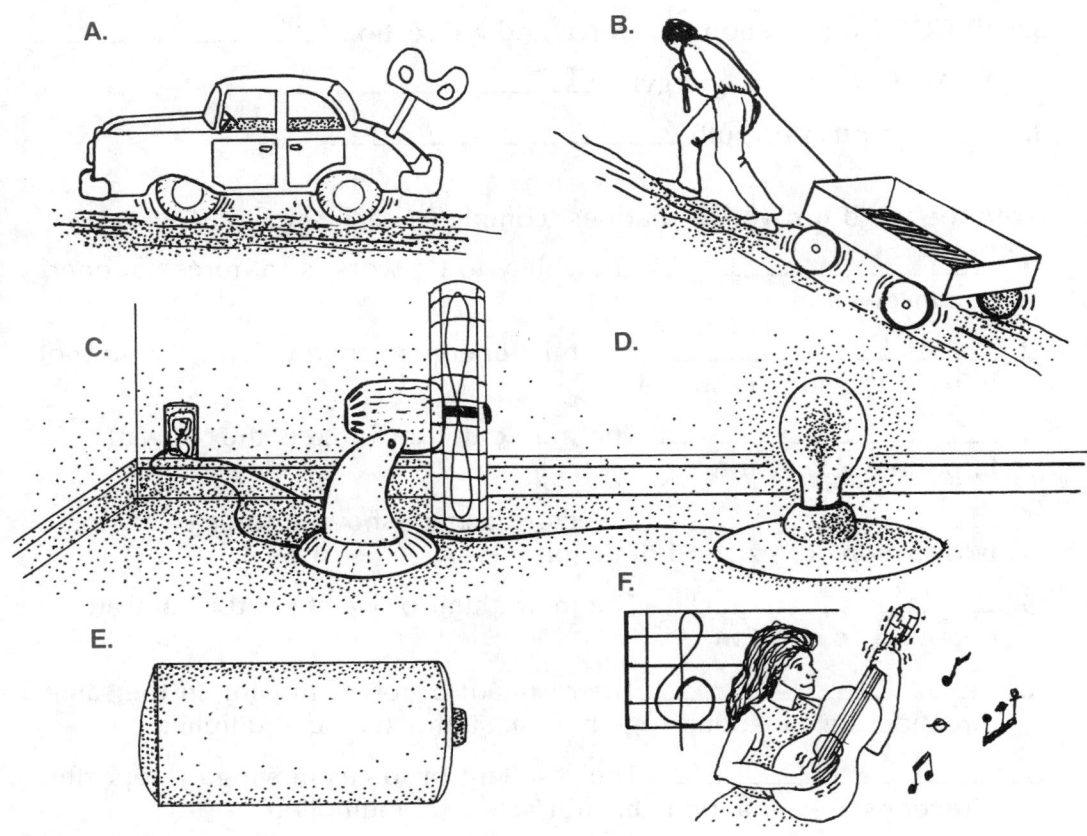

a. Electrical energy is changed to mechanical energy in drawing _____.

b. Potential chemical energy is shown in drawing _____.

c. Electrical energy is changed to heat and light energy in drawing _____.

d. Mechanical energy is changed to sound energy in drawing _____.

e. Mechanical energy is shown in drawing _____.

f. Kinetic energy is being changed to potential energy in drawing _____.

2. Study the drawing below. Then use the following words to fill in the blanks.

    vector diagram   **resultant**   **5 cm**   **4 cm**   **2 cm**   **3 cm**   **400 N**

    500 N   **300 N**   **SE**   **SW**   **scale**

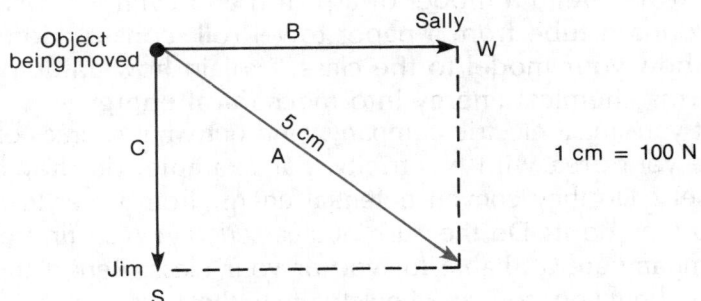

Sally is moving an object to the west with a force of 400 N. Jim is moving the same object to the south with a force of 300 N.

a. The drawing above is called a _____ .

b. 1 cm = 100 N is the _____ .

c. The object is moving _____ .

d. Vector B is _____ long.

e. Vector C is _____ long.

f. Vector A is the _____ .

g. The size of the force shown by vector A is _____ .

3. Sam is moving an object to the east with a force of 40 N. Sarah is moving the same object to the north with a force of 30 N. Draw a vector diagram that shows the movement of the object. Use 1 cm = 10 N for your scale.

D. Find Out More

1. Make a data table showing your weight on the moon and each of the planets in the solar system. Reference books in your library will help you find the gravitational forces for each of these bodies.
2. Use references to find out how the pistons and cylinders in an automobile engine work. Build a model of a piston and cylinder. Use materials such as the cardboard tube from a paper towel roll, construction paper, and paper clips. Show your model to the class. Explain how an automobile engine transforms chemical energy into mechanical energy.
3. Contact your local electric company. Find out what sources of energy they use to provide your area with electricity. For example, do they burn fuels, and, if so, what fuels? Do they convert potential energy from water to electrical energy? If so, how do they do it? Do they use nuclear energy? Also find out if tours of the electric company are available for you or your class. Report the information that you discover about your sources of electricity to the class.

Summing Up
Review What You Have Learned So Far

A. Study the illustrations. Then identify each illustration on the line below it. Choose from the items below.

atomic nucleus   vector diagram   chemical symbol   molecule

negative ion   change of phase   chemical formula   chemical equation

crystal   positive ion

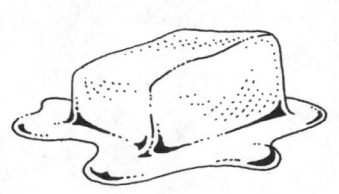

1. _____

2. _____

3. _____

Ca

Fe_2Cl_3

$Zn + H_2SO_4 \rightarrow ZnSO_4 + H_2$

6p
6n

4. _____

5. _____

6. _____

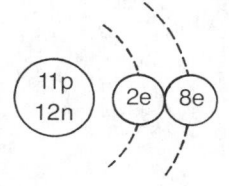

11p
12n 2e 8e

O C O

7. _____

8. _____

B. Each statement below refers to the illustration having the same number as the statement. Circle the underlined word or phrase that makes each statement true.

1. The physical change shown here is (freezing/melting).
2. Arrow Z in this drawing is called the (resultant/product).
3. This illustration represents the element (carbon/calcium).
4. One molecule of the compound represented here is made up of (two/five) atoms.
5. The H_2 in this illustration is a (reactant/product).
6. A neutral atom of this element will have (six/twelve) electrons.
7. The particle shown here has (lost/gained) an electron.
8. The compound illustrated here is (calcium hydroxide/carbon dioxide).

UNIT

8

WORK AND MACHINES

Lesson One

What Is Work?

Exploring Science

Pumping Iron. At one time, lifting weights was a rare activity. However, the recent concern for physical fitness has brought about a big change. The terms "pumping iron" and "moving mass" have become common. Americans can be found lifting weights at health spas, on beaches, and in their homes.

Some sports medicine experts are concerned about the number of people lifting weights. They warn weight-lifters of a hidden danger—high blood pressure.

The normal blood pressure of humans is $^{120}/_{80}$. A reading of $^{150}/_{100}$ is considered high. But readings as high as $^{400}/_{300}$ have been recorded in weight-lifters. As athletes try to lift very heavy weights, their blood vessels are pinched closed by tensed muscles. When this happens, the blood cannot flow properly and their blood pressure becomes dangerously high.

Doctors believe that athletes should not struggle to lift the greatest possible mass. Instead, smaller weights should be lifted repeatedly. This type of work-out helps to strengthen muscles without pinching off the blood vessels.

● What is one danger of weight-lifting?

What is one advantage of lifting weights?

Work and Power

A scientist defines **work** as a force that moves an object over a distance. Suppose an athlete applied a force of 980 N but no weights were moved. He has not done any work. Even if he struggled and became exhausted, no work was done because no object was moved.

Now suppose the amount of weight on the bar is reduced. Again the athlete applies a force of 980 N. This time he lifts the weights 1 meter. Work has been done.

The amount of work the athlete did can be found by using an equation. Let's find out how much work was done by the athlete.

work = force applied × distance moved
(W)   (f)   (d)

$$W = f \times d$$
$$W = 980 \text{ newtons} \times 1 \text{ meter}$$
$$W = 980 \text{ newton-meters}$$

The newton-meter (Nm) may be used as a unit for work. But a more convenient unit is the joule (JOOL). A **joule** is the metric unit for work. One joule is equal to one newton-meter. The symbol for the joule is **J**. So,

$$1 \text{ J} = 1 \text{ Nm}$$

Converting from newton-meters to joules is easy because the ratio is 1:1. For example, 10 Nm = 10 J. 40 Nm = 40 J. So, the amount of work done by the athlete must be 980 J since he did 980 Nm of work.

Let's look at another example. Suppose you used a force of 10 N to move a book 100 meters. Let's find out how much work you did using the equation for work.

$$W = f \times d$$
$$W = 10 \text{ N} \times 100 \text{ m}$$
$$W = 1000 \text{ Nm}$$
$$W = 1000 \text{ J}$$

In other words, you did more work than the weight-lifter!

In the examples we've looked at, the amount of time it took to do the work was not considered. However, if the amount of time it takes to do the work is known, then power can be measured. **Power** is the rate at which work is done.

Recall that the joule is the unit for work. Power also has its own unit. The unit for power is the **watt**. One watt is equal to 1 joule of work per second.

Which illustration shows work being done? How do you know?

Let's find out how much power the athlete used if it took him 1 second to lift the weights.

$$\text{Power} = \frac{\text{Work}}{\text{Time}}$$

$$P = \frac{W}{t}$$

$$P = \frac{980 \text{ J}}{1 \text{ sec}}$$

$$P = 980 \text{ watts}$$

Suppose you took 10 seconds to move the textbook. How much power did you use?

$$P = \frac{W}{t}$$

$$P = \frac{1000 \text{ J}}{10 \text{ sec}}$$

$$P = \frac{100 \text{ J}}{1 \text{ sec}}$$

$$P = 100 \text{ watts}$$

Compare the amount of work you did with the amount of work done by the athlete. You will find that you did more work but the athlete used more power.

To Do Yourself   How Is Work Measured?

You will need:

Spring scale, meter stick, book

1. Attach a book to a spring scale.
2. Use the spring scale to lift the book one meter off the floor. Find out how much work was done. Your answer should be given in joules.
3. Repeat step **2** lifting the book two meters. Again find out how much work was done.
4. Use the spring scale to drag the book along the floor for a distance of one meter. Find out how much work you did.
5. Repeat step **4** dragging the book two meters.
6. Hold the book steady and pull on the spring scale to the highest reading. Find out how much work was done?

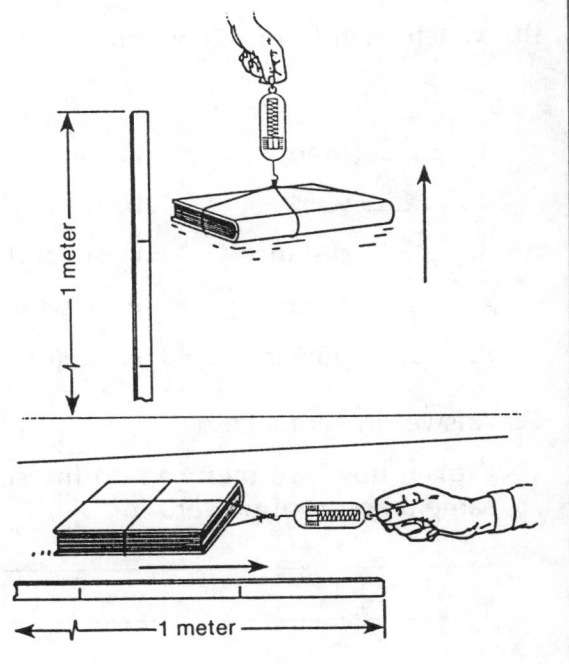

Questions

1. Was the work done in steps 2 and 3 equal? Explain. _____

_____

2. Which step required the most work 4 or 5? _____

_____

3. How much work was done in step 6? Explain your answer. _____

_____

Review

I. Fill in each blank with the word that fits best. Choose from the words below.

**force   distance   work   power   joule   watt   multiply   divide
newton**

If a _____ causes an object to move, _____
has been done. The _____ is the unit for work. To find out
how much work has been done, we must _____ the
strength of the force by the _____ the object moved.
_____ is the rate at which work is done. The
_____ is the unit for power.

II. Which statement seems more likely to be true?

A. _____ Work is accomplished when a force is applied.

B. _____ Sometimes a force is applied but no work is done.

III. Match each term in column **A** with its unit in column **B**.

A	B
1. _____ time	**a.** joule
2. _____ work	**b.** watt
3. _____ distance	**c.** second
4. _____ force	**d.** newton
5. _____ power	**e.** meter

IV. Answer in sentences.

Explain how two men can do the same amount of work without using the
same amount of power.

_____

_____

_____

_____

Lesson Two

What Is Friction?

Exploring Science

Modern Mississippi Adventure. Who were the two lads on the Mississippi River? Tom Sawyer and Huckleberry Finn? No, Dean Pollee and Terry Chapman.

In July 1981, these two seventeen-year olds and Bob Windt set a new world's record. In only 13 days, these men "flew" from Illinois to Louisiana and back. In this trip, the word "flew" took on a new meaning. They were not in an airplane. They were in an ACV—an air-cushioned vehicle. This ACV was designed by Bob Windt.

An ACV does not travel on land or in water. It travels about 30 centimeters above the surface. The ACV has a very special design and motor. The motor produces a cushion of air under the ACV. The cushion keeps the bottom of the ACV from rubbing against the surface of the water. This allows the ACV to travel at very unusual speeds.

The top speed for an ocean liner is about 65 kilometers per hour. An ACV can easily travel at 90 kilometers an hour. The ACV also has another advantage. When traveling down river, Windt and his crew reached a traffic jam. Boats were lined up to by-pass a dam in the river. To

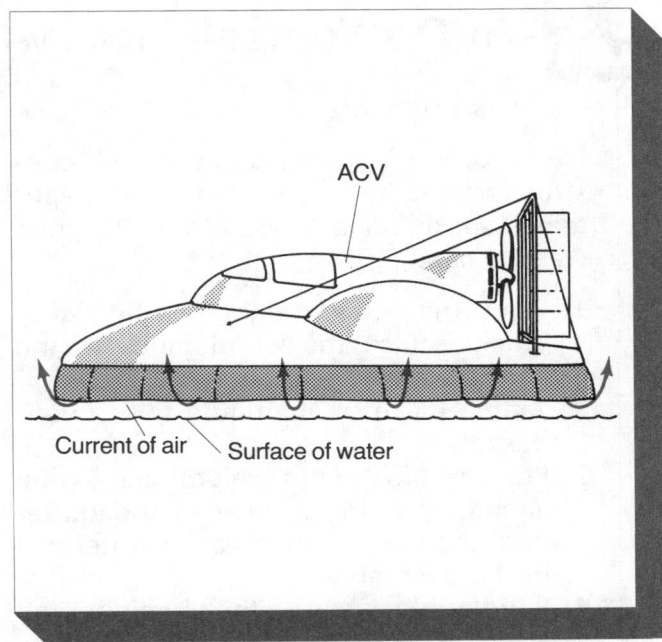

Current of air   Surface of water

everyone's amazement, Windt took the ACV right over the dam.

● Why can a powerful ocean liner only travel at 65 kilometers an hour and an ACV travel at 90 kilometers an hour?

Friction

Why can an ACV travel faster than a high-powered ocean liner? One answer is because of friction (FRIK-shun). **Friction** is the rubbing of one body against another body.

Friction is always present when work is done. Remember, work *always* involves a force moving an object. Friction opposes, or works against, this force causing the movement of the object to be slowed.

Let's look at how friction affects the movement of the ocean liner and the ACV. An ocean liner travels in water—a liquid. However, the ACV travels above the surface of the water on a cushion of air. Air is a mixture of gases. Recall that the molecules of a liquid are closer to-

gether than the molecules of a gas. This means that the ocean liner comes into contact with more molecules than the ACV. There is more friction (more opposing force) between the ocean liner and the water than there is between the ACV and the air. Therefore, the ocean liner moves more slowly. The bottom of the ACV rubs against air molecules. The friction from the air is less than the friction from the water. There is less opposing force.

Because of friction, more energy is required to do work. Friction is *always* present when work is done. Recall the law of conservation of energy. It states that energy cannot be created or destroyed. But it *can* be changed. Friction usually changes mechanical energy to heat energy.

Let's look at an example of this. Rub the palms of your hands together, slowly. Notice that the palms of your hands begin to feel warm. This is because friction has changed some of the mechanical energy to heat energy. Now rub your hands together more quickly. Your hands

To Do Yourself   How Can Friction Be Measured?

You will need:

Pieces of sandpaper, aluminum foil, construction paper, freezer wrap, and paper towel, spring scale, wood block, 500 gram weight, tape

1. Place the 500-g weight on the wood block. Attach the wood block to the spring scale.
2. Tape a piece of aluminum foil to your desk.
3. Pull the block (and weight) across the aluminum foil at a constant speed. Record the force to the nearest .1 newton in the data table.
4. Pull the block across the aluminum foil again. Record the force in the data table.
5. Repeat steps 2–4 for each of the other surfaces. Record the force required to drag the block across each surface in the data table.

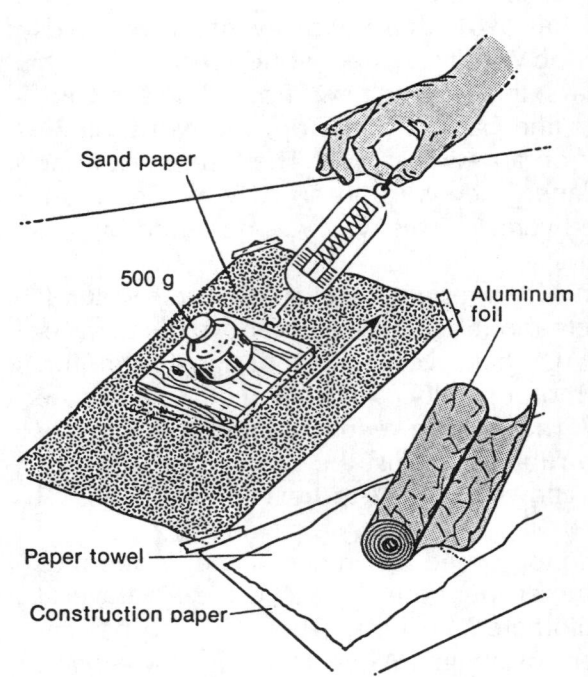

Data Table

	Aluminum Foil	Sandpaper	Paper Towel	Freezer Wrap	Desk Top	Construction Paper
1st Trial	N	N	N	N	N	N
2nd Trial	N	N	N	N	N	N

Questions

1. Which surface had the least amount of friction? _____
2. Which surface had the most friction? _____
3. Which surface probably increased the most in temperature because of friction?

_____

4. What could you have done to reduce the friction of each surface? _____

_____

will begin to feel even warmer. In fact, the more quickly you rub your hands together, the warmer they will become. This is because you are increasing the amount of friction between your hands.

Every time we do work, two forces work against us. The first is friction. The second is resistance (rih-ZIH-stunts). **Resistance** is anything that slows down or prevents motion. The weight of an object is one cause of resistance. The more an object weighs, the more resistance it has.

It is difficult to change the weight of an object. For example, an object with a weight of 100 N will always weigh 100 N. However, we can reduce the amount of resistance of the object. If we reduce the effect of friction on the object, its resistance will be less.

Imagine a 200-N object being dragged along a rough surface. The amount of friction between the surface and the object is very high. We cannot change the weight of the object. But, if we make the surface smoother, the resistance will be less. There will be less friction between the object and the surface. This is why mechanical surfaces need to be kept clean and polished. A lubricating (LOOB-rih-kay-ting) oil can also be used to reduce friction.

In Lesson 3, we will begin to study the work of machines. To make our calculations easier, we will pretend that there is no friction. But remember, no such surface exists.

Review

I. Fill in each blank with the word that fits best. Choose from the words below.

work   power   friction   smooth   rough   heat   light

The movement of an object is always slowed by _____ .

The effect of friction is greater on a _____ surface.

_____ is made more difficult because of friction. Some

energy is changed to _____ energy.

II. Which statement seems more likely to be true?

A. _____ Friction is greater on gravel than on ice.

B. _____ Friction is less in water than in air.

III. Place a check (✔) next to each example where friction has caused a change.

a. _____ a runner stops on a gym floor

b. _____ a man lights a match

c. _____ water becomes ice

d. _____ sandpaper is used to smooth wood

e. _____ a car slides on an icy road

f. _____ a student erases an error

IV. Answer in sentences.

How can friction be both helpful and harmful to the driver of an automobile?

Lesson Three

How Do Machines Help Us?

Exploring Science

Space Mechanics. Many young people dream about becoming astronauts. They see themselves in the pilot's seat at lift-off. They see themselves discovering a new moon or comet. They don't see themselves as space mechanics. Yet, this may be the most important role for astronauts in the twenty-first century.

The equipment we put in space often costs billions of dollars. Sometimes, after use, a part has to be replaced. One satellite needed a new fuse box. To do the space repairs, special tools had to be designed. One space-wrench cost more than a million dollars. Then the astronauts had to be trained because using tools in space causes new problems. For example, a wrench on Earth can be used to loosen bolts. If the handle of the wrench is turned, the bolt will turn. This is not the case in space. In space, if an astronaut turns the handle of a wrench, the astronaut revolves.

Sometimes the equipment we put in space does not work properly from its first day in orbit. This was the case with the Hubble Space Telescope. It was placed in space in April 1990. Many scientists waited for the first Hubble images. With this telescope we expected to answer many questions about the universe. Instead, the pictures were fuzzy! Could billions of dollars have been wasted? Did astronomers have to wait for their answers?

The original 2.4-meter mirror had a tiny flaw that had to be corrected. The space mechanics

Kathryn Thornton, a woman astronaut, making repairs to the Hubble Telescope

came to the rescue. In December 1993 seven astronauts worked for five days to correct the problem. Using their space tools, they turned a space failure into a space victory.

● What would happen if an astronaut in space tightened a screw with an ordinary screwdriver?

Simple Machines

A **simple machine** is a mechanical device that changes an applied force. It can increase the force. It can change the direction of the force. Or, it can do both. But *a machine cannot increase the amount of work that is done.*

The million-dollar wrench is not a simple machine. But its design is based on the wrench that is a simple machine on Earth. In fact, most motor-driven equipment uses technology based on our knowledge of simple machines.

Humans are limited in the amount of force they can push or pull. This applied force is called the **effort** (EF-urt). Most simple machines multiply our effort. They do not lessen our work, but they do allow us to work easier or faster.

Suppose your greatest pushing effort was 200 N and you needed to move a large object. If a device could increase your applied force to 400 N, it would be a simple machine. It would multiply your effort by 2. The amount that any machine multiplies a force is called the **mechanical advantage** (muh-KAN-ih-kul ad-VAN-taj) of the machine. Mechanical advantage is abbreviated **MA.**

Now, suppose your greatest pulling force is 200 N. If you had to move a 1000 N object, you would need help. You could use a simple machine with an MA of 5. This machine would multiply your effort by 5. If the tool had a smaller MA it could not move the 1000 N resistance.

How can we determine what MA is needed to move a resistance? An equation can be used to find the needed mechanical advantage. Let's see how the mechanical advantage for the example above was found.

$$\text{Mechanical Advantage} = \frac{\text{Resistance (weight of object)}}{\text{Effort (applied force)}}$$

$$MA = \frac{R}{E}$$

$$MA = \frac{1000 \text{ N}}{200 \text{ N}}$$

$$MA = 5$$

There is no unit for mechanical advantage. Mechanical advantage is only a number that multiplies the effort, or applied force.

There are six kinds of simple machines. They are listed in the table below. We will discuss several of these machines in future lessons.

The Six Kinds of Simple Machines

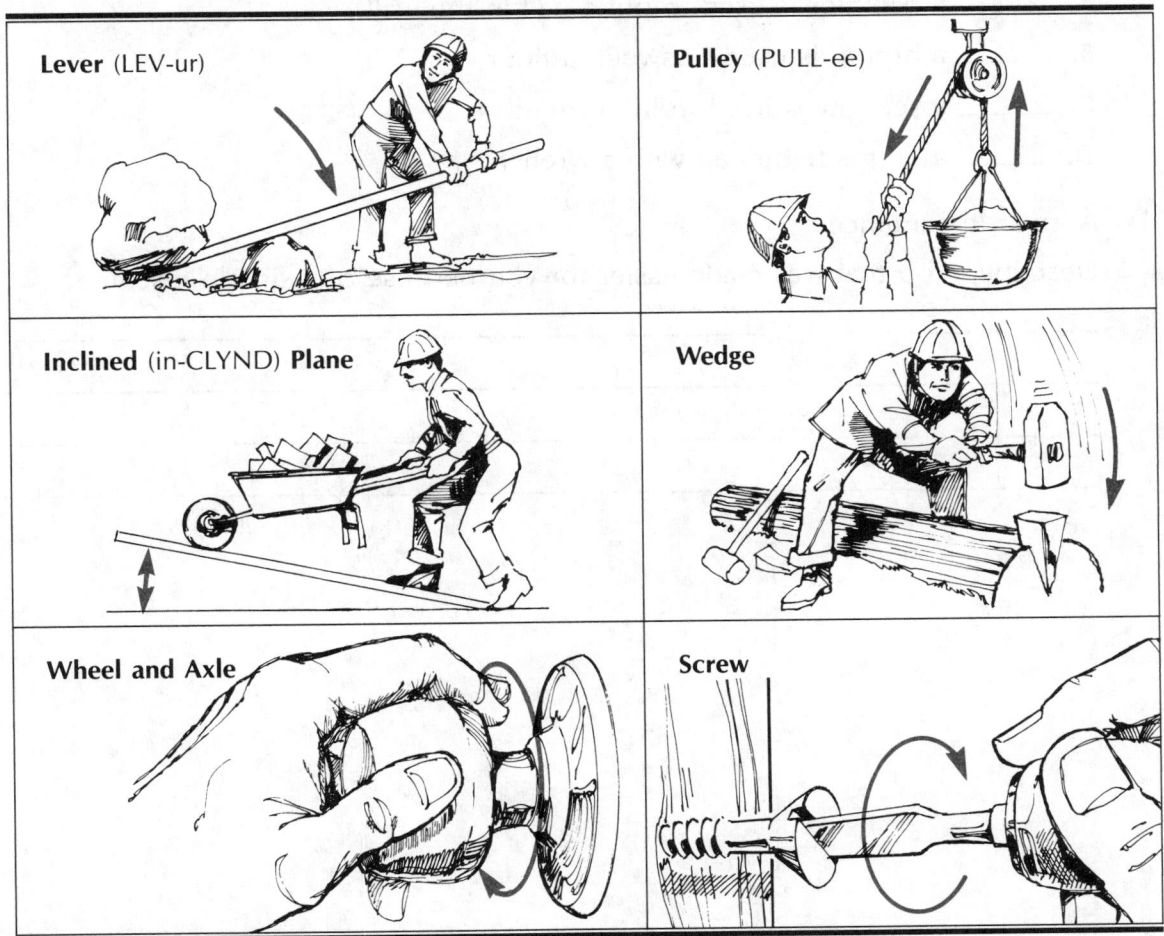

Lever (LEV-ur)

Pulley (PULL-ee)

Inclined (in-CLYND) Plane

Wedge

Wheel and Axle

Screw

Review

I. Fill in each blank with the word that fits best. Choose from the words below.

**mechanical advantage   work   resistance   simple machine   effort
energy**

A pulley is a _____ . A simple machine allows us to

multiply our applied force, or _____ . The number that

multiplies our applied force is called the _____ . A simple

machine does not lessen the amount of _____ that is
done.

II. Which statement seems more likely to be true?

A. _____ A simple machine with an MA of 2 can lift a 300 N resistance with an effort of 150 N.

B. _____ A simple machine with an MA of 2 can lift a 300 N resistance with an effort of 600 N.

III. Place a check (✔) next to each device that changes the direction of the force.

A. _____ a hammer is used to put a nail in the wall

B. _____ a broom is used to sweep a floor

C. _____ a see-saw is used to lift a friend

D. _____ a bolt is tightened with a wrench

IV. Answer in sentences.

Describe a job that was made easier for you because of a simple machine.

_____

_____

_____

_____

What Is a Lever?

Exploring Science

Future Workers. The first industrial robot was developed in the 1950's. It was little more than a motorized arm controlled by computers. These arms can pick up a small part and place it in a new position. Such "pick and place" robots still work on automobile assembly lines.

Today, more than 13,000 systems using robots can be found in the United States. Heavy duty robots have joined the smaller models. These heavy duty robots have arms that may be as long as 2 meters. Many can lift loads as heavy as 200 kilograms.

Robotic (roh-BOT-ik) arms have been used in space to capture damaged satellites. They have been used in nuclear plants to handle dangerous materials. They have even been used in hospitals to help with patient care.

What can we expect of robots in the future? Humans of the future will surely have some androids (AN-droydz) as co-workers. Scientists are developing robots that can rescue fire victims. Others will explore the ocean floor. Several robots are already available for use in the home. These robots can serve drinks or carry objects. Some can even paint or play an organ.

● What features would you expect most robots to have?

Robots serve humans in many ways. If you had a robot, what would you want it to be able to do?

Levers

A **lever** (LEV-ur) is a simple machine made of a bar or board that is supported at one place. The support of a lever allows it to **pivot,** or move, at that place. This support, or pivot, is called a **fulcrum** (FUL-krum).

The first step in the design of a robot is to identify what the robot will be doing. The force it will have must be known. The length of its arms (levers) and the placement of its joints (fulcrums) affect the effort of the robot. A study of the lever will help you understand why this is so.

One example of a lever is a see-saw. The fulcrum of a see-saw is the place where the see-saw is attached to the ground. A triangle (△) is often used to show the location of the fulcrum in a diagram. Look at the diagram below.

Fulcrum

A see-saw is a lever. Where is the fulcrum on this see-saw located?

Suppose a force of 100 N is applied to each end of a see-saw with equal arms. As long as the board is identical on both sides, the see-saw will balance, or be at **equilibrium** (ee-kwul-LIB-ree-um).

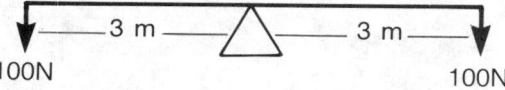

Now, suppose you wanted to make a see-saw for two people of different weights. One of the people weighs 100 N. The other weighs 200 N. The lever shown above would not work properly. The 100-N effort would not be able to lift the 200-N effort. We need to increase the mechanical advantage of the person weighing 100 N.

If you experimented, you would find that a lever like the one shown below would work best.

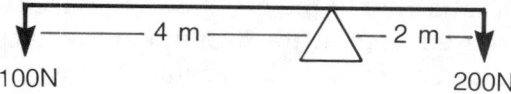

The lighter person needs a longer arm to have an equal advantage. The short arm of a lever is called the resistance arm. The **resistance arm** is the arm the resistance is resting on. The longer arm is the effort arm. The **effort arm** is the arm that is doing the lifting. The mechanical advan-

tage of a lever can be found by dividing the length of the effort arm by the length of the resistance arm.

Let's find the mechanical advantage for the lever above.

$$MA = \frac{\text{length of effort arm (EA)}}{\text{length of resistance arm (RA)}}$$

$$MA = \frac{EA}{RA}$$

$$MA = \frac{4m}{2m}$$

$$MA = 2$$

This means the effort needed to lift the 200-N person must be multiplied by 2. Thus, the distance between the effort and the fulcrum must be multiplied by 2.

Think about a 100-cm lever that is being used to lift a rock weighing 4000 N. Look at the diagram below.

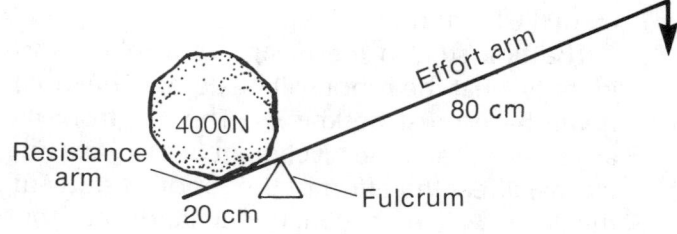

196

The mechanical advantage for this lever can be found using our equation.

$$MA = \frac{EA}{RA}$$

$$MA = \frac{80 \text{ cm}}{20 \text{ cm}}$$

$$MA = 4$$

This means that the lever will multiply the effort by 4. To find how little force is needed, *divide* by the MA. An effort of 1000 N is needed to lift a rock that weighs 4000 N.

Of course, this MA is ideal. It ignores friction. In a real situation, slightly more effort would be needed to overcome the affect of friction.

When using levers, the fulcrum is not always between the resistance and the effort. Look at the diagram below.

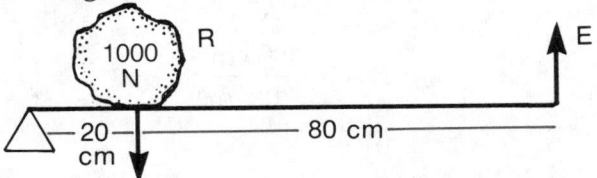

This type of lever is used in machines such as the wheelbarrow. Let's identify each factor that helps to find the mechanical advantage.

R – the load to be lifted = 1000 N

RA – the distance from R to △ = 20 cm

EA – the distance from E to △
= 20 cm + 80 cm = 100 cm

To Do Yourself   How Can a Lever Reduce Effort?

You will need:

Spring scale, wood block, meter stick, 500-g weight

1. Balance the meter stick on the wood block at the center mark.
2. Place the 500-g weight on one end of the meter stick. Attach the spring scale to the other end of the meter stick.
3. Measure the force needed to lift the weight by pulling down on the spring scale.
4. Repeat step **4** with the wood block 5 cm closer to the 500 g weight.
5. Repeat step **4** with the wood block 5 cm farther from the weight.
6. Lay the meter stick flat on your desk. Place the 500-g weight in the center of the meter stick. Lift one end of the meter stick with the spring scale.
7. Move the weight 5 cm to either side of the center mark. Again lift the meter stick with the spring scale.

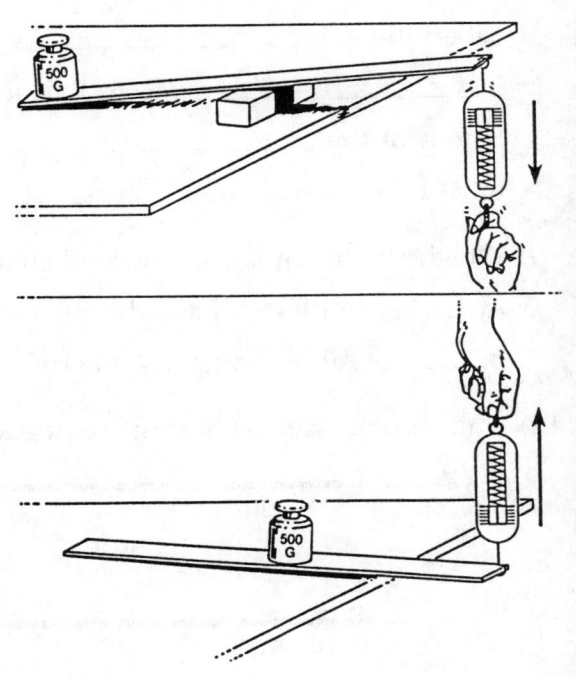

Questions

1. In steps **3, 4,** and **5,** when was it easiest to lift the weight? _____ When was it
   hardest? _____ Why? _____
2. In step **7,** when was it easiest to lift the weight? Why?

_____

Now, let's find the MA of this lever.

$$MA = \frac{EA}{RA}$$

$$MA = \frac{100 \text{ cm}}{20 \text{ cm}}$$

$$MA = 5$$

In this case, the effort will be multiplied by 5. The 1000-N load could be lifted with an effort of 200 N (1000 N ÷ 5). The formula for MA could have helped us arrive at this answer.

$$MA = \frac{\text{Resistance}}{\text{Effort}}$$

$$5 = \frac{1000 \text{ N}}{?}$$

Effort = 200 N

Now recall the description of robots at the beginning of this lesson. Their arms are often built so that the elbow is the fulcrum. For an arm of 2 meters, where must the elbow be to give the arm a mechanical advantage of 3?

Review

I. Fill in each blank with the word that fits best. Choose from the words below.

**fulcrum   lever   arms   pivots   pulley   longer   shorter
equilibrium**

A _____ is a board that pivots at one point. That point is

called the _____ . When the board is balanced, it is at

_____ . To balance a board with unequal weights, the

length of the _____ must be different. The heavier weight

must be placed on the _____ arm.

II. Which statement seems more likely to be true?

A. _____ All levers have the fulcrum between the resistance and the effort.

B. _____ All levers have a fulcrum.

III. Find the mechanical advantage of each lever.

A. _____   E |————————————————————————————| R
　　　　　　　　20 cm　　　　　△　　　　20 cm

B. _____   R |————————————————————————————| E
　　　　　　　10 cm △　　　　　　　100 cm

IV. Answer in sentences.

A box contains many different objects. Two of the objects have the same weight. How could you make a lever to identify the two objects?

_____

_____

What Is An Inclined Plane?

Exploring Science

Pyramid Wonders. The Egyptian (ee-JIP-shun) pyramids (PEER-uh-midz) were built more than 4000 years ago. Travelers have been curious about the pyramids for centuries. The pyramids brought to mind dreams of hidden treasures and mummy's curses.

The Great Pyramid was originally 146.7 meters tall. It was made from 2,300,000 blocks of stone. Each block weighed about 2272.5 kilograms. Imagine the amount of work that was involved in the building of this pyramid.

Every man in Egypt helped to build the pyramids. Each work gang had 100,000 men. These men moved stone for three months. Then another 100,000 men replaced them. It took ten years to lay down the stone for the roadbed that leads to the pyramid. Then, fine limestone was brought from across the river. Beautiful granite was carried from more than 1800 kilometers away.

The Great Pyramid took more than twenty years to build. Some experts think that four ramps were built around the pyramid to move the heavy stones to the top. Others believe only one ramp was used. By lengthening this ramp, workers could raise stones to higher levels of the pyramid.

● How are ramps used today?

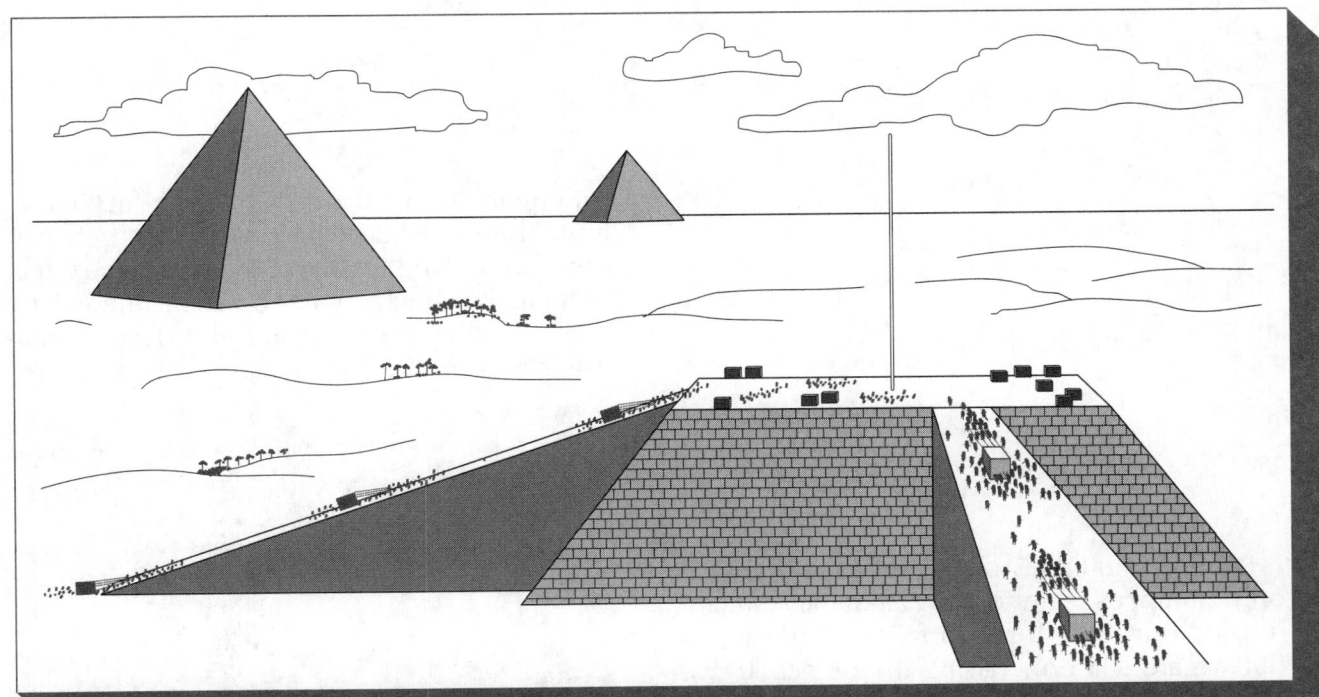

Inclined Planes

A ramp is an inclined (in-CLYND) plane. An **inclined plane** is a simple machine made by a sloping surface. It is probably the simplest ma-chine. Think about two ramps that could be used to lift a heavy load. Look at the illustrations of the two ramps on the next page.

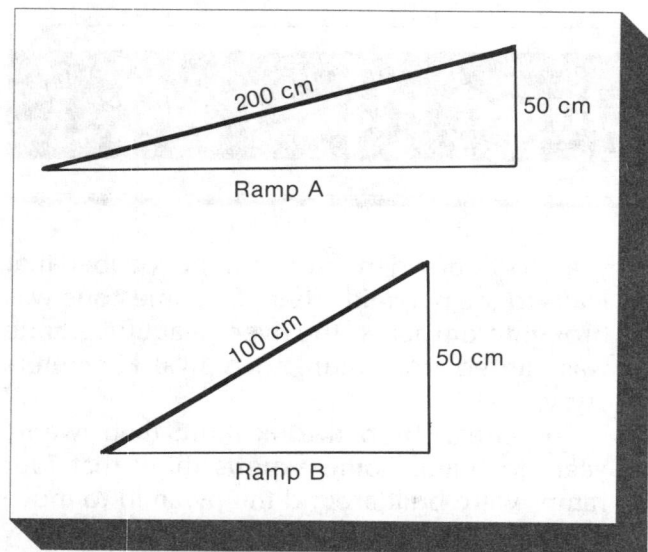

Ramp A

200 cm

50 cm

Ramp B

100 cm

50 cm

It is always helpful to make a diagram of the problem. Look at the diagram below.

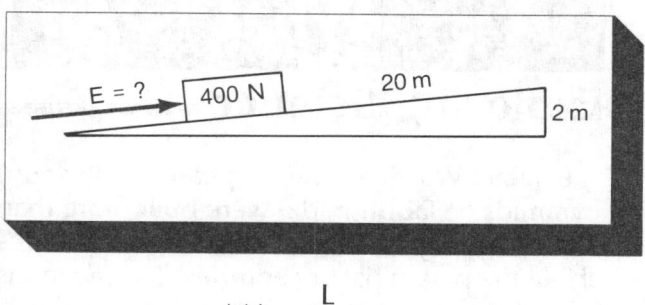

E = ?   400 N   20 m   2 m

$$MA = \frac{L}{H}$$

$$MA = \frac{20 \text{ m}}{2 \text{ m}}$$

$$MA = 10$$

From your experience walking up hills, you probably know that ramp A requires less effort to walk up than ramp B. Let's find the MA of ramps A and B. To find the MA of an inclined plane, the length of the inclined plane is divided by the height the object is to be lifted.

Ramp A $MA = \dfrac{\text{Length}}{\text{Height}}$

$$MA = \frac{200 \text{ cm}}{50 \text{ cm}}$$

$$MA = 4$$

Ramp B $MA = \dfrac{\text{Length}}{\text{Height}}$

$$MA = \frac{100 \text{ cm}}{50 \text{ cm}}$$

$$MA = 2$$

Ramp A multiplies our effort by 4. Ramp B only multiplies our effort by 2. In ramp A we are moving our force through twice the distance, but we are still only raising the object 50 cm.

This illustrates a principle stated in an earlier lesson. Simple machines do not lessen the amount of work we do. They just make the work easier.

Let's try a problem. Suppose a person is pushing a 400-newton crate up a 20-meter ramp. The top of the ramp is 2 meters off the ground. How much force will have to be applied?

This inclined plane multiplies the effort by 10. But how much effort is required to move the crate up the ramp? To find out, we must use our formula for finding mechanical advantage.

$$MA = \frac{R}{E}$$

$$10 = \frac{400 \text{ N}}{?} \text{ or } (400 \div 10)$$

$$E = 40 \text{ N}$$

Remember, we are ignoring the effect of friction. There would be a lot of friction between the crate and the surface of the ramp. The effect of friction is much greater on an inclined plane than it is on a simple machine like a lever. Scientists call a mechanical advantage that ignores

How is an inclined plane helpful to a person in a wheel chair?

the effect of friction an **ideal mechanical advantage,** or **IMA.**

Now imagine the building of the pyramids. The pyramids could not have been built without the use of inclined planes. Even 100,000 men could not have carried the blocks of stone straight to the top. Do you think there was one big ramp or a smaller ramp at each side?

To Do Yourself   How Does an Inclined Plane Reduce Effort?

You will need:

Two meter board, spring scale, book, chair, protractor, meter stick

1. Set the board at a 30° angle from the floor. Rest the board on the seat of the chair.
2. Attach the book to the spring scale. Measure the force required to pull the book up the board (from the floor to the chair).
3. Find the IMA by dividing the effort distance (from the floor to the chair) by the resistance distance (height of the chair seat). Record your results in the data table.
4. Repeat steps **2** and **3** with the board at a 50° angle.
5. Repeat steps **2** and **3** with the board at a 70° angle.

Data Table

	30° angle	50° angle	70° angle
Effort Distance	cm	cm	cm
Resistance Distance	cm	cm	cm
Mechanical Advantage			

Questions

1. At what angle was it easier to pull the book? _____
2. Which angle had the greatest IMA? _____
3. Describe the kind of ramp that is best for a wheelchair. _____
_____
_____

Review

I. Fill in each blank with the word that fits best.

ramp   friction   incline   machine   plane   rough   smooth

A flat surface is called a _____ . A tilt is called an

_____ . So a _____ is more properly

called an inclined plane. For such a simple _____ to be

efficient, its surface should be _____ . This will reduce

the effect of _____ .

II. Which statement seems more likely to be true?

A. _____ The longer the inclined plane the higher the MA.

B. _____ The shorter the inclined plane the higher the MA.

III. Find the MA of each inclined plane.

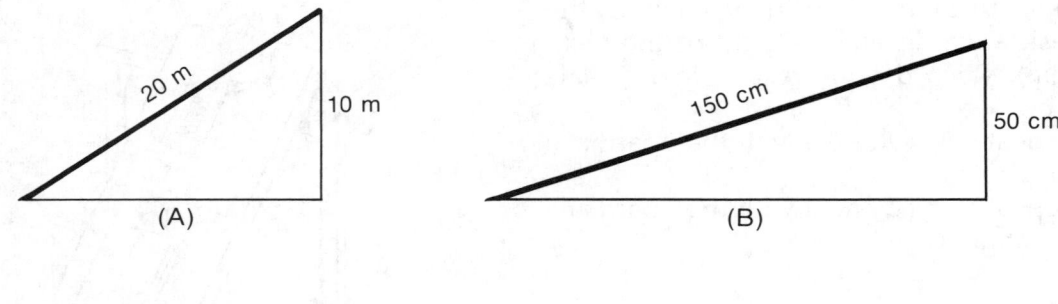

(A) 20 m 10 m

(B) 150 cm 50 cm

_____   _____

IV. Answer in sentences.

What are two disadvantages of a long inclined plane?

_____

_____

_____

What Is a Wheel and Axle?

Exploring Science

The Original 4-Wheel Drive. How was the wheel invented? Researchers think that the first wheels were used between 5000 and 6000 years ago. The wheel was developed in an area called Mesopotamia (mes-uh-puh-TAY-nee-uh). Wheels were not used in North America until Columbus reached the New World.

One theory explains that the wheel developed from a fast-moving "lazy Susan" used to shape pots from clay. The wet clay was turned round and round by the potter as she shaped it. Potters found it easier to turn a heavy pot on a tray balanced on a rock. The pot also could be turned steadily. Spinning the rim of the tray at a fast speed moved the pot slowly around. Soon a circular tray was used, mounted on a short rod in place of the rock.

Scientists have found such "Potter's Wheels" in Mesopotamia. They date from just *before* the first wheeled cart. Pictures of the first carts show what appear to be sleds that have wheels like Potter's Wheels. The wheels are connected by a long rod instead of the short rod used on the Potter's Wheel.

These first carts had a surprising use. The best roads were the entrances to temples. When an important person, such as a king, died, the body was taken to the temple. Wheeled carts demonstrated the importance of the body.

The earliest roads outside of towns were for sleds to use in carrying stone to temples. Soon

A potter's wheel is based on the wheel and axle concept

builders found that carts could carry stones better than sleds.

● Why is the wheel considered to be a great invention?

Wheel and Axle

How would you describe a wheel and axle? Most people would describe a large wheel that turns freely around a small axle that does not move. This kind of wheel and axle makes work easier by reducing friction. It is *not* the wheel and axle that is a simple machine.

The simple machine called the **wheel and axle** is made up of two wheels of different sizes. The smaller wheel is called the axle. One wheel is used to turn the other. Usually, the larger wheel is used to turn the smaller wheel.

Think about the steering system of a car. It is a wheel (steering wheel) and axle (steering column). By exerting a small force (effort) on the steering wheel, we can make the steering column turn. This turns the front wheels of the car. It is much easier to move the steering wheel than it would be to move the steering column. How much easier?

To find the mechanical advantage of a wheel and axle, we need to know the **diameter** (DY-am-ih-ter), or distance across, the wheel and the diameter of the axle. Now, suppose the diameter of the steering wheel is 35 centimeters and the diameter of the steering column (axle) is 7 centimeters. What is the mechanical advantage of this wheel and axle?

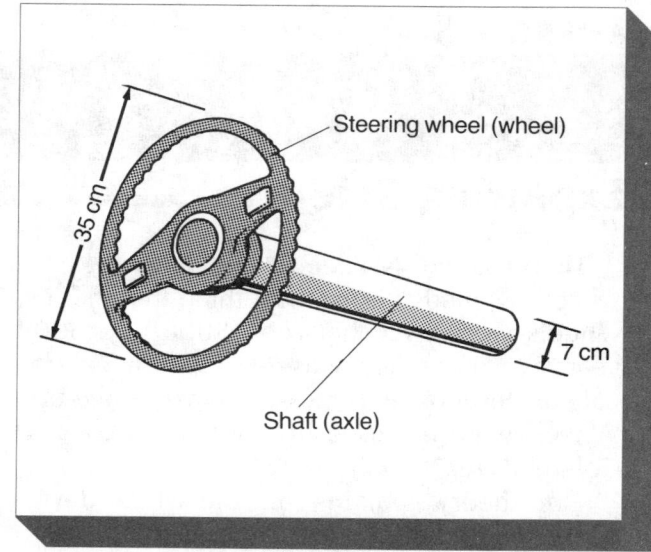

Steering wheel (wheel)

35 cm

7 cm

Shaft (axle)

$$MA = \frac{\text{Diameter of Wheel}}{\text{Diameter of Axle}}$$

$$MA = \frac{DW}{DA}$$

$$MA = \frac{35 \text{ cm}}{7 \text{ cm}}$$

$$MA = 5$$

Sometimes the wheel and axle are not so obvious. Consider a screwdriver. When you turn the hand grip, you are actually turning the wheel. The connected shaft is really the axle.

Suppose the handle of a screwdriver is 3 cm in diameter and its shaft is 1 cm in diameter. What would its mechanical advantage be?

$$MA = \frac{DW}{DA}$$

$$MA = \frac{3 \text{ cm}}{1 \text{ cm}}$$

$$MA = 3$$

How are the gears on the bicycle (right) different from the gears on the watch (left)?

If the handle were removed from the screwdriver, three times as much effort would be required to do the work. What would happen if a screwdriver with a bigger handle was used?

An invention similar to the wheel is the gear. A gear is a toothed wheel. Gears can connect with other gears to change the direction of the force.

Take a look at a bicycle. Notice how the gears connect. How many other things can you think of that use gears? How would vehicles be different if gears had not been invented?

Review

I. Fill in each blank with the word that fits best. Choose from the words below.

rotate   simple machine   wheel   axle   once   twice   greater   smaller

A doorknob is a _____ . The knob is the

_____ . The shaft is the _____ . If the

knob turns once, the shaft turns _____ . The diameter of

the wheel is always _____ than the diameter of the axle.

II. Which statement seems more likely to be true?

A. _____ A tire is a simple machine.

B. _____ A hand mixer is a simple machine.

III. A doorknob has a diameter of 6 cm. The shaft has a diameter of 2 cm. Find the mechanical advantage of the doorknob.

MA = _____

IV. Answer in sentences.

A simple Potter's Wheel is 50 cm in diameter. When it turns once, a point on its rim moves a little more than 150 cm. A pot 20 cm in diameter is centered on the Wheel. When the Wheel turns once, so does the pot. As the pot turns once, a point on its rim moves a little more than 60 cm. Explain how the mechanical advantage of the wheel and axle make it easier to shape the pot.

_____

_____

_____

What Is a Pulley?

Exploring Science

The Sky's the Limit. Paintings of cities of the early 1800's can be fascinating. Most of the paintings show very low buildings. There may be a single tall monument such as the Eiffel Tower, but there are no tall office buildings with many stairs to climb.

Simple lifts or dumbwaiters were used to carry heavy materials. But they were not safe enough to move people. If a rope snapped, the load fell to the ground.

Elisha Otis solved this problem. In 1852, he invented an elevator with a safety feature. If a rope broke or slipped, jagged teeth would grasp and hold the platform. The ride was slow and jerky, but there was less risk.

Today's elevators are much better. Engineers are designing elevators for even taller buildings. Elevators that will move people to their building, and then lift them to their floor are being developed. Future elevators may even be double-deckered so that they can service two floors at a time.

● What features would you install in elevators of the future?

Pulleys

A **pulley** is a wheel-and-axle turned by a rope. There are three types of pulleys. A **fixed pulley** is the simplest. The axle of the wheel is attached to something so that it can rotate, but not move otherwise. The ideal mechanical advantage (IMA) of a fixed pulley is *always* 1.

A single fixed pulley is used to change the direction of a force. It allows you to pull down, instead of lift up, on a resistance. You can raise an object all the way up to the pulley, while you remain in place below it or at the side.

Most elevators use a cable and pulley system. Some use a single fixed pulley. When a force pulls down on one end of the cable, the elevator car on the other end is lifted.

A **movable pulley** is a pulley that is not attached to a surface. When using a movable pulley, the pulley moves up as the resistance moves up. A single movable pulley *always* has an IMA of 2. Look at the illustration of the movable pulley below. How does this pulley differ from the single fixed pulley on the left?

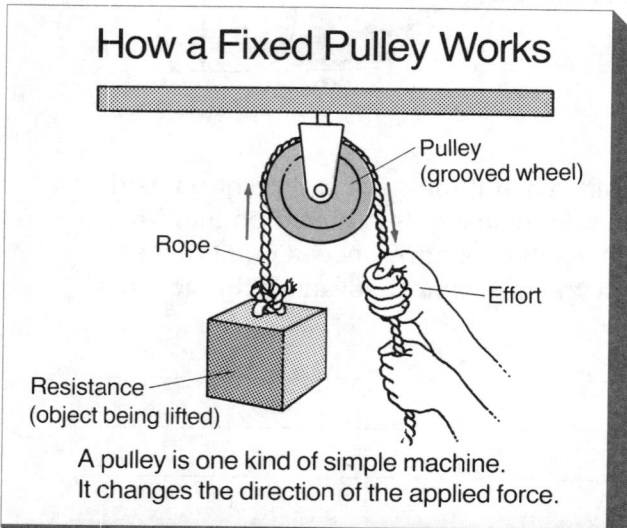

How a Fixed Pulley Works

Pulley (grooved wheel)

Rope

Effort

Resistance (object being lifted)

A pulley is one kind of simple machine. It changes the direction of the applied force.

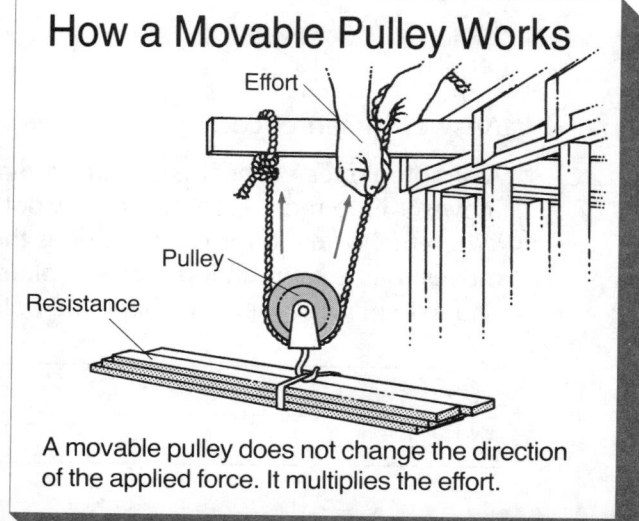

How a Movable Pulley Works

Effort

Pulley

Resistance

A movable pulley does not change the direction of the applied force. It multiplies the effort.

The third type of pulley is the pulley system. A **pulley system** is a combination of pulleys connected by a supporting rope or cable. A block and tackle is an example of a pulley system.

A simple method can be used to find the IMA of a pulley system. The IMA is *equal to the number of supporting ropes.* Only the supporting ropes affect the IMA of a pulley system. The rope being pulled is not counted. What is the IMA of the pulley system in the illustration? The IMA of this pulley system is 4. There are four ropes supporting the resistance. Remember, the rope that is being pulled on is not counted in the IMA.

By using a combination of movable and fixed pulleys, many different machines can be made. The more pulleys and ropes the system has, the higher the IMA will be.

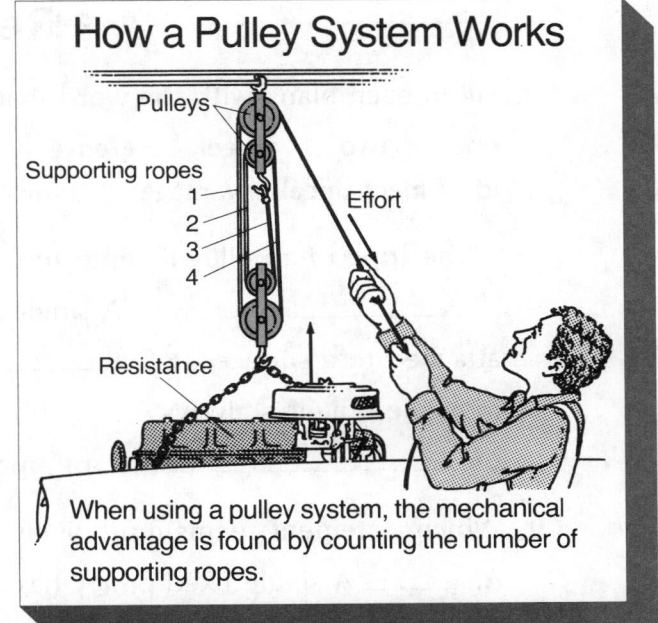

How a Pulley System Works

When using a pulley system, the mechanical advantage is found by counting the number of supporting ropes.

To Do Yourself What Kind of Pulley Works Best?

You will need:

Ringstand with clamp, single wheel pulley, spring scale, 500-g weight, rope

1. Attach the weight to the spring scale. Pick up the weight with the spring scale. Record the force needed to lift the weight in newtons.
2. Set up the fixed pulley as shown in sketch **A.**
3. Pull down on the spring scale. Record the force needed to lift the weight in newtons.
4. Set up the pulley as shown in sketch **B.**
5. Pull up on the spring scale. Record the force needed to lift the weight in newtons.

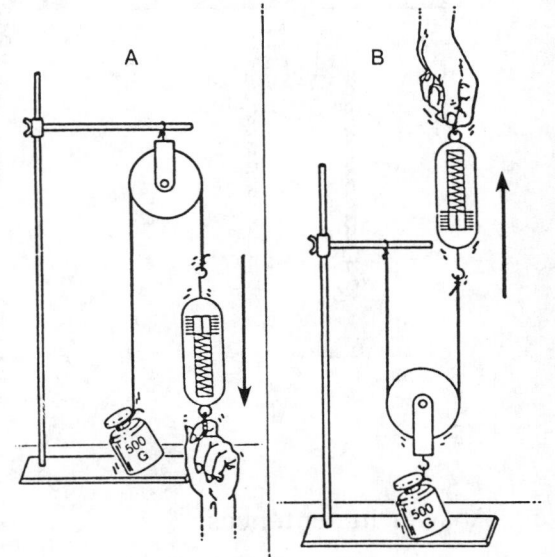

Questions

1. Which pulley lifted the weight more easily? _____

2. What is the IMA of the pulley used in step **3**? _____

3. What is the IMA of the pulley used in step **5**? _____

4. When you lifted up on the weight in step **5,** what did the pulley do? _____

Review

I. Fill in each blank with the word that fits best. Choose from the words below.

one   two   wheel   groove   direction   fixed   resistance ideal mechanical advantage   movable

The rope of a pulley fits into the _____ of its

_____ . A single _____ pulley is

attached to a surface. Its _____ is always one. The IMA of a

movable pulley is always _____ . A pulley is used to change

the _____ of an applied force.

II. Which statement seems more likely to be true?

A. _____ A single fixed pulley has a higher IMA than a single movable pulley.

B. _____ A single movable pulley has a higher IMA than a single fixed pulley.

III. What is the IMA of each of the pulleys shown below?

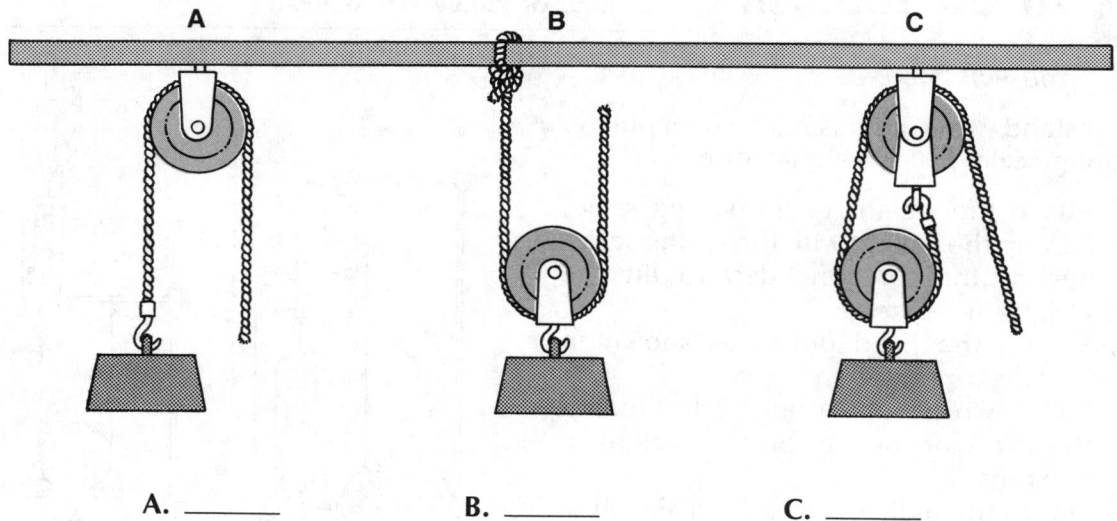

A. _____   B. _____   C. _____

IV. Answer in sentences.

Explain why a pulley is a simple machine.

_____

_____

_____

UNIT 8

Review What You Know

A. Use the clues below to complete the crossword.

Across

1. Caused by rubbing surfaces
4. An opposition to a force
6. A push or pull
7. The *M* in IMA
9. Metric unit of work
10. The movement of an object over a distance
12. Simple machine that changes the direction of a force
14. The *A* in IMA
15. Applied force

Down

1. The pivot of a lever
2. Simple machine similar to the wheel and axle
3. Metric unit of force
5. A ramp is one
8. Work ÷ Time
11. A see-saw is one
13. Metric unit of power

B. Write the word (or words) that best completes each statement.

1. _____ One newton-meter of work is equal to one
   a. watt **b.** kilometer **c.** joule

2. _____ The rate at which work is done is **a.** power
   b. effort **c.** resistance

3. _____ Friction usually changes mechanical energy to
   a. effort **b.** heat **c.** resistance

4. _____ The amount of friction between two objects can
   be reduced by adding **a.** lubrication **b.** resistance **c.** more mass

5. _____ A device that changes an applied force is a
   a. lubricant **b.** simple machine **c.** resistance

6. _____ Another term for applied force is **a.** resistance
   b. power **c.** effort

7. _____ The number that multiplies the force is a simple
   machine's **a.** resistance **b.** mechanical advantage **c.** effort

8. _____ If an object is moved a distance of 5 meters using a
   force of 20 newtons, the work that has been done is **a.** 20 joules
   b. 4 joules   **c.** 100 joules

9. _____ If 50 joules of work was done in 10 seconds, the
   amount of power was **a.** 50 watts **b.** 500 watts **c.** 5 watts

10. _____ Of the following, the machine which would
   require the least effort is the machine with the IMA of **a.** 1 **b.** 2 **c.** 5

11. _____ A playground see-saw is an example of **a.** an
   inclined plane **b.** a pulley **c.** a lever

12. _____ A ramp is an example of **a.** an inclined plane
   b. a lever **c.** a pulley

13. _____ A lever can reduce our effort if the fulcrum is
   placed **a.** closer to the load **b.** closer to the effort **c.** farther from the
   resistance

14. _____ An example of a tool that is really a lever is a
   a. shovel **b.** ramp **c.** screwdriver

15. _____ A 10-meter ramp with a height of 5 meters has an
   IMA of **a.** 2 **b.** 50 **c.** 5

16. _____ A mechanical advantage that ignores the effect of
   friction is **a.** a resistance **b.** an ideal mechanical advantage **c.** an effort

17. _____ An inclined plane can lessen our effort if it is
   made **a.** longer **b.** higher **c.** rougher

18. _____ The mechanical advantage of a pulley system is
   equal to the number of **a.** pulleys **b.** supporting ropes **c.** ropes and
   pulleys

19. _____ A device that is very similar to the wheel and axle is the **a.** pulley **b.** gear **c.** fulcrum

20. _____ All of the following are examples of a wheel and axle except **a.** a screwdriver **b.** a shovel **c.** a door knob

21. _____ A single fixed pulley **a.** multiplies our effort by 2 **b.** changes the direction of the applied force **c.** multiplies our effort by 5

22. _____ If the shaft of a screwdriver is 1 cm and the handle of the screwdriver is 3 cm, the mechanical advantage of the screwdriver is **a.** 1 **b.** 3 **c.** 6

23. _____ An opposition to a force is **a.** an effort **b.** a resistance **c.** a fulcrum

24. _____ If the effort arm of a see-saw is equal in length to the resistance arm, the see-saw is at **a.** resistance **b.** equilibrium **c.** effort

25. _____ The pivot of a lever is called the **a.** fulcrum **b.** friction **c.** effort

C. Apply What You Know

1. Study the drawing of the construction site. Then use the words below to fill in the blanks on page 212. Some words may be used more than once.

direction **amount** **effort** **resistance** **friction** **wheel and axle**
inclined plane **pulley system** **lever** **fixed** **moveable**
increased **decreased** **1** **2** **4**

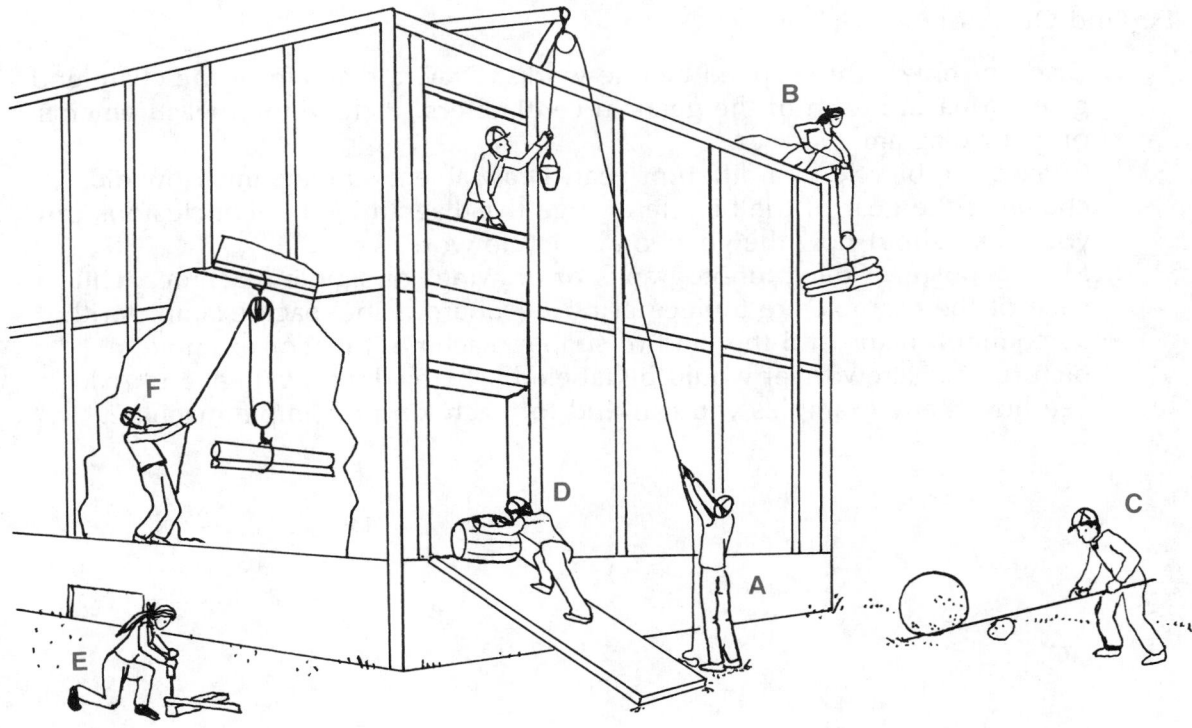

a. Worker A is using a single _____ pulley. It changes the _____ of his effort. It has an IMA of _____ .

b. Worker B is using a single _____ pulley. This pulley doubles her effort. It has an IMA of _____ .

c. Worker C is using a _____ to move a rock. The rock is called the _____ . This simple machine can change the _____ and _____ of his effort.

d. Worker D is using a(n) _____ . If the ramp is smooth, the amount of _____ between the ramp and the box will be _____ .

e. Worker E is using a(n) _____ . If the handle on the screwdriver were larger, the mechanical advantage of the screwdriver would be _____ .

f. Worker F is using a block and tackle. A block and tackle is an example of a _____ . The mechanical advantage of the block and tackle is _____ . This simple machine changes the _____ and _____ of the worker's applied force.

D. Find Out More

1. Discover how a three-speed bicycle works. Draw a diagram of the chain and gears. Measure each of the gears in centimeters. Include the measurements on your diagram.

   Place the bicycle in a different gear. Draw a second diagram. How did changing the gear of the bicycle change the diagram? What conclusions can you make about how the gears on a bicycle work?

2. Make a poster. Collect photographs or drawings of simple machines. Glue each of the examples to a piece of poster board. Label each example with its common name and the kind of simple machine it is. For example, a picture of a screwdriver would be labeled: "Screwdriver, Wheel and Axle." See how many examples you can find for each kind of simple machine.

Careers in Physical Science

Under Construction. During the 1980's, about two million American homes were built each year. Taller skyscrapers, longer bridges, and larger sports arenas were also built. Construction projects provide thousands of new jobs each year.

The construction business is very complicated. Buildings and bridges come in many shapes and sizes. Many new materials are required to meet the needs of today's buildings. New building techniques are also required.

Structural Engineers. Structural engineers make sure that a building will meet the needs of its occupants. They also make certain that construction projects are built properly and safe for their intended use. A structural engineer must know how strong each floor of a building must be. This person must also know what materials are best for a specific job. Structural engineers are construction specialists.

Structural engineers complete at least four years of college. Many go on for additional college training. But practical experience is also important. This job combines work in the office with work in the field.

Building contractors. Architects and engineers plan and design building projects. Building contractors (kon-TRAK-tors) supervise the actual construction. Large construction projects may require more than one contractor. So, it is not uncommon to have several contractors at one building site.

Some building contractors work for large companies. Others, work for themself. All building contractors have one thing in common. They must understand, and know how to complete, all parts of a construction project. These people must use many different skills in their work.

Some building contractors have college degrees. These specialists take courses in the areas of materials, structure, and management. Many building contractors do not have a college education. Their knowledge of the construction business comes from years of work in the field. These contractors learn the business through other trades such as bricklaying, carpentry, and electrical work. ·

This construction engineer and contractor are looking over the blueprints for the structure in the background.

UNIT

9

MAGNETISM AND ELECTRICITY

Lesson One

What Is Magnetism?

Exploring Science

High-Tech Archeology. Scientists from New York's Museum of Natural History went to the island. They were archeologists (ar-KEE-ol-oh-jists), scientists who study the past. For ten years, they looked for signs of the old Spanish mission. The mission was built in the 16th century.

The island is St. Catherine's Island. It is named after the old mission. The island is located off the coast of Georgia. Today, it is a wildlife preserve.

No signs of the old mission could be found. Then a high-tech research team came to the rescue. The team carried a magnetometer (mag-nih-TOM-ih-tur). This device detects changes in the Earth's magnetism (MAG-nuh-tiz-um). Such changes are caused by buried objects.

After studying their data, the scientists chose a location that showed great disturbance. The area was cleared, and a dig was begun. One year later, the walls of the mission were uncovered. Pieces of bone, medals, and teacups were found. These objects revealed much information about the Indians and Spaniards who once lived there.

● Why didn't the museum team just pick a spot and start digging?

The site of this mission was found because of data provided by a magnetometer.

Magnetism

Magnetism is a force created by the attraction of unlike electrical charges between two substances. Although all matter has electrical charges, not all matter is magnetic (mag-NET-ik). In fact, only materials containing the elements iron, nickel, or cobalt (COH-bawlt) are magnetic. But, what makes a material magnetic?

The atoms of magnetic materials are arranged in groups called **domains** (doh-MAYNS). Each domain is made up of one positive charge and one negative charge. Look at the diagram of the iron bar below. Each domain is shown as a block.

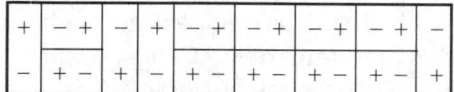

Look at how the domains are arranged. Recall that like charges *repel,* or push each other apart. Unlike charges *attract,* or pull toward each other. Thus, the domains are arranged so that each positive charge is next to a negative charge. But is the iron bar magnetic? No.

In order for a material to be magnetic, its domains must be arranged in a special pattern. The domains must be lined up. When the domains are lined up, they are called **magnetic domains.**

Look at the diagram of the iron bar below.

−	+	−	+	−	+	−	+	−	+	−	+	−	+
−	+	−	+	−	+	−	+	−	+	−	+	−	+

The domains in this bar are lined up. They are magnetized (mag-nuh-TYZED). The iron bar will produce a magnetic force. The bar is now called a **magnet.**

The Earth acts as if it contained a giant magnet. In fact, the Earth produces a weak magnetic force. If you have ever used a compass to find your way, you have made use of the Earth's magnetic force.

A **compass** is a magnet that responds to the magnetic pull of the Earth. The needle of the compass points to where the magnetic pull of the Earth is strongest—the Earth's magnetic pole. The needle of the compass points to the Earth's **North Magnetic Pole.** This pole is located in Canada.

A compass also has magnetic poles. A **magnetic pole** is that part of the magnet where the force is strongest. If you are using a bar magnet, such as the iron bar, the magnetic poles are located at each end of the magnet.

The poles of a magnet are labeled **North (N)** and **South (S).** The North pole of the magnet is the pole that would point to the Magnetic North Pole of the Earth if the magnet was hanging freely in the air.

If two N poles of a magnet are placed near one another, they repel each other. However, if an N pole and an S pole are placed near one another, the poles pull together. This illustrates the **Law of Magnetic Poles.** This law states that unlike poles attract, and like poles repel. This is similar to the reactions of positive and negative electrical charges.

When using a compass, the North pole of the compass points toward the Earth's North Magnetic Pole. How can this be? The Law of Magnetic Poles states that like poles repel, or push each other apart. This occurs because the Earth's North Magnetic Pole is magnetically a South pole. It is really a "north-seeking" pole.

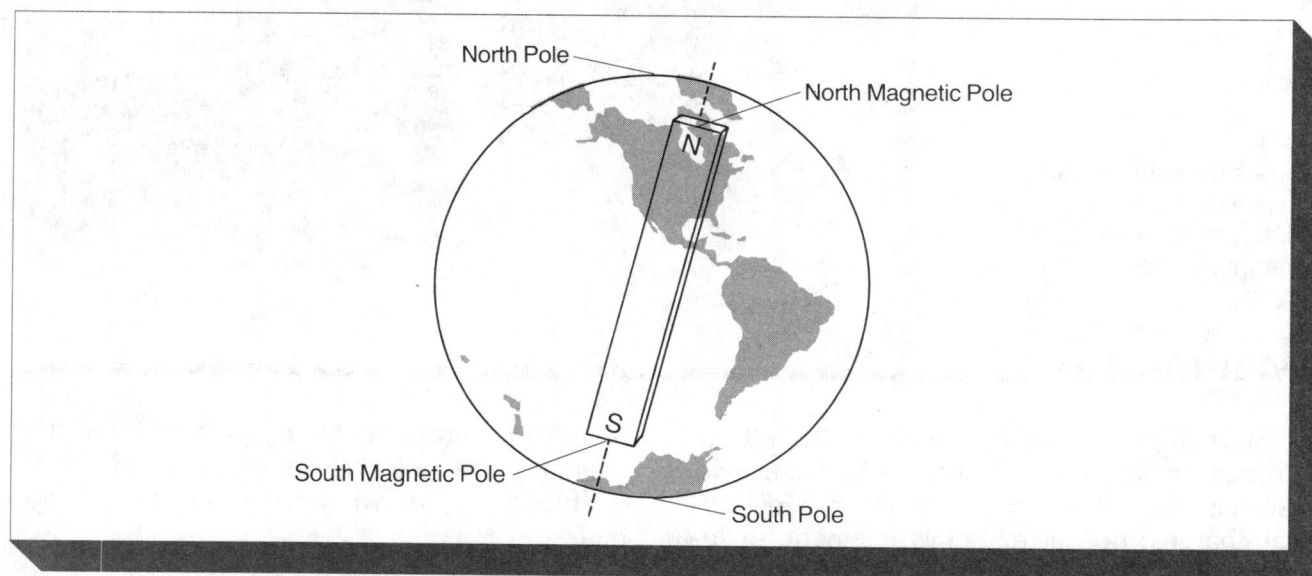

People use compasses to find direction. The needle of a compass always points to the north. How is knowing where north is helpful in finding direction?

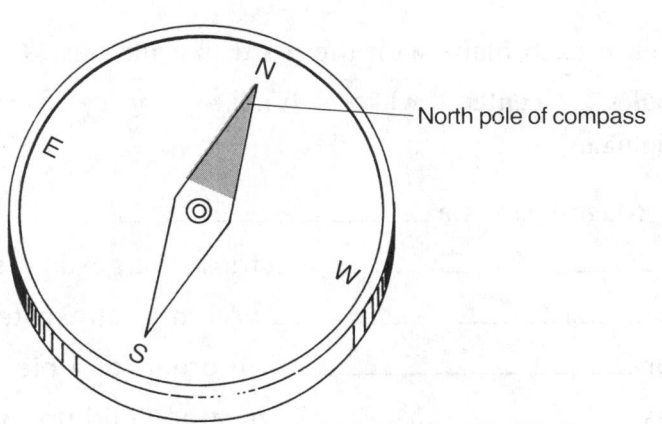

North pole of compass

To Do Yourself   How Can the Poles of a Magnet Be Determined?

You will need:

Two bar magnets, two small iron nails, steel paper clip

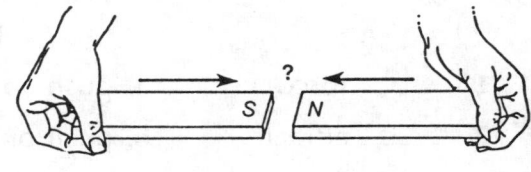

1. Place the N poles of the magnets near each other. Observe what happens.
2. Place one N pole and one S pole near each other. Observe what happens.
3. Hold the nails in your hand. Stroke both nails with the N end of the magnet. Stroke each nail slowly, in one direction, 25 times.
4. Place the ends of the nails close to the paper clip. Observe what happens.
5. Place the two ends of the nails close together. Observe what happens.

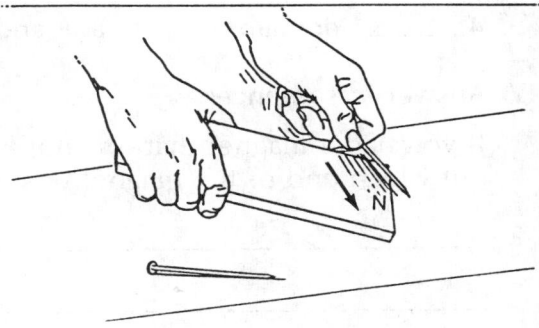

Questions

1. How do like poles react when placed near each other? _____
_____

2. How do unlike poles react when placed near each other? _____
_____

3. What happened to the nails in step 3? _____
How do you know something happened? _____

4. Did the nails have north and south poles? _____

Review

I. Fill in each blank with the word that fits best. Choose from the words below.

**poles   center   like   unlike   force   magnetic   lead   nickel
domains**

Magnetism is a _____ . It is created by the attraction of

_____ electrical charges in a substance. Not all materials

are _____ . All magnetic materials contain iron, cobalt,

or _____ . In order for a piece of iron to be a magnet,

its _____ must be lined up. A magnet is strongest

at its _____ .

II. Which statement seems more likely to be true?

A. _____ Some magnets are made of lead.

B. _____ Some magnets are made of cobalt.

III. Match each term in column **A** with its description in column **B**.

	A		B
1.	_____ force	a.	push apart
2.	_____ attract	b.	push or pull
3.	_____ repel	c.	pull together
4.	_____ domain	d.	a + and a − charge

IV. Answer in sentences.

If you had a magnet that was not labeled north and south, how could you find out which end of the magnet was the north pole?

_____

_____

_____

_____

Lesson Two

What Is a Magnetic Field?

Exploring Science

Sense of Direction. Migrations (my-GRAY-shuns) usually mark a change in seasons. A migration is the movement of a group of animals from one place to another.

People have been watching birds migrate for years. Many birds fly south for the winter. But birds are not the only animals that migrate. Monarch butterflies travel from Canada to Mexico. Caribou (KAR-ih-boo) move south when the Alaskan winter nears. They return in the spring when the weather is warmer.

Much research about migration has been done. Some animals find their way by using the sun's position. Others use star patterns. However, recent research shows that some birds have a magnetic sense.

An experiment using three sets of nests and magnets was performed. The nests contain eggs of birds that always fly south for the winter. Magnets were placed in the first set of nests. The N pole of each magnet pointed west. Magnets pointing east were placed in the second set of nests. The third set of nests was left alone.

Birds hatched in all three nests. As winter approached, the birds began to migrate. Birds from the first set of nests flew west. Birds from the second set of nests flew east. Those from the third set flew south.

● Why was the third set of nests included in the experiment?

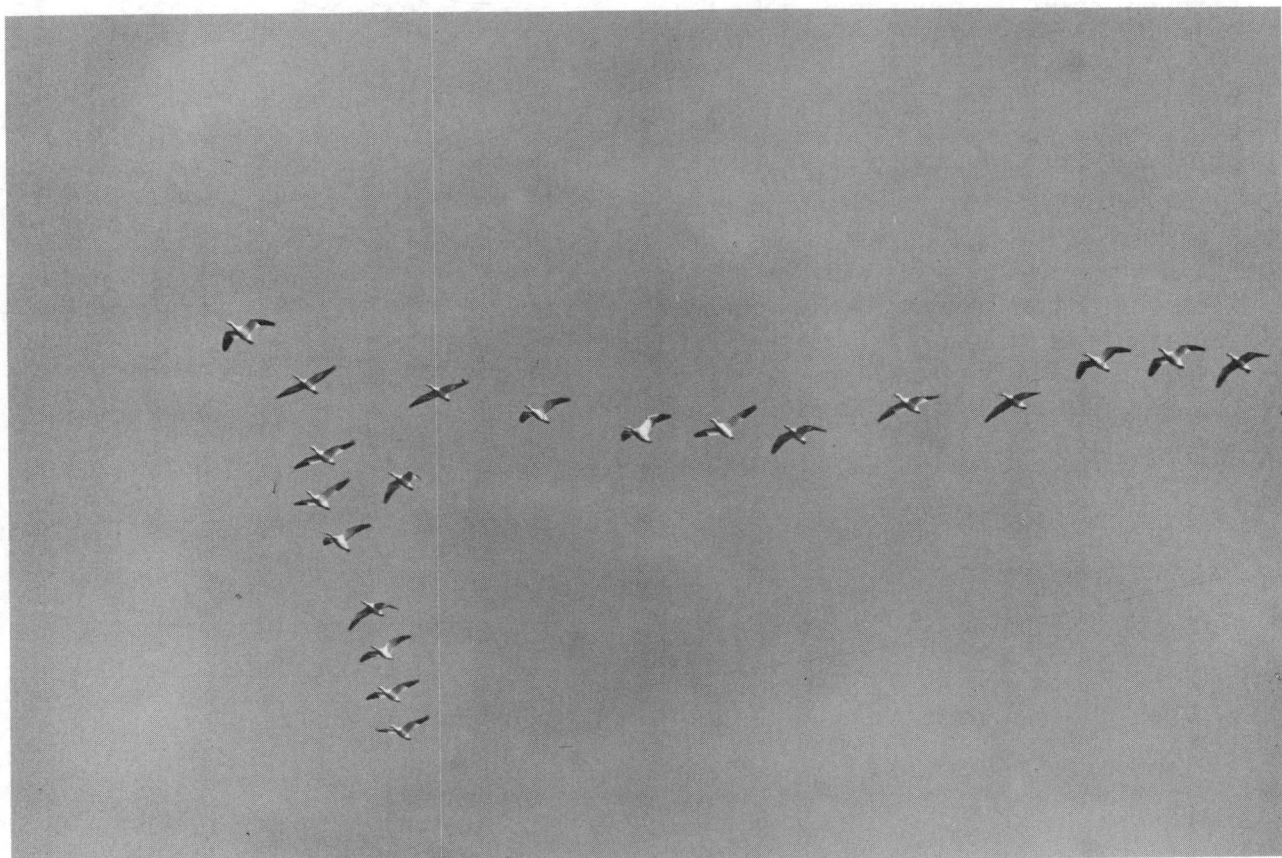

People have been fascinated by the annual migration of geese for many years.

Magnetic Fields

The Earth, like all magnets, has a magnetic field. A **magnetic field** is the area around a magnet where the magnetic force is found. We cannot see this field. We cannot feel this field. But, perhaps birds and other animals can.

If iron filings are sprinkled around a single bar magnet, the lines of magnetic force become visible. The filings are close together at the poles. This is because the magnetic field is strongest at the poles. They get farther apart as their dis-

To Do Yourself   What Does the Field Around a Magnet Look Like?

You will need:

Two bar magnets, iron filings, newspaper, one sheet of white paper, small cup

1. Place a sheet of newspaper on your desk. Place one bar magnet in the center of the newspaper.
2. Place the sheet of white paper on top of the bar magnet so that the magnet is centered underneath.
3. Slowly sprinkle iron filings onto the sheet of white paper. Observe what happens.
4. Collect the iron filings and place them in a small cup.
5. Put the second bar magnet on the piece of newspaper. Place the second magnet as close to the first magnet as possible.
6. Cover the two magnets with the sheet of white paper so that the magnets are in center of the paper.
7. Again sprinkle the iron filings onto the paper. Observe what happens.

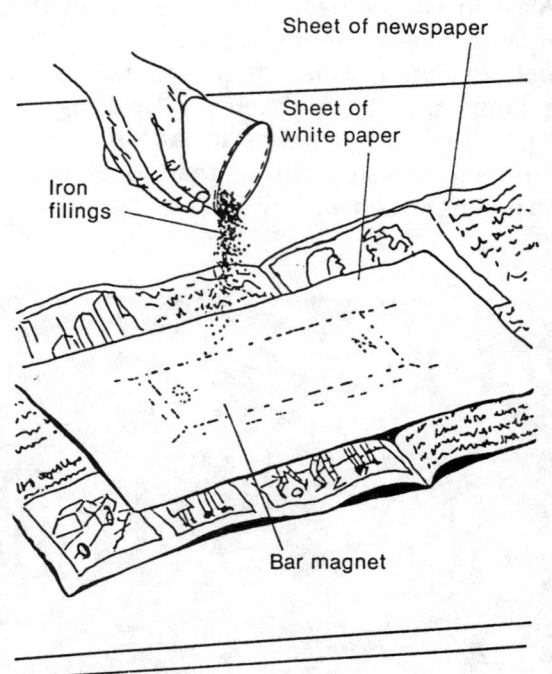

Sheet of newspaper

Sheet of white paper

Iron filings

Bar magnet

Questions

1. What happened as the filings fell on the sheet of paper in step **3**? _____

_____

2. Where were the filings closest to the magnet? _____

_____

3. Would the pattern be the same if you turned the magnet around? _____

_____

4. Was the pattern of the iron filings different in step **7** than it was in step **3**? How?

_____

tance from the magnet increases. The magnetic field is not as strong farther from the magnet.

If another magnet is placed in the magnetic field of the first magnet, the lines of force change. The pattern is disturbed. If the poles are different, the new pattern shows attraction. If the poles are alike, the new pattern shows repulsion (ree-PUL-zhun).

This explains why some of the birds did not fly south. The magnets in the nests disturbed the birds' magnetic sense. The hatching birds oriented themselves to the field of the magnets. Only the birds in the control nest flew south.

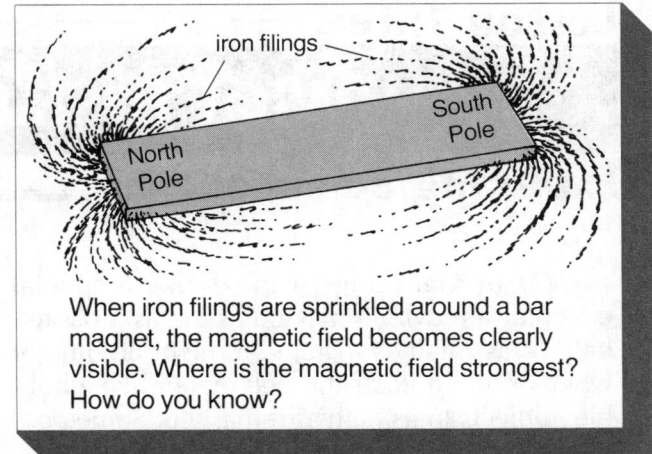

When iron filings are sprinkled around a bar magnet, the magnetic field becomes clearly visible. Where is the magnetic field strongest? How do you know?

Review

I. Fill in each blank with the word that fits best. Choose from the words below.

pattern   **field**   **center**   **poles**   **filings**   **disturbed**

The area around a magnet where the magnetic force is located is called the magnetic _____ . The magnetic force is strongest at the magnetic _____ . Iron filings can be used to show the _____ of the magnetic field. This field can be _____ by another magnet.

II. Which statement seems more likely to be true?

A. _____ Lines of force run from pole to pole.

B. _____ Lines of force are strongest at the center of a magnet.

III. Predict the lines of force for each situation.

A. | N | | N | **B.** | N | | S |

IV. Answer in sentences.

If you had one strong magnet and one weak magnet, how could you find the stronger magnet using iron filings?

Lesson Three

What Is Static Electricity?

Exploring Science

UFO or Lightning? It looks like a special effect for a movie. It may be as big as a basketball, or as small as a grape. It might be blue or white or red. It floats through the air like a bubble. Objects in its path are smashed. Some people call it ball lightning.

Witnesses claim that ball lighting appears during or just after thunderstorms. Scientists have tried to study ball lighting, but it is difficult. Ball lightning is ten thousand times as rare as normal lightning.

A physicist named David Turner is studying ball lightning in England. He believes that a bolt of lightning traps a lump of a charged gaslike state of matter called plasma.

In plasma, the electrons in become separated from the rest of the atom. The plasma forms a ball. As the ball floats, its chemicals interact and change. The surface of the ball becomes cool and watery. The temperature inside the ball is so high that it can do great damage. The energy that escapes is high enough to push the ball forward.

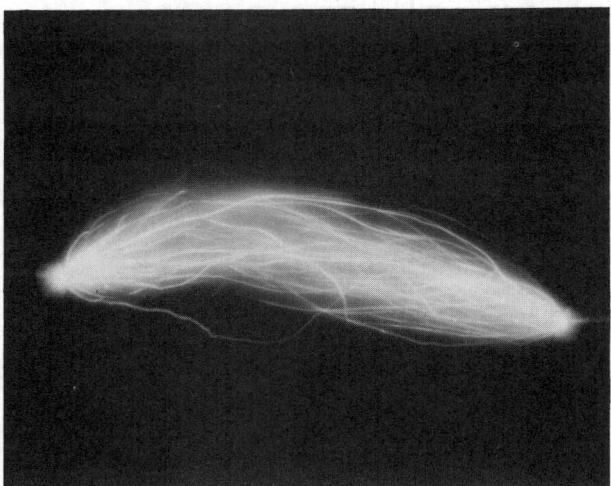

Scientists can make lightning in the laboratory. This is one way that scientists can study lightning.

Dr. Turner has not created ball lighting in his laboratory, or even seen it. Yet, many other scientists will use his theory to develop experiments that will test his ideas.

● Why is Dr. Turner's idea called a theory?

Static Electricity

Lightning is caused by the build-up of electrical charges in a cloud. The molecules of a cloud are electrically neutral. But wind, rain, and friction can change these molecules. Electrical charges begin to form in the cloud. This build-up of electrical charges is called **static electricity.**

Recall that electrical charges can be positive or negative. If electrons are removed from molecules, there are more protons than electrons. Thus, the substance becomes positively charged.

If you rub wool against rubber, the wool is left with a positive charge. The rubber has a negative charge. If you looked at the rubber, you could not see the charges. But, a device called an **electroscope** (EE-lek-troh-skohp) can show that the charges are there.

The top of an electroscope is metal. Recall that many metals are good conductors. A **con-**

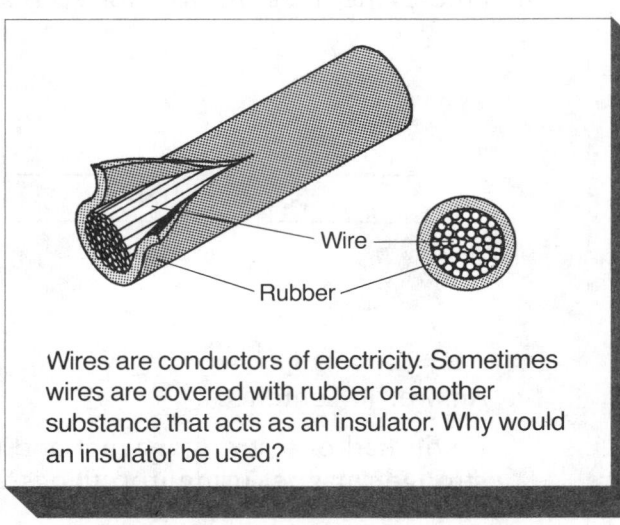

Wires are conductors of electricity. Sometimes wires are covered with rubber or another substance that acts as an insulator. Why would an insulator be used?

ductor (kon-DUK-tohr) is a material that electricity can pass through easily. Let's look at how an electroscope works.

If you touch a charged rubber rod to an electroscope, the extra electrons leave the rubber. They travel into two very thin metal leaves of the electroscope. The leaves become negatively charged. The negative charges from the electrons cause the leaves to push apart. This occurs because like charges repel each other.

If the rubber was charged, why wouldn't you be able to feel the charge? Because rubber is an insulator (IN-suh-LAY-tohr). An **insulator** does not allow electrical charges to pass through it easily. It is the opposite of a conductor.

Silver, gold, and copper are good conductors. Rubber, glass, and plastic are good insulators. Can you think of other substances that are either good insulators or good conductors?

Think about the lightning. The static charges continue to build up in the cloud. The more it rains, the more charges build up. Eventually, the charges must be released. A **discharge** occurs. This causes the electrons to move. This movement of electrons emits light. The result is lightning.

Review

I. Fill in each blank with the word that fits best. Choose from the words below.

protons   electrons   static electricity   like   unlike   positively negatively

If you rubbed a balloon against your hair, the rubber would strip

_____ from your hair. The balloon would become

_____ charged. Your hair would become

_____ charged. Your hair will build up _____

_____ . This will cause your hair to stand on end because

_____ charges repel each other.

II. Which statement seems more likely to be true?

A. _____ Leaves of an electroscope are conductors.

B. _____ Electrical discharges are harmless.

III. Complete each statement by writing a word in the blank.

1. An instrument that detects static electricity is a(n) _____ .

2. A substance that allows the movement of an electrical charge is called

   a(n) _____ .

3. The charge that is caused by a loss of electrons is

   a(n) _____ charge.

4. Most _____ are good conductors.

IV. Answer in sentences.

What would happen if a positively charged object touched the ball of the electroscope?

What Is Current Electricity?

Exploring Science

All Sciences Contribute. If you compared a medical text from 1935 with one from today, you would see many differences. The causes and cures of many diseases have been found. New medicines have been developed. The tests and treatments for many diseases have changed. Patients have new hope for cures.

Many people believe that advances in medicine are the result of our knowledge of biology (BY-ol-oh-gee). **Biology** is the science that deals with the study of life. But all of the sciences help in making medical discoveries.

In 1910, a new disease was named. The disease caused the deaths of many children. A doctor studied the blood of the sick children. He saw that many of their red blood cells looked strange. They looked like sickles. So, the disease was named sickle-cell anemia (uh-NEEM-ee-uh).

Medical help for this disease was limited. No cause for the disease was known. But, in 1949, a chemist named Linus Pauling provided new information. Pauling knew about a special technique using electricity. The technique involved passing electricity through a solution of proteins. When this was done, the proteins separated. The process is called electrophoresis (ee-LEK-troh-FOR-ee-sis).

Using this process, Pauling could identify a special molecule in the sickle-shaped cells. The molecule is called the S molecule. This S molecule is the cause of sickle-cell anemia. Because of Pauling's work, the cause of this disease could be separated from the other red blood cells.

Much has been learned about the S molecule that causes sickle-cell anemia. In the 1990s, doctors developed new ways to treat the disease based on what they had learned.

● Why is this story called "All Sciences Contribute?"

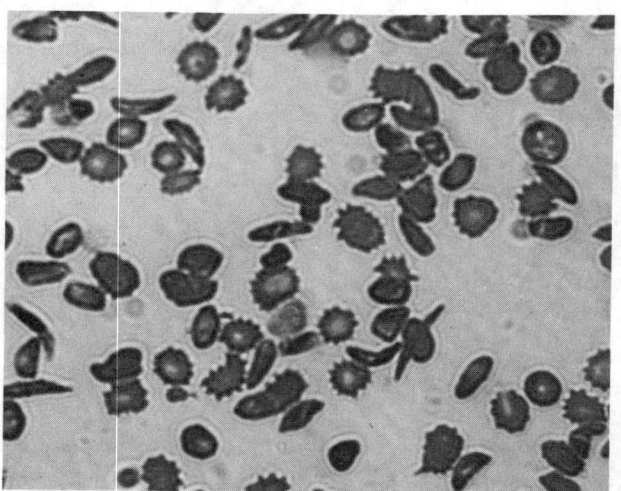 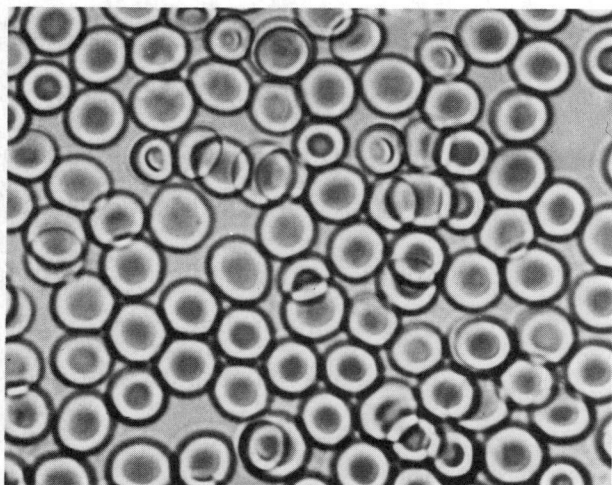

The blood cells on the left are from a person with sickle cell anemia. Compare these blood cells with the normal (round) blood cells on the right.

Current Electricity

Electrical current is the flow of electrons from one point to another. These points, or poles, are called **electrodes** (ee-LEK-trohds). Current *always* travels from the negative electrode to the positive electrode. Since electrons are moving, they are attracted to the opposite charge.

The rate at which electrons pass through a conductor determines the strength of the current. For example, if few electrons pass through a conductor per second, the current is weak. If many electrons pass through the conductor, the current is strong. An **ammeter** (AM-mee-tur) is used to measure the strength of electrical currents. An ammeter counts the number of electrons that pass through one point of a conductor in one second.

The **ampere** (AM-peer) is the unit for measuring electrical current. A current of 1 ampere, or 1 **amp,** is equal to 624,000,000,000,000,000 electrons per second. The ampere is a much more convenient unit than an electron count.

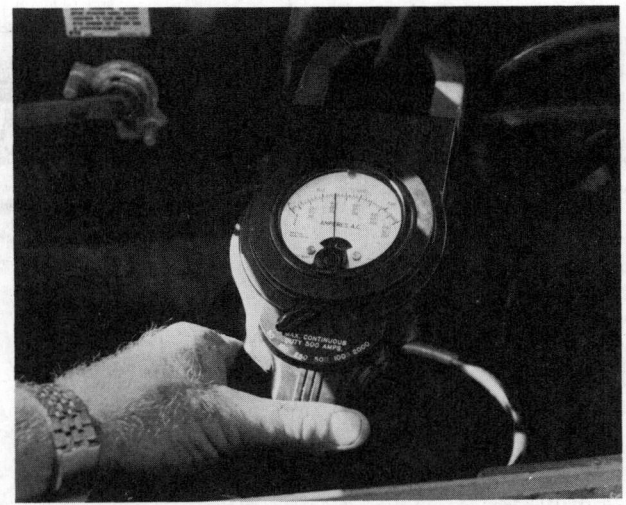

An ammeter is used to measure electrical current. The unit for electrical current is the ampere, or amp.

Review

I. Fill in each blank with the word that fits best. Choose from the words below.

**ammeter   ampere   current   electrode   electrons   conductor
insulator**

The flow of electrical charges is called _____ . It is

measured with a(n) _____ . A current is strong if many

_____ flow per second. The _____ is

the unit for current. A current flows through a(n) _____

from one _____ to the other.

II. Which statement seems more likely to be true?

A. _____ Electrical current flows from the positive electrode to the negative electrode.

B. _____ Electrical current flows from the negative electrode to the positive electrode.

III. Answer in sentences.

What is meant by the term "strong" current?

_____

_____

_____

What Is Electromotive Force?

Exploring Science

Emissions Zero. Many states have passed laws to provide clearer air. California has led this movement. A deadline was set for when at least 2 percent of the cars sold in California must be emission-free. Auto makers knew that they must find an alternative to a gasoline-powered car. They must meet this deadline.

Auto makers developed improved designs for electric vehicles. These cars are quiet and produce almost no emissions. They emit no toxic gases. They seem to be the best solution to the California problem.

However, there are two major difficulties. Electric cars run on batteries—think how long a cord they would need to be plugged in to for a trip! A battery will power a car for about 100 miles. Then the battery needs to be recharged. Also, the common automobile battery contains acid and lead. These materials are dangerous pollutants when the battery is discarded.

The auto makers have recently invested over $18 million in developing a new battery. The nickel-hydride battery almost doubles the distance a car can travel before having to be recharged. With it, a compact car can travel 200 miles. Also, nickel is a less dangerous pollutant than lead.

Although 200 miles between charges is still less than acceptable, the progress is hopeful.

● In your opinion, is the California law a reasonable solution to the pollution problem?

The Impact is one of the new electric-powered cars.

Many states have facilities for collectiong car batteries.

Electromotive Force

Gasoline-powered motors use chemical reactions that release heat and cause the expansion of gases. Electric-powered motors, however, run on electric current, the flow of electrons in a conductor.

A battery provides **electromotive** (ee-LEK-troh-moh-tiv) **force.** This is the push that moves electrons. Without electromotive force, or EMF, electrical current could not flow.

There are many types of batteries. The most common type is the dry cell. This is the type of battery used in most flashlights. A dry cell works by moving electrons from a negative pole to a positive pole.

Look at the diagram of the battery. The outside case of this battery is made of zinc. This zinc case is the negative electrode. A carbon rod in the center of the battery is the positive electrode. A chemical reaction takes place between the zinc and the carbon. Electrons begin to collect on the zinc. It becomes negatively charged. The carbon loses electrons, so it gains a positive charge.

When a wire is connected to each of the poles, electrons will flow from the negative pole to the positive pole. The negative pole of a battery is marked −. The positive pole is marked +. As electrons move from one pole to the other, each pole builds up charges. The more charges gained by the poles, the stronger the EMF will become.

The **volt** is the unit for measuring electrical pressure, or EMF. Electromotive force is therefore called **voltage.** A flashlight battery has an EMF of 1.5 volts. Many car batteries are 12 volts. Household electricity is usually 110 volts. The **voltmeter** is used to measure voltage.

Think about water flowing through a hose. The flow of the water can help us understand the difference between current and EMF. The amount of water moving through the hose each second is the current. The force, or pressure, pushing the water is similar to the EMF.

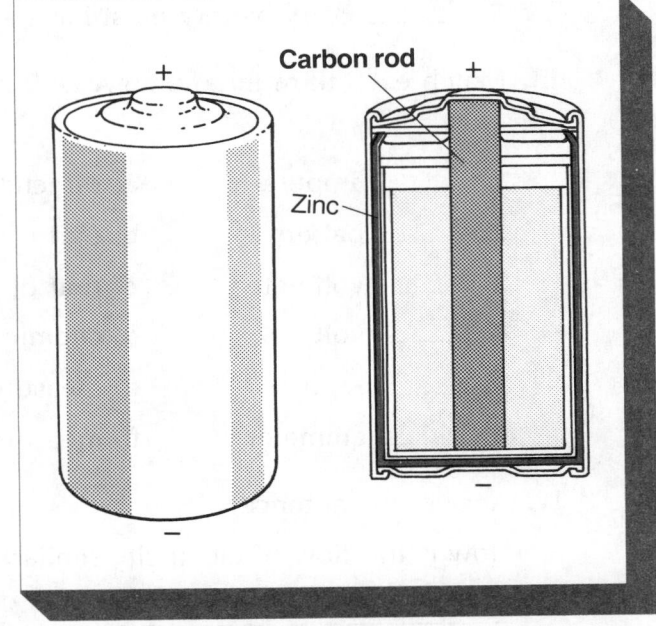

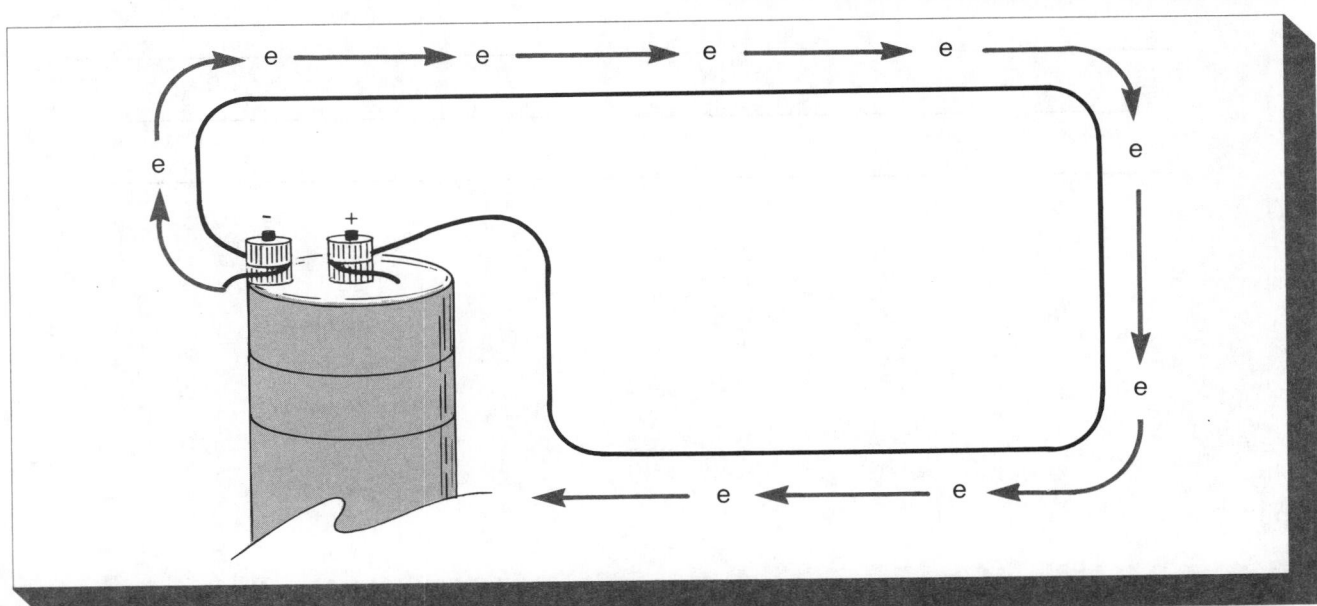

Review

I. Fill in each blank with the word that fits best. Choose from the words below.

voltage   **volt**   **voltmeter**   **force**   **ammeter**   **chemical**   **electrons**

Electromotive force is often called _____ . The unit for

EMF is the _____ . EMF is measured with a(n)

_____ . This device measures the amount of

_____ pushing the _____ .

II. Which statement seems more likely to be true?

A. _____ Every battery must be made of zinc.

B. _____ Every battery must have a + and a − pole.

III. Match each term in column **A** with its description in column **B**.

A	B
1. _____ ampere	a. direction of flow
2. _____ battery	b. unit of current
3. _____ voltmeter	c. unit of EMF
4. _____ volt	d. chemical to electrical
5. _____ − to +	e. measures EMF
6. _____ ammeter	f. measures strength of the current

IV. Answer in sentences.

How is the flow of electricity similar to the flow of water?

_____

_____

_____

_____

What Is a Circuit?

Exploring Science

Off and On. Computer languages are very complicated. All of the letters, numbers, and digits in a computer are in codes using 0 and 1. The 0 is *off*. The 1 is *on*. This is called an on-off pulse of electricity.

Suppose the letter "B" has the code 01100010. These 8 numbers are called a **byte**. When "B" is touched on the keyboard, the electricity flowing through a part of the computer makes the same pattern as the byte: off-on-on-off-off-off-on-off. The computer understands the byte. "B" means nothing.

The computer's information is stored in memory chips. Each on-off pulse involves 500,000 electrons. So the memory storage is limited. Today's scientists are working to create a pulse of a single electron. If the electron is present, = 0. If it is not, = 1. This change could increase memory a million times. Think about the video or computer games that would be possible with this new technology.

Many computer users no longer use their keyboards. An electronic device called a mouse is often used to send signals to the computer. Other computers have touch-screens. The

touch of a finger on a part of the screen sends a signal to the computer.

No matter how the command is given, it must be changed into bytes. The computer does not understand anything else.

● What code would you use if you wanted to write your name in bytes?

Electric Circuits

The idea of codes using on and off patterns is not new. The technology used by computers is based on a knowledge of electric circuits.

A **circuit** (sur-CUT) is the path taken by a current. A path with no breaks is called a **closed circuit.** When there is a closed circuit, the electricity is on.

If there is a break in the path the current follows, the circuit is incomplete. A break in the path is called an **open circuit.** The electricity is off if the circuit is open.

Recall that electrons flow through a conductor. Look at the diagram at the right. The wires that connect the battery to the lamp are the conductors.

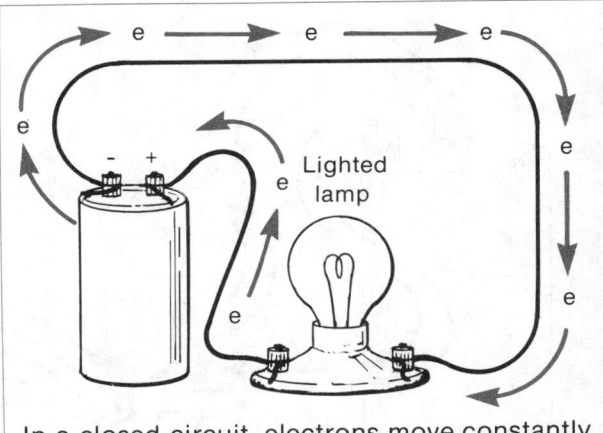

In a closed circuit, electrons move constantly from the negative pole to the positive pole.

Arrows in the diagram on page 229, show the direction that the electrons are moving. The current flows from the negative pole through the wire. Then it returns through the wire to the positive pole. "Positive" and "Negative" are the opposite of what you might expect.

This movement of electrons will continue because the circuit is closed. Unless the circuit is broken or opened, the lamp will remain lighted. A switch can be put into the circuit. A **switch** is a simple device that opens or closes an electric circuit.

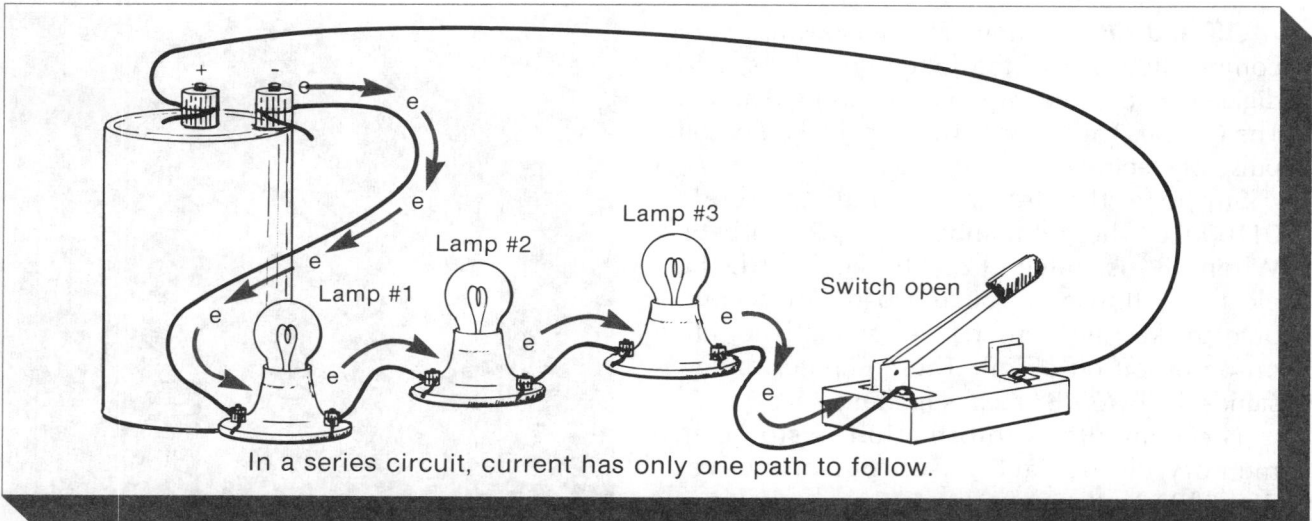

In a series circuit, current has only one path to follow.

Let's add a switch and two more lamps to our example. Look at the diagram above.

When the switch is open, the lamps are not lit. The circuit is open. The current cannot jump across the gap made by the switch. There is only one path for the current. So, a break at any point stops the current from flowing.

This type of circuit is called a **series circuit**. A series circuit is an electric circuit with only one path for a current to follow. Many of the first homes to get electricity were wired in this way.

When one light was turned off, or a bulb burned out, all the lights in the house went off. This system was not very convenient, so a new wiring method was developed.

A **parallel** (PAR-uh-lel) **circuit** is a circuit with more than one path for a current to follow. The diagram below shows how our lamps would be connected in a parallel circuit.

The arrows in the diagram show the directions the electrons can travel. If the switch near the third lamp is opened, only the third lamp.

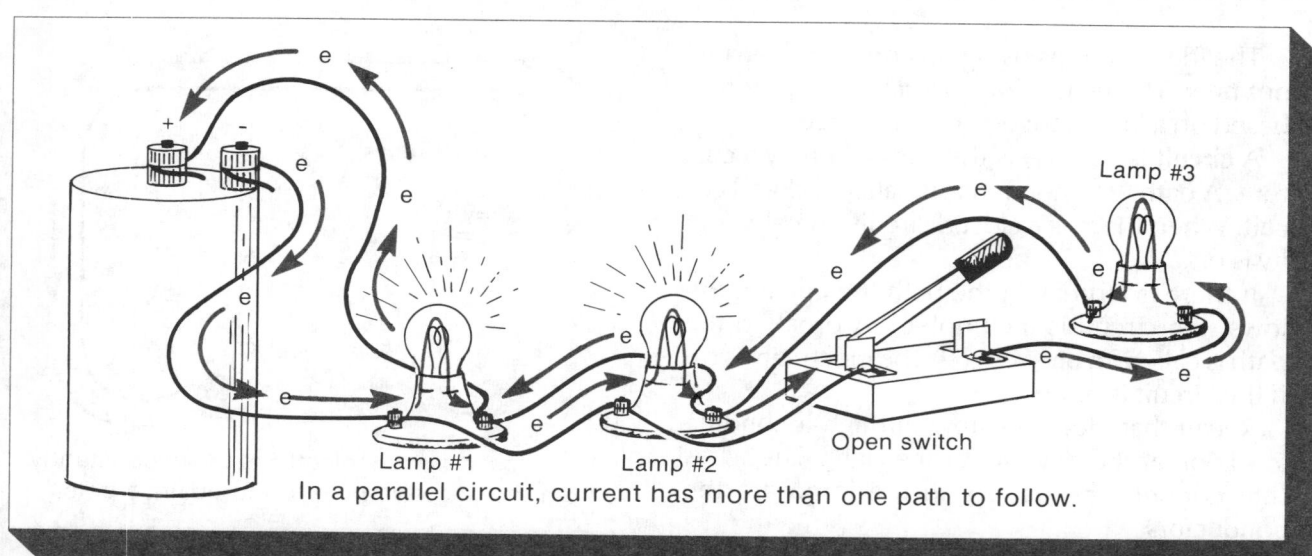

In a parallel circuit, current has more than one path to follow.

will go off. The other lamps will remain lighted. The current will follow another path.

Today, most homes are wired in parallel circuits. If we looked at the wiring, we would find that each branch of the circuit would have its own switch. Look at the diagram to the right.

This type of diagram is called a schematic (skee-MAT-ik). A **schematic** is a diagram of an electrical system. The symbols on a schematic tell where switches and other features are located. Look at the schematic. Can you figure out what each symbol stands for?

This schematic has three symbols. ⊣⊢ is the symbol for a dry cell. A lamp or light bulb is shown by this symbol ⎯Ⓠ⎯. ⟍ is the symbol for a switch. The solid lines (⎯) in the schematic represent the wires.

The schematic for our series circuit with the three lamps would look like this. Label each of the symbols on the schematic.

Most electrical appliances have schematics. A schematic is helpful when repairs on the appliance are needed. By using a schematic, the repair-person can find out exactly where the wires and switches in an appliance are located.

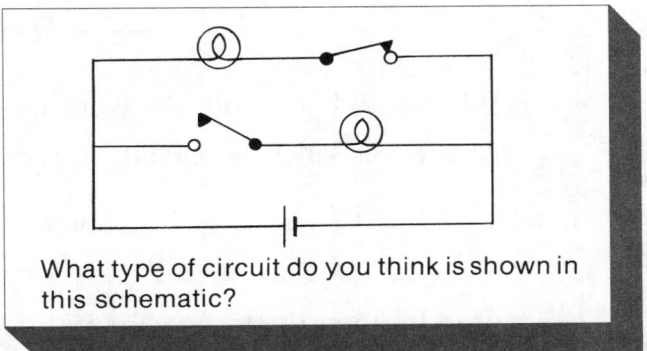

What type of circuit do you think is shown in this schematic?

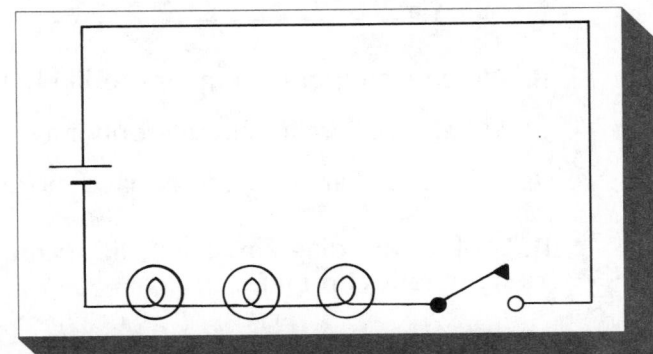

To Do Yourself   How Can a Series Circuit Be Made and Drawn?

You will need:

Two batteries, bulb and socket, knife switch, ammeter, wire, tape, screwdriver

1. Tape the batteries together (+ to −).
2. Connect the negative end of the batteries to the negative terminal of the ammeter.
3. Connect the positive terminal of the ammeter to the bulb and socket.
4. Connect the other side of the bulb and socket to an open knife switch. Then connect the switch to the positive end of the batteries.
5. Close the switch and observe.
6. Use symbols to draw this circuit.

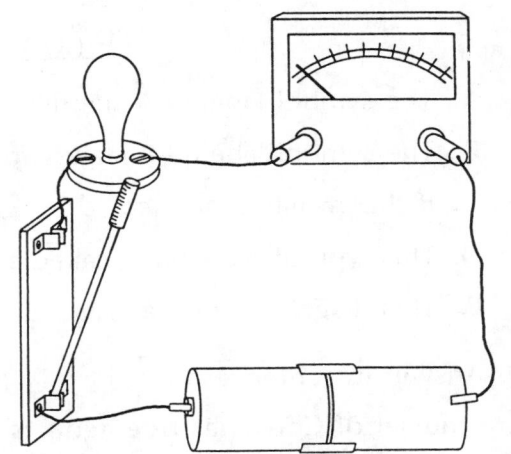

Questions

1. What happens when the switch is open? ⎯⎯⎯⎯⎯⎯⎯

2. What happens when the switch is closed? ⎯⎯⎯⎯⎯

3. How many paths can the current follow? ⎯⎯⎯⎯⎯

Review

I. Fill in each blank with the word that fits best. Choose from the words below.

parallel   **series**   **circuit**   **schematic**   **switch**   **opened**   **closed**

A _____ is a path for an electric current. When the path is _____ , it is complete. A _____ can be used to open or close a circuit. In a _____ circuit, there is only one path for the current to follow. A sketch of a circuit and its features is called a _____ .

II. Which statement seems more likely to be true?

A. _____ Parallel circuits only have one path.

B. _____ Parallel circuits have more than one path.

III. Look at the diagram. Then, fill in the word (or words) that make each statement complete.

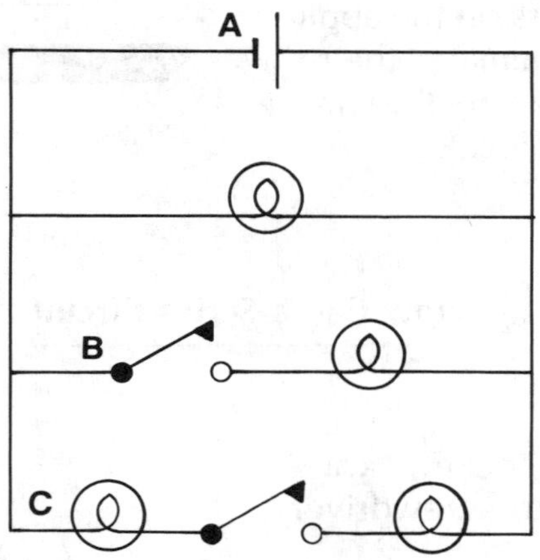

A. The symbol labeled **A** stands for a _____ .

B. The symbol labeled **B** stands for a _____ .

C. If the switch is opened, _____ lamp(s) will go out.

D. This type of diagram is called a _____ .

E. This diagram shows a _____ circuit.

IV. Answer in sentences.

One set of Christmas tree lights is connected by a series circuit. Another set is connected by a parallel circuit. Which set of lights would you buy? Explain your answer.

_____

_____

_____

What Can Affect Currents?

Exploring Science

Unaffected Current. A lead ring was cooled using liquid helium. The temperature was lowered to −265°C. Then, several hundred amperes of current were fed into the metal.

The current was switched off. Scientists watched as the current made circuit after circuit around the ring. The ring was kept at the very low temperature. The current never changed.

The electricity in the ring flowed for three years. There was almost no energy loss. Scientists had made the first superconductor. A **superconductor** is a material that does not oppose a current. It can only exist at extremely low temperatures.

Since the invention of the superconductor, scientists have worked to improve the technology. Until recently −140°C was the warmest temperature for this effect. In 1993, however, French scientists reported a success. They recorded superconductivity at −23°C. Additional tests are being done to confirm this report.

At −23°C, this technology could be used for home appliances. Imagine having a home freezer with no cost for electricity. Today, however, machines that use superconductors must be kept at very low temperature. If they are not, the superconductor will not superconduct. Scientists use superconductors only when the benefits make up for the inconvenience.

● Why are metals used for most of these types of experiments?

Increasing and Decreasing Current

The opposition to current is called **resistance.** Until the discovery of superconductors, all conductors had some resistance.

Suppose you had two wires of the same thickness and length. One of the wires is steel. The other wire is silver. Which wire do you think would be a better conductor? If you guessed the silver wire, you were correct. In fact, the silver wire can carry almost ten times more current than the steel wire. Silver has much less resistance than steel.

The unit for resistance is the **ohm.** This unit is written using the symbol Ω. The symbol is called an **omega** (OH-may-guh). The omega is the Greek symbol for the letter O. This symbol is used so that the letter O will not be confused with a zero.

There is no meter for measuring resistance directly. The amount of resistance can be found by using **Ohm's Law.** This law states that current is affected by electromotive force and resistance. If the EMF is high, the current is high. But if the resistance is high, the current is low.

Let's look at the equation for Ohm's Law. When using this equation, the answer is given in amperes.

$$\text{Current} = \frac{\text{Electromotive Force}}{\text{Resistance}}$$

Suppose a 6-volt battery was used to provide electrical energy. If the battery was connected to a circuit with 3 ohms of resistance, how much current would flow through the circuit? Let's use the equation for Ohm's Law to find out.

$$\text{Current} = \frac{\text{Electromotive Force}}{\text{Resistance}}$$

$$\text{Current} = \frac{\text{EMF}}{\text{ohms}}$$

$$\text{Current} = \frac{6 \text{ volts}}{3 \ \Omega} = 2$$

Recall that the answer for this equation is given in amperes. Thus, our answer must be 2 amperes.

Now, suppose the resistance of the conductors was doubled. $2 \times 3\ \Omega = 6\ \Omega$. How much current would flow through the wire now?

$$\text{Current} = \frac{\text{EMF}}{\text{ohms}}$$

$$\text{Current} = \frac{6\ \text{volts}}{6\ \Omega}$$

$$\text{Current} = 1\ \text{ampere}$$

As you can see, doubling the resistance lessens the flow of the current.

Suppose we went back to our original resistance of 3 ohms. Let's use 2 batteries to give us 12 volts. How much current will travel through the wire?

$$\text{Current} = \frac{\text{EMF}}{\text{ohms}}$$

$$\text{Current} = \frac{12\ \text{volts}}{3\ \Omega}$$

$$\text{Current} = 4\ \text{amperes}$$

Notice that if we increase the voltage, the flow of the current also increases. Changing the length or thickness of the wire is another way to change the amount of current that can pass through a conductor.

Suppose you were using a hose and wanted to get a lot of water from it as fast as possible. You would choose a wide hose. A wider hose would have less opposition to the flow of the water. You could also use a shorter hose. This hose would also have less opposition to the current.

The same rules apply to conductors. A thin wire offers great resistance. A thicker wire offers less resistance. A long wire has greater resistance than a shorter wire.

What happens to an electrical current when it meets high resistance? Think about a toaster. Current enters the toaster through a short, thick wire. The current then passes to the heating element of the toaster. The element is a long thin wire that is twisted into coils. A heating element has great resistance. The electrical energy is

To Do Yourself How are Resistance, Current, and Voltage Related?

You will need:

Two batteries, bulb and socket, ammeter, wire, tape, screwdriver

1. Use one battery. Connect the negative end of the battery to the negative terminal of the ammeter.
2. Connect the positive terminal of the ammeter to a bulb and socket.
3. Connect the other side of the bulb and socket to the positive end of the battery.
4. Take the reading on the ammeter and the voltage on the battery. Use Ohm's Law to find the resistance in the circuit.
5. Now repeat steps 1–4 using two batteries.

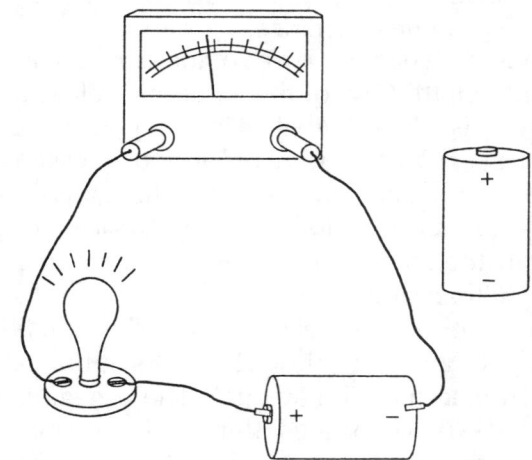

Questions

1. What was the resistance using one battery? _____
2. What was the resistance using two batteries? _____
3. How did two batteries affect the current? _____

transformed to heat energy when it flows through the element. The element gets hot and toasts your bread.

A light bulb works in a similar way. A high-resistance metal is used inside the light bulb. The metal is twisted into a long thin wire. The electrical energy is transformed into light energy when it meets the high resistance of the wire. This causes the wire in the bulb to glow.

Temperature also affects resistance. A cold wire has much less resistance than a hot wire. Recall the lead ring. It was kept at a very low temperature. Current flowed through the ring without meeting any resistance. The super-low temperature made the ring a superconductor. This superconductor had no resistance at all.

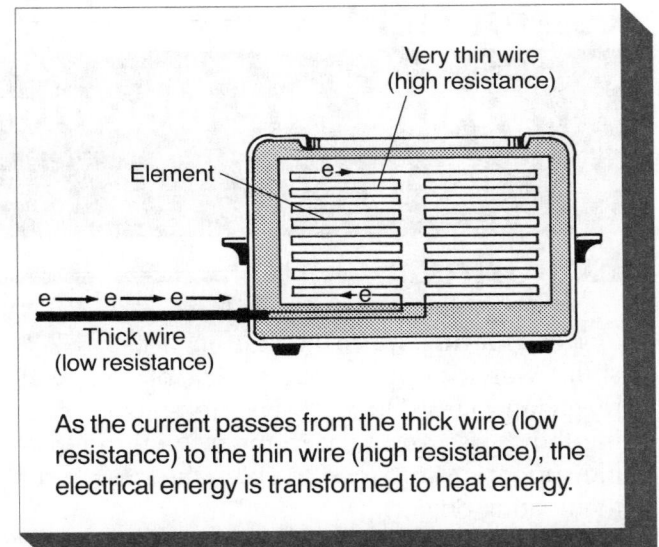

As the current passes from the thick wire (low resistance) to the thin wire (high resistance), the electrical energy is transformed to heat energy.

Review

I. Fill in each blank with the word that fits best. Choose from the words below.

current   ohm   ampere   resistance   increased   decreased

The opposition to the flow of current is called _____ .

The _____ is the unit for resistance. Ohm's law states

that if resistance is _____ , then the _____
is decreased.

II. Which statement seems more likely to be true?

A. _____ If a 12 volt battery pushed a current through a resistance of 2 ohms, the current would be 24 amperes.

B. _____ If a 12 volt battery pushed a current through a resistance of 2 ohms, the current would be 6 amperes.

III. Place the following wires in order from greatest resistance to least resistance. Use the numbers **1** (for greatest) through **4** (least).

A. _____ a 15 cm long, 3 cm thick silver wire

B. _____ a 15 cm long, 5 cm thick silver wire

C. _____ a 10 cm long, 5 cm thick silver wire

D. _____ a 15 cm long, 3 cm thick steel wire

IV. Answer in sentences.

Why is a high resistance wire sometimes better than a low resistance wire?

235

How Are Electricity and Magnetism Related?

Exploring Science

The Floating Train. Levitation (lev-uh-TAY-shun) was a popular magic trick of the great magician Harry Houdini. People enjoyed seeing things rise and float in the air with no visible support. Many tried to solve the mystery of the floating objects.

Today, levitation is a more serious term. Instead of a magician making a table float, an entire train is levitated. The train is called Maglev. This name is short for magnetic levitation.

Researchers describe Maglev as a floating train. The rails for Maglev are T-shaped. The bottom of the car wraps around the rail. This design may keep the train from running off the track. It also allows the train to travel safely at speeds up to 400 kilometers per hour.

Maglev runs on electricity and magnetism. The car does not touch the rails. When the train is running, the bottom of the train repels the top of the rail. So, the car floats about one centimeter above the rail as it moves forward. Because there is no friction between the rail and the car, the train also runs very quietly.

Germany and Japan have Maglev test tracks in operation. The first operating American Maglev is planned for Orlando, Florida.

● What are three benefits of Maglev?

Maglev uses both electricity and magnetism.

Electromagnets

Maglev is one example of technology that uses both electricity and magnetism. An electromagnet (ee-LEK-troh-MAG-net) is another example. An **electromagnet** is a device that uses electricity to make magnetism. Let's look at how an electromagnet works.

Current flowing through a wire produces a small magnetic field. However, if the wire is wrapped into a coil, the magnetic field becomes stronger. As long as a current exists, a magnetic field exists. But if the circuit is broken, the magnetic field disappears.

If an iron core is placed into the coil, the magnetic field increases further. The iron core also becomes a strong magnet. As long as the current flows, the core is a magnet. If the current stops, however, the magnetic force also stops. This creates a useful tool. The tool is the electromagnet.

A large electromagnet can lift heavy metal objects. It can move these objects to a new location. Then, with the flick of a switch, the current can be stopped. This stops the magnetism. The object is dropped.

Recall Ohm's Law. It states that increasing the EMF, increases the current. This means that if we increase the EMF in the coil, we will also increase the magnetic force. A stronger magnet is made.

The magnetic force of an object can be increased in another way. Increasing the number of coils around the core increases the magnetic force. So the more turns of wire in the coil, the greater the magnetic force.

If an electric current can make a magnet, can a magnet be used to make an electric current? Yes. When a magnet is placed inside a coil of wire, there is no current. But, if the magnet is moved back and forth inside the coil, a current

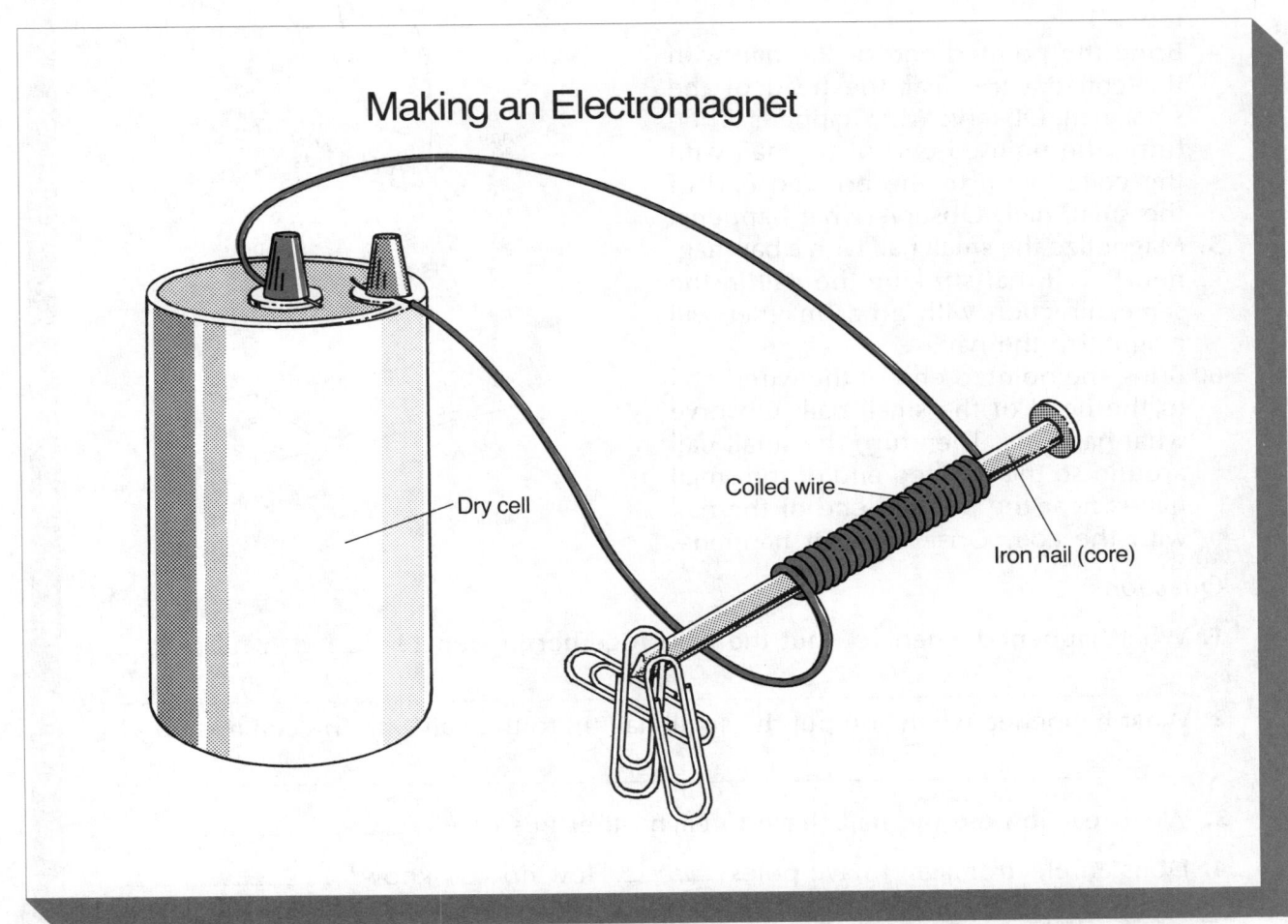

Making an Electromagnet

Dry cell

Coiled wire

Iron nail (core)

is produced in the wire. This type of current is called an induced (in-DOOSD) current. An **induced current** is a current made by the forced flow of electrons.

If the magnet is moved faster, the current increases. The more turns in the coil, the greater the current becomes. A greater current can also be made with a stronger magnet.

A machine that induces current in this way is called a **generator** or dynamo. An electric generator converts mechanical energy into electric current. Most generators are powered by steam produced by burning fuel or by nuclear fission. Some get energy from the motion of falling water. A Maglev changes electrical energy back into mechanical energy.

To Do Yourself How Are Electricity and Magnetism Related?

You will need:

Two iron nails (one large and one small), wire, battery, bar magnet, tape

1. Bring the large nail near the small nail. Observe what happens.
2. Coil the wire around the large nail from the head to the point.
3. Tape both ends of the wire to the battery.
4. Bring the pointed end of the nail with the coiled wire near the head of the small nail. Observe what happens. Then bring the pointed end of the nail with the coiled wire to the pointed end of the small nail. Observe what happens.
5. Magnetize the small nail with a bar magnet. Recall that stroking the nail in the same direction with a bar magnet will magnetize the nail.
6. Bring the pointed end of the wired nail to the head of the small nail. Observe what happens. Then turn the small nail around so the pointed end of the small nail is near the pointed end of the nail with the coil. Observe what happens.

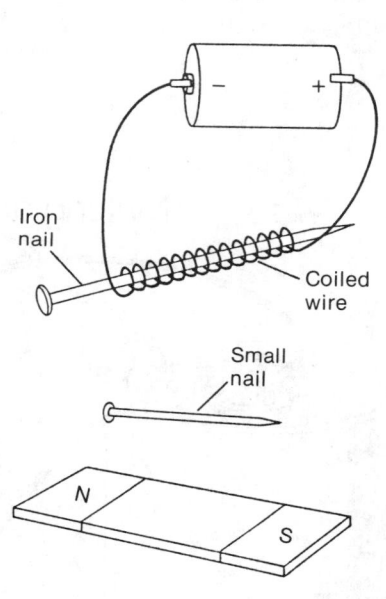

Questions

1. What happened when you put the nails together in step **1**? _____

_____

2. What happened when you put the small nail up to the nail with the coil in step **4**?

_____

3. What reaction did the nails have to each other in step **6**? _____

4. Does an electromagnet have poles? _____ How do you know? _____

Review

I. Fill in each blank with the word (or words) that fits best. Choose from the words below.

magnetic field   electromagnet   coil   compass   induced current

power source

Wire twisted into a series of loops is called a _____ . If the wire is connected to a _____ , a _____ is produced. If an iron core is added to the coil, a(n) _____ is made. If a magnet is moved back and forth inside a coil a(n) _____ is made.

II. Which statement seems more likely to be true?

A. _____ An electromagnet can easily be turned off.

B. _____ An electromagnet must have a soft iron core.

III. Place the electromagnets in order from strongest to weakest. Use the numbers **1** (strongest) through **4** (weakest).

A. _____ 6 volt battery, 30 turns of wire, iron core

B. _____ 3 volt battery, 15 turns of wire, iron core

C. _____ 3 volt battery, 30 turns of wire, iron core

D. _____ 9 volt battery, 30 turns of wire, iron core

IV. Answer in sentences.

How could you force a piece of wire to carry a current without using a power source? Explain your answer.

_____

_____

_____

_____

Review What You Know

A. Hidden in the puzzle below are ten science words related to magnetism and electricity. Use the clues to help you find the terms. Circle each word you find in the puzzle. Then write each term on the line next to its clue.

```
G C O B A L T S
I L F P O E S C
N A I G P C T U
S A M M E T E R
U F O T N R I R
L A H E P O L E
A I M W A S O N
T S W I T C H T
O E D R I O P L
R E P E L P S I
T S E R I E S T
```

Clues:

1. A magnetic material. _____

2. Place where the magnetic force is strongest. _____

3. To push away from. _____

4. Device used to measure current. _____

5. A circuit where no electricity flows. _____

6. A device used to open or close a circuit. _____

7. Circuit that has only one path. _____

8. Unit of resistance. _____

9. Flow of electrons through a conductor. _____

10. Material that electricity cannot pass through. _____

B. Write the word (or words) that best completes each statement.

1. _____ The atoms of a magnet are arranged in groups called **a.** poles **b.** currents **c.** domains

2. _____ All magnets have **a.** nickel **b.** a bar shape **c.** poles

3. _____ Sprinkling iron filings around a magnet will show the magnet's lines of **a.** cobalt **b.** force **c.** poles

4. _____ The magnetic field of a bar magnet is strongest at the **a.** center **b.** poles **c.** north pole

5. _____ A buildup of electrical charges is called **a.** static electricity **b.** a discharge **c.** current electricity

6. _____ The flow of electrons through a conductor is called **a.** resistance **b.** current **c.** discharge

7. _____ The ampere is the unit for **a.** resistance **b.** electromotive force **c.** current

8. _____ A battery may be used in a circuit to provide **a.** resistance **b.** a discharge **c.** electromotive force

9. _____ The poles on a battery are **a.** electrodes **b.** insulators **c.** domains

10. _____ A voltmeter is used to measure **a.** current **b.** resistance **c.** electromotive force

11. _____ Current will not flow if a circuit is **a.** open **b.** closed **c.** connected to a battery

12. _____ A circuit with only one path for electrons to follow is **a.** a parallel circuit **b.** a series circuit **c.** an open circuit

13. _____ A circuit with more than one path for electrons to follow is **a.** a parallel circuit **b.** a series circuit **c.** an electromagnet

14. _____ A diagram of an electrical circuit with symbols showing where each part of the circuit is located is a **a.** schematic **b.** dry cell **c.** circuit board

15. _____ Opposition to a current is **a.** amperage **b.** voltage **c.** resistance

16. _____ A good conductor of electricity is **a.** silver **b.** glass **c.** plastic

17. _____ A material that does not allow electricity to pass through it easily is **a.** a conductor **b.** an insulator **c.** copper

18. _____ A low-resistance wire would be **a.** short and thick **b.** long and thin **c.** short and thin

19. _____ A coil of wire wrapped around an iron core can be used to make **a.** a dry cell **b.** an electromagnet **c.** a voltmeter

20. _____ Ohm's Law states that there is a relationship between **a.** EMF and voltage **b.** EMF and resistance **c.** magnetism and like charges

21. _____ If more voltage is added to a circuit, the current will **a.** increase **b.** decrease **c.** remain the same

22. _____ Moving a magnet through a coiled wire will produce **a.** resistance **b.** electricity **c.** a switch

23. _____ The Law of Magnetic Poles states that **a.** like poles attract **b.** like poles repel **c.** all magnets must be made of iron

24. _____ Of the following, the material least likely to be found in a magnet is **a.** iron **b.** cobalt **c.** carbon

25. _____ If a magnet were suspended freely in air, the north pole of the magnet would point toward the earth's **a.** geographical north pole **b.** magnetic south pole **c.** magnetic north pole

C. Apply What You Know

Study the diagram below. Then answer the questions.

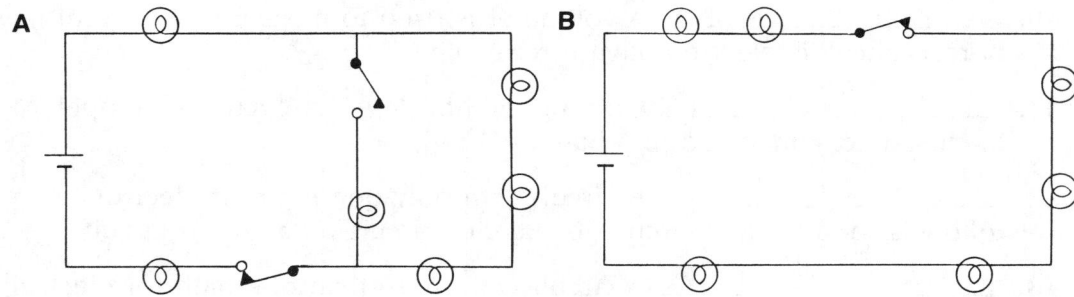

a. Drawing A shows a (series/parallel) circuit.
b. Drawing B shows a (series/parallel) circuit.
c. There are (1/2) switches in drawing A.
d. The power source in drawing B is a (motor/dry cell).
e. There are (1/2) switches in drawing B.
f. The switches in drawing B are (open/closed).
g. If the power is on, (5/6) lights are lit in drawing B.
h. If the power is on, (5/6) lights are lit in drawing A.
i. If one bulb goes out in drawing A, the other lights will (remain lit/ go out).
j. If one bulb goes out in drawing B, the other lights will (remain lit/ go out).
k. The current in diagram A can follow (1 path/2 paths).
l. Diagrams A and B are called (circuit boards/schematics).

D. Find Out More

1. Find out how a flashlight works. Examine a flashlight to find the conductors. Then trace the path of electricity. Draw a diagram of the flashlight. Label the conductors and the source of power. Include arrows in your diagram to show the path of the current.
2. Practice using a compass. Plan a treasure hunt. Hide objects in various locations. Instead of a map, list instructions that use a compass and a meterstick. Each instruction should state the distance and compass reading for where an object can be found. See how many objects are found by following the instructions.

Summing Up
Review What You Have Learned So Far

A. Study the illustrations. Then identify each illustration on the line below it. Choose from the items below.

parallel circuit   pulley system   balanced equation   isotopes
unbalanced equation   lever   atomic nucleus   vector diagram
inclined plane   series circuit   electromagnet   positive ion

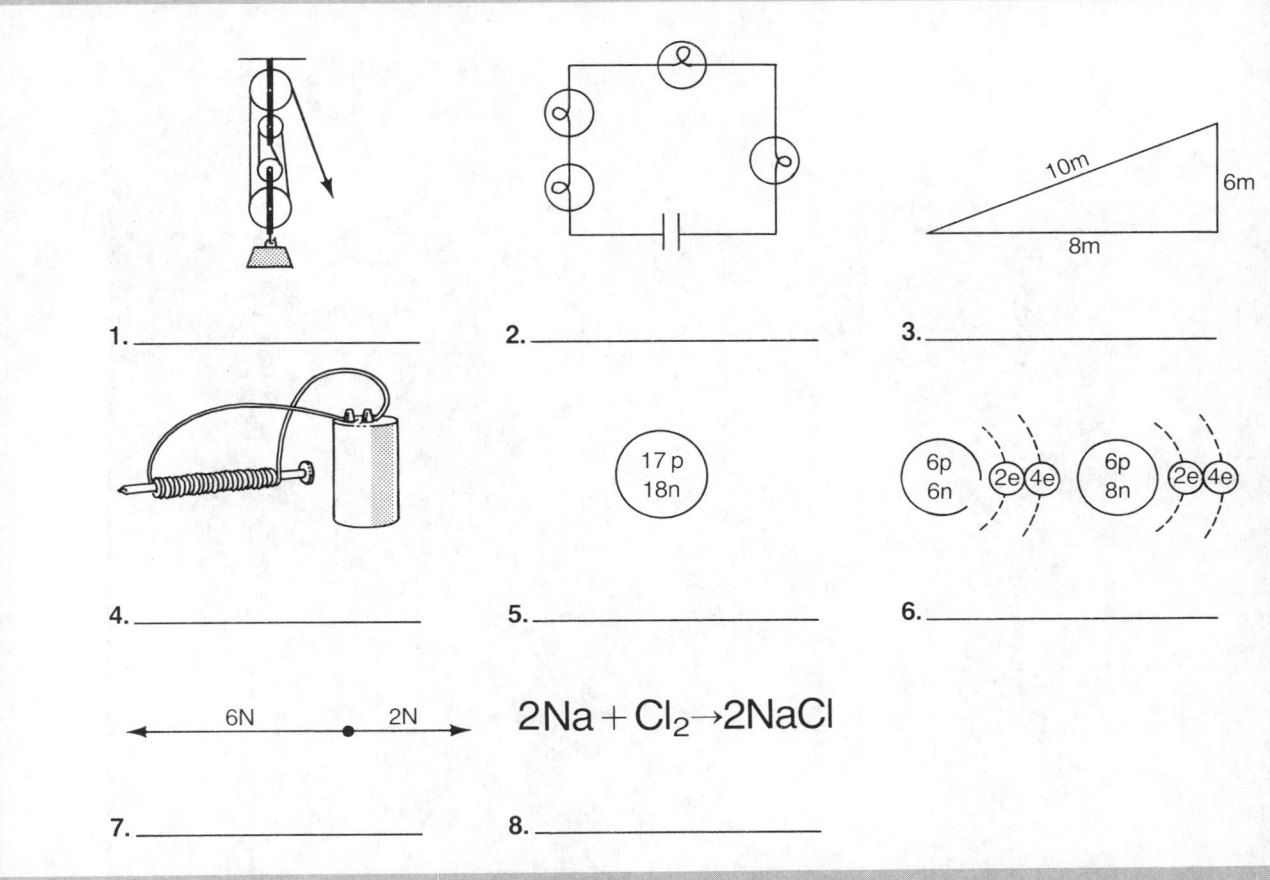

1. _____
2. _____
3. _____
4. _____
5. _____
6. _____
7. _____
8. _____

B. Each statement below refers to the illustration having the same number as the statement. Circle the underlined word or phrase that makes each statement true.

1. The mechanical advantage of this device is (two/four).
2. If one bulb in this circuit goes out, the rest of the bulbs will (remain burning/go out).
3. The length of this machine is (6 meters/10 meters).
4. If you add another dry cell to this device, it will become (stronger/weaker).
5. The atomic number of the element shown here is (17/35).
6. The two atoms represented in this illustration have (the same/different) atomic masses.
7. The resultant of the forces shown here will be to the (left/right).
8. The 2 in front of the NaCl is called a (subscript/coefficient).

10

HEAT, LIGHT, AND SOUND

How Is Heat Measured?

Exploring Science

Correct 98.6°F. Thousands of students have memorized as a fact that humans have a body temperature of 98.6°F. This number sounds very accurate, but, in truth, it is not.

A recent study recorded the temperatures of 148 healthy men and women several times a day. The data showed a normal range to be 97.2°F to 100.4°F. Also, morning temperatures were lower than night temperatures. The scientists concluded that a healthy human has a range of temperatures. The *average* is not even 98.6°F, but 98.2°F.

These scientists used electronic thermometers. Over 100 years ago, a German doctor and his team used less accurate equipment. They mea-sured the temperatures of over 25,000 people. They found the average to be 37°C to the nearest degree. A temperature of 37°C converts *exactly* to 98.6°F. This set the 98.6°F standard for health. If a thermometer showed 100.4°F, the patient "had a fever." The patient was probably sick.

The new results indicate that 100.4°F is normal for some people at some time. Perhaps, each of us should repeat the experiment. Only if we know our range is a temperature reading very helpful.

● Suggest reasons for the differences in data when we compare the new experiment to that of the 19th century.

A person's body temperature changes throughout the day

Measuring Heat and Temperature

Even an accurate thermometer has limitations. A thermometer can only measure temperature. It cannot measure heat. Heat and temperature are not the same.

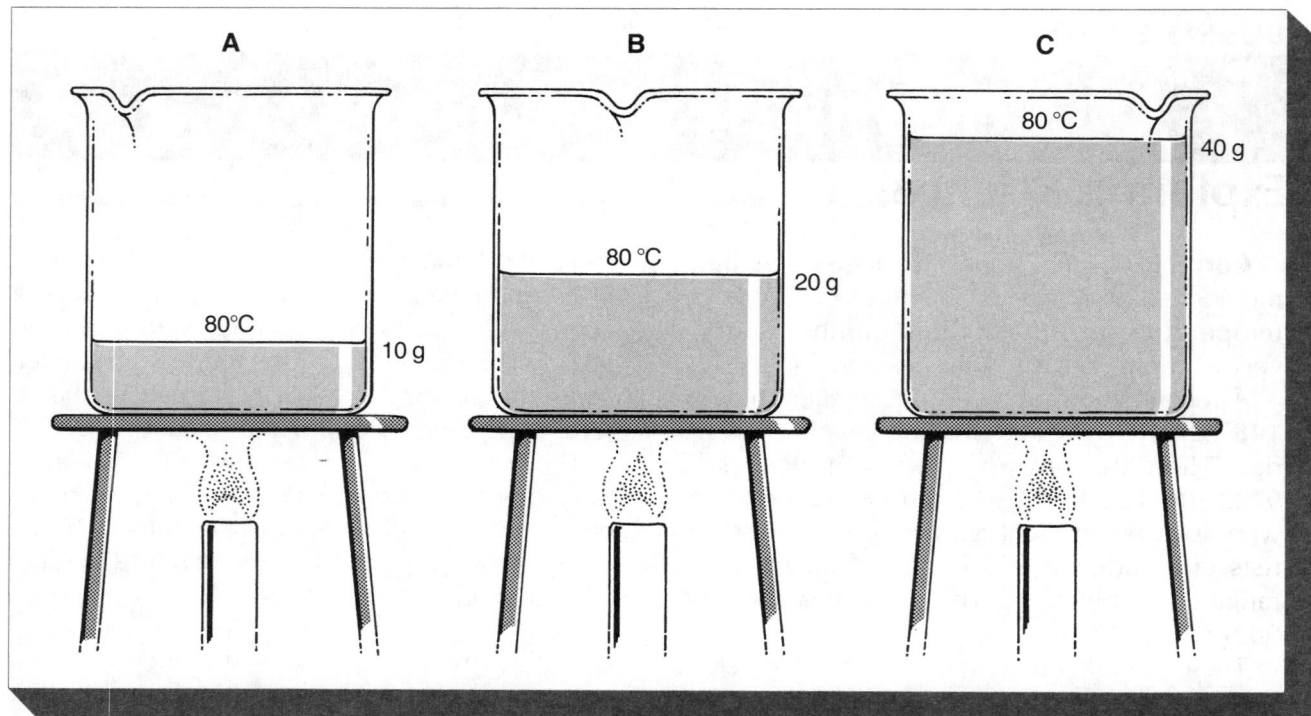

Temperature is a measure of the *average kinetic energy* of the molecules in a substance. Temperature is measured in **degrees Celsius.** This unit is abbreviated **°C.**

Heat is a measure of the *total kinetic energy* of all the molecules in a substance. Recall the kinetic-molecular theory. It states that molecules are always in motion and bumping into each other. The faster the molecules move, the greater the kinetic energy. Much of this energy is released as heat energy. In fact, the heat energy increases as the kinetic energy increases.

What happens when a flame is placed under a pan of water? The water molecules speed up. They gain kinetic energy. As the flame gets hotter, the molecules move faster. The kinetic energy increases. Much of this energy is in the form of heat energy.

Let's look more closely at heat and temperature. Think about three water samples. Each sample is being heated by a flame. Look at the diagram above.

Each water sample has the same temperature—80°C. But the mass of each sample of water is different. Sample **B** has twice the mass of sample **A.** So, sample **B** has twice as many molecules. Sample **C** has four times the mass of sample **A.** So, sample **C** has four times as many

molecules as sample **A.**

The number of molecules in each sample is different. Thus, the total amount of kinetic energy in each sample is different. The amount of heat is also different. Recall that heat is the *total kinetic energy* of all of the molecules in a sample. Sample **C** has more molecules than the other samples. So, sample **C** must have more kinetic energy and more heat.

Sample **B** has twice as many molecules as sample **A.** It has twice as much kinetic energy. So, it also has twice as much heat.

The calorie (KAL-uh-ree) is the unit for heat. A **calorie** is the amount of heat needed to raise the temperature of one gram of water one degree Celsius. The symbol for a calorie is a small letter **c.**

Suppose the temperature of our 10 gram sample was raised from 80°C to 81°C. The temperature changed 1°C. How can we find out how much heat was needed to raise the temperature of the water? We must multiply the change in temperature by the number of grams in our water sample.

$$1°C \times 10g = 10c$$

Suppose we raised the temperature in our 40 gram sample from 80°C to 82°C. How much heat was used? The temperature increased 2 degrees. The mass of the water heated was 40 grams. To find the heat used, we must multiply the two numbers. This will tell us how many calories were used.

$$2°C \times 40g = 80 \text{ c}$$

We used 80 calories of heat. This heat raised the temperature of 40 grams of water 2 degrees Celsius.

Most people think of food when they think of calories. Some foods, like candy, are said to be high in Calories. Notice the capital letter C. This is the symbol for kilocalories. Recall that the prefix kilo- means 1000. Thus, a **kilocalorie** is really 1000 calories.

One jelly bean contains about 6 Calories. This is the same as 6000 calories. This energy could raise the temperature of 6000 grams of water 1°C!

To Do Yourself   How Can Heat Be Measured?

You will need:

Beaker, ringstand, clay triangle, candle and holder, matches, thermometer, clock, water, heat-resistant mitt, heat-resistant pad.

1. Set up the ringstand, triangle, and candle as shown in the diagram.
2. Put 100 mL of water into the beaker. 100 mL of water weighs 100 g. Place the beaker of water on the ringstand.
3. Record the temperature of the water in °C.
4. Heat the water with the candle. **Caution: Use great care when heating substances. Be very careful when working near an open flame. Keep hair and clothing away from the flame. Always wear a heat-resistant mitt when handling hot objects. Never set hot objects directly on a desk or table top. Place them on a heat-resistant pad.**
5. After 5 minutes, blow out the candle. Record the temperature of the water.
6. Find out how much heat was added to the water by the candle. Your answer should be in calories.

Questions

1. Was heat added to the water? How do you know? _____

_____

2. How much heat was added to the water? Your answer should be in calories.

_____

Review

I. Fill in each blank with the word that fits best. Choose from the words below.

**degrees Celsius   calorie   kilocalorie   total   average   heat
temperature   gram   milliliter**

A thermometer measures _____ . Scientists use the unit

_____ to measure temperature. Temperature is the

_____ kinetic energy of all of the molecules in a substance.

Heat is the _____ kinetic energy of all of the molecules in

a substance. The unit for heat is the _____ . One calorie

can raise the temperature of one _____ of water one degree Celsius.

II. Which statement seems more likely to be true?

A. _____ Temperature is the total kinetic energy of all of the molecules in a
substance.

B. _____ Temperature is the average kinetic energy of all of the molecules in
a substance.

III. How much heat is needed in each of the following?

A. _____ To raise the temperature of 1 gram of water from 11°C to 12°C.

B. _____ To raise the temperature of 2 grams of water from 30°C to 35°C.

C. _____ To raise the temperature of 50 grams of water from 80°C to 81°C.

D. _____ To raise the temperature of 10 grams of water from 23°C to 33°C.

IV. Answer in sentences.

Why would a food chart use the unit Calories instead of calories? Explain your
answer.

_____

_____

_____

_____

Lesson Two

How Does Heat Travel?

Exploring Science

Friend and Enemy. On April 14, 1912, the *Titanic* struck an iceburg. The ship began to sink. More than 2000 passengers were forced into the icy water.

Many of these people did not have any injuries. But they had to wait in the water until help arrived. The wind and icy water caused their body temperatures to drop very low. Doctors call this condition **hypothermia** (HY-poh-THUR-mee-uh). Hypothermia is often fatal.

In some cases, hypothermia has been helpful. People with serious injuries have lived *because* their body temperature was low. The low body temperature makes the heart slow down. This reduces the blood flow. All body activities slow down. So, the victim needs less oxygen.

A study of hypothermia victims led to an important medical hypothesis. Doctors observed that hypothermia might allow them to do surgery that is not possible at normal body temperature. This hypothesis proved to be correct.

Open-heart surgery is one procedure that makes use of hypothermia. The patient is placed in an ice bath. The patient's blood is directed out of the body. It flows through tubes over cooling coils. The cooled blood is then directed back into the body. When the body temperature drops from 37°C to 26°C, surgery can begin. The drop in temperature makes the surgery possible.

Hypothermia can still cause death. A sudden loss of body heat is very dangerous. But under the right conditions, hypothermia can be a "friend" instead of an "enemy."

● What do you think the prefix hypo- means?

When the Titanic sank, more than 1000 people were forced into the icy water.

Conduction, Convection, and Radiation

Heat always moves from warmer areas to cooler areas. In solids, heat travels by conduction (kon-DUK-shun). **Conduction** is the movement of heat caused by atoms or molecules bumping into each other. Only the molecules of a solid are close enough for conduction to take place.

Some solids are very good conductors of heat. For example, many metals are good heat conductors. Stainless steel, copper, and aluminum conduct heat easily. This is why so many pots and pans are made from these metals.

Let's look at how conduction works. Suppose a silver spoon is placed in a cup of hot tea. Look at the illustration.

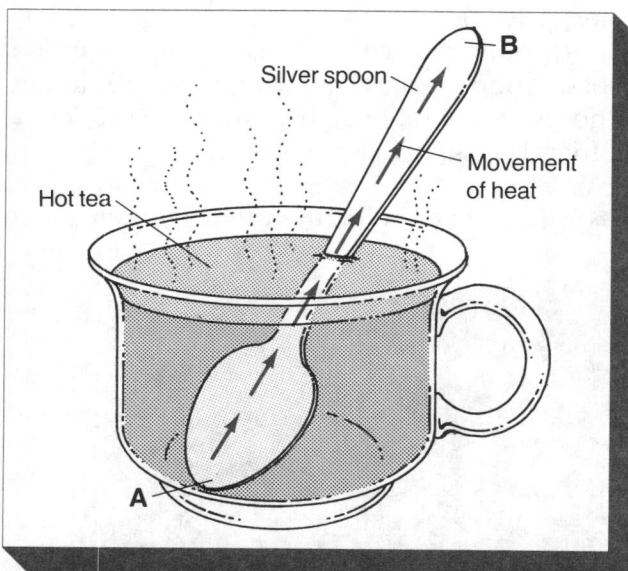

The silver atoms touching the hot tea gain kinetic energy from the tea. These atoms bump into the other silver atoms. Much of this kinetic energy is transformed to heat energy. This bumping of atoms continues all the way up the handle of the spoon. In a short time, the heat energy moves from point **A** to point **B**.

Some solids do not conduct heat well. Such substances are called **heat insulators.** Wood and wax are good heat insulators.

Heat does not travel by conduction through liquids and gases. The molecules are too far apart. How does heat travel in liquids and gases? Heat travels in liquids and gases by **convection** (kon-VEK-shun).

To understand convection, you must understand density. Recall that a cork floats on water. This is because the cork has a lower density than the water.

Hot water is less dense than cold water. This is because the molecules of hot water are farther apart than the molecules of cold water.

Let's look at how convection works. Suppose heat is applied to a pan of water at point **A.** As the water at this point is heated, it becomes less dense. Cooler, denser water pushes under the warm water, forcing it to rise. The cooler water takes its place. The rising water carries its heat to water in other parts of the pan.

This pattern continues as long as heating continues. The result of this moving air is a circular movement of heat called a **convection current**.

The same type of movement carries heat in gases. Imagine a hot stove in one corner of a room. The air near the stove is heated. Denser cool air moves in and pushes the hot air up. The result is a convection current. In a short time, the whole room is warm.

Both convection and conduction require matter. A third type of heat movement occurs only in the absence of matter. This movement of heat is called radiation (RAY-dee-ay-shun). **Radiation** is the movement of heat without

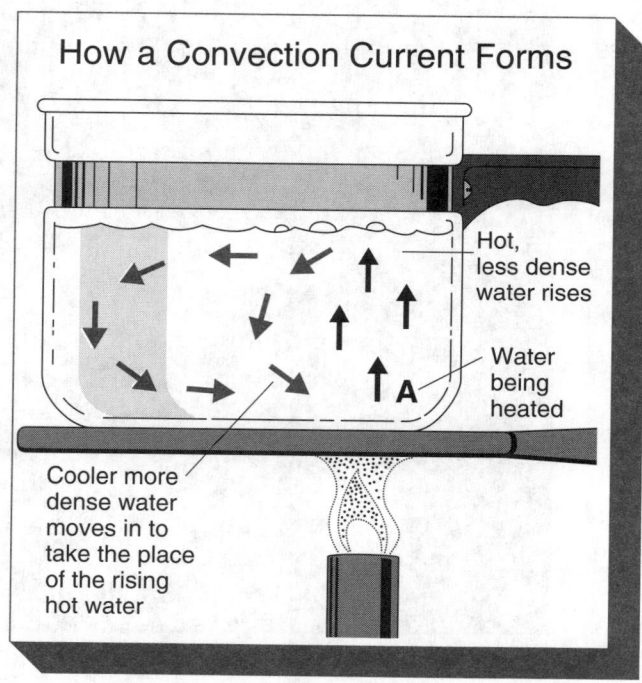

How a Convection Current Forms

Hot, less dense water rises

Water being heated

Cooler more dense water moves in to take the place of the rising hot water the involvement of molecules. When radiation takes place, heat is transferred through a **vacuum** (VAK-yew-wum). Recall that a vacuum is the absence of matter.

Most of the heat on Earth comes from the sun. This heat travels as **radiant** (RAY-dee-unt) **energy**. Space is a vacuum. The heat that we receive from the sun travels through space to the earth. Thus, the heat that we get from the sun travels by radiation.

Heat energy is carried through empty space in various ways. One way is by light. You know that you feel warmer in sunlight than in shade. Some of the warmth comes from the light itself. More warmth comes from "invisible light." You can feel it, but you cannot see it.

To Do Yourself   How Does Heat Move In a Beaker of Water?

You will need:

Beaker, ringstand, triangle, heat source, eye dropper, ink, water, heat-resistant mitt, heat-resistant pad

1. Set up the ringstand and triangle as shown in the diagram.
2. Fill a beaker three quarters full with water. Place the beaker of water on the ringstand.
3. Half-fill the eye dropper with ink.
4. Carefully lower the dropper into the beaker of water. Do not squeeze the dropper bulb. Position the dropper to one side of the beaker near the bottom. Slowly squeeze the ink into the water.
5. Add heat to the beaker directly below the ink. Add the heat very slowly. **Caution: Use great care when heating substances. Be very careful when working near an open flame. Keep hair and clothing away from the flame. Always wear a heat-resistant mitt when handling hot objects. Never set hot objects directly on a desk or table top. Place them on a heat-resistant pad.**
6. Observe what happens to the ink as the water is heated. Stop the heating when a change occurs.

Questions

1. How did the ink act when heated? _____
_____

2. What happened to the water at the other side of the beaker? _____
_____

3. How was the heat transferred? _____

Review

I. Fill in each blank with the word that fits best. Choose from the words below.

conduction   convection   radiation   vacuum   transfer

The movement of heat from place to place is an energy _____ .

A _____ is the absence of matter. The movement of heat

through a solid is called _____ . The only way heat can

move through a vacuum is by _____ . Heat moves through

liquids and gases by _____ .

II. Which statement seems more likely to be true?

   A. _____ Water at 70°C is less dense than water at 10°C.

   B. _____ Water at 70°C is more dense than water at 10°C.

III. Complete each statement with the word **conduction, convection,** or **radiation.**

   A. A person floating in icy water loses heat by _____ .

   B. Warm currents of air transfer heat by _____ .

   C. Heat travels through many solids by _____ .

   D. Heat travels from the sun to the Earth by _____ .

IV. Answer in sentences.

   Why are pots and pans often made with metal bottoms and wooden handles?
   Explain your answer.

   _____

   _____

   _____

   _____

Lesson Three

What Are Waves?

Exploring Science

New Technology Can Bring New Problems. A flight attendant reads a new warning: "Passengers may not use video games or personal computers until we reach cruising altitude." Pilots had reported that instruments were affected by signals from the cabin.

A hotel in the Carribean is showing cable movies. The hotel does not pay for this service. An illegal receiving dish has been set up on the roof of the hotel. The hotel is pirating this satellite signal.

An automobile engineer is testing an experimental car. When the driver uses the telephone, the power windows move. When the car passes a radio tower, the safety air bags inflate. Stray signals have ruined the tests.

These situations are the result of advances in technology. Soon there will be more than a hundred communications satellites orbiting the earth. It has been calculated that one tenth of the total cost of a new car is for the electronic equipment.

Many new problems can be caused by advances in technology. These problems can get very serious. New laws and regulations are necessary to make sure that technology is safe.

● List 6 products mentioned in this story that involve technology.

Characteristics of Waves

Many forms of energy travel as waves. **Waves** are repeated patterns of motion that carry energy from one place to another. There are different types of waves. The diagrams show examples of the two most common types of

Waves have definite patterns. But the pattern of wave A is very different from the patterns of the other waves. Wave A is an example of a **longitudinal** (LAWN-jih-tood-ih-nul) **wave**. The particles of a longitudinal wave move back and

forth in the direction of the wave motion. Sound energy travels in longitudinal waves. We will study sound energy and longitudinal waves in later lessons.

Waves **B, C,** and **D** are examples of **transverse waves**. The particles in a transverse wave move at right angles to the direction in which the wave moves. Light energy travels in transverse waves. Radio and television signals also travel in transverse waves.

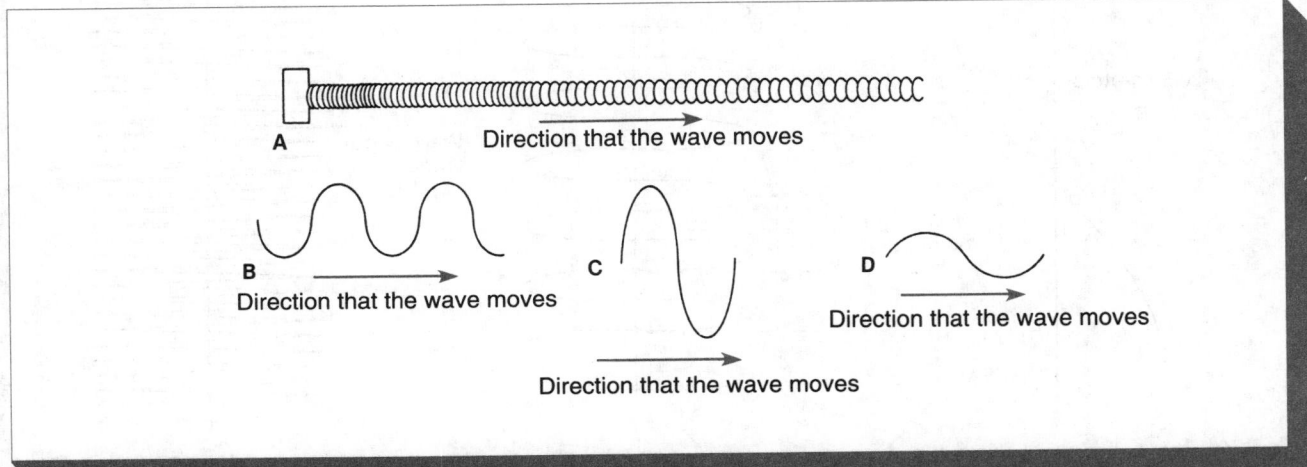

A — Direction that the wave moves

B — Direction that the wave moves

C — Direction that the wave moves

D — Direction that the wave moves

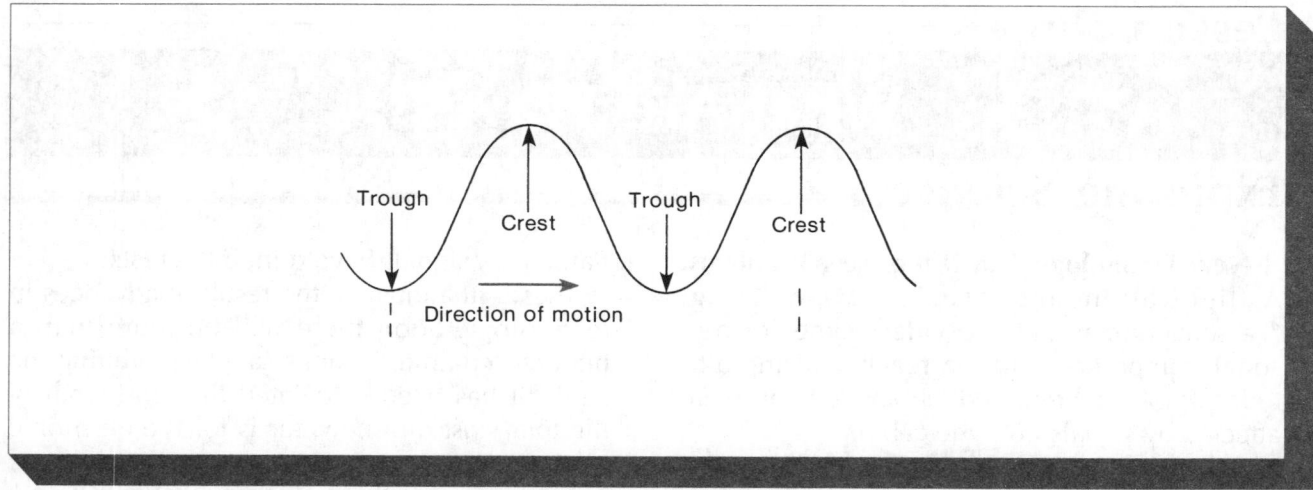

A transverse wave has several parts. The **crest** is the top of the wave. The bottom of the wave is called the **trough** (TROF). The entire pattern of one crest and one trough is called a **cycle.**

Look at the diagrams of the television, radio, and radar waves. Each of these waves has crests and troughs. But the distance from crest to crest is very different for each wave. This distance is called the wavelength. A **wavelength** is the distance from a point on one cycle to the same point on the next cycle.

Wavelengths are usually measured from one crest to the next. They can also be measured in another way. Look at the middle diagram below—the one for FM and television waves. The two points marked **W** and **L** are each the

same part of a wave cycle. The distance between them is the same as the distance from crest to crest, one wavelength.

The height of a wave is also important. The height of a wave is called its **amplitude.** Amplitude can be measured using an imaginary line drawn through the center of the wave. This line is drawn in blue in the diagrams below. The distance from this line to the top of the crest is the amplitude of the wave. Look at the diagram on page 255. Waves X, Q, and Z all have the same wavelengths, but a different amplitude.

There is one other way that waves differ from each other—frequency (FREE-kwun-see). **Frequency** is a measure of the number of cycles of a wave that pass through a point in one second.

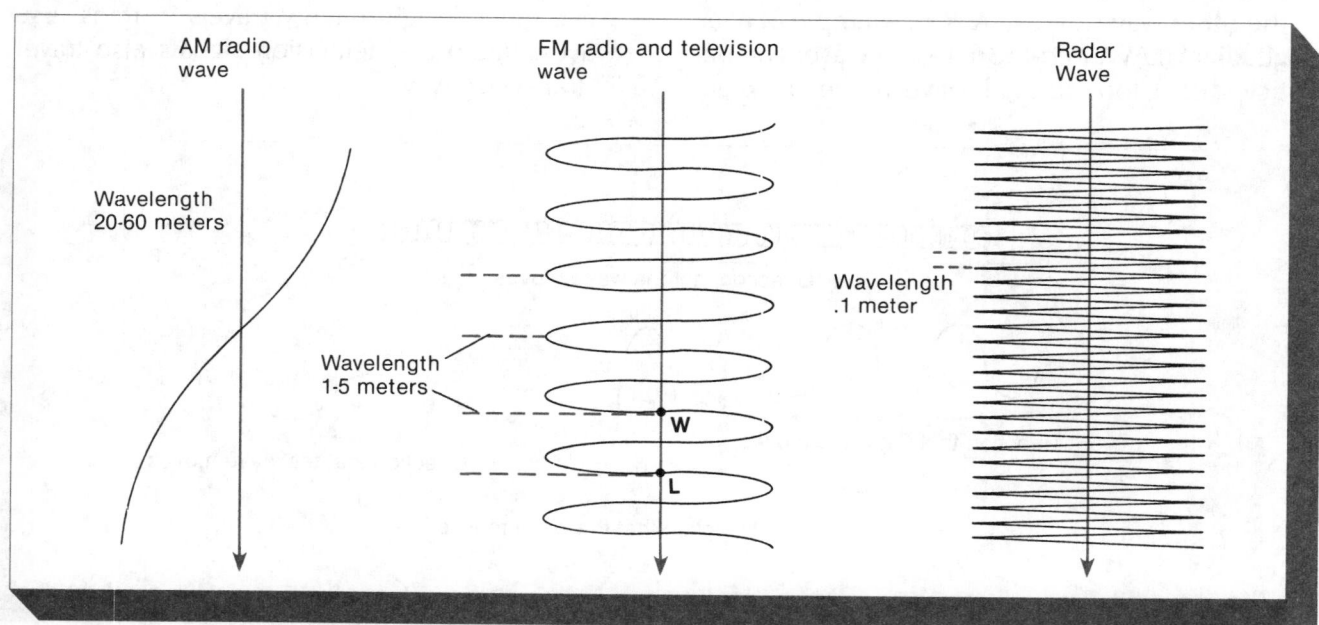

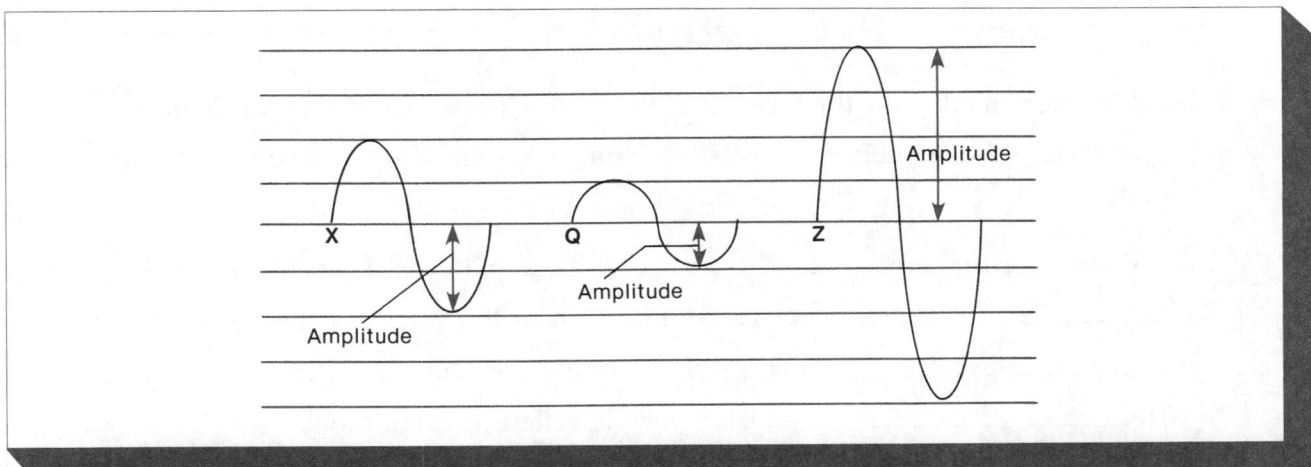

Imagine that three wave cycles pass through a point in one second. This means that one crest and one trough will pass through this point three times. The frequency of this wave would be 3 cycles per second.

Radio and television stations are assigned special frequencies. If two waves from two different stations have the same frequency, interference occurs. This is what causes signals to become mixed up.

To Do Yourself   How Can We Observe Moving Waves?

You will need:

6 meters of rope

1. Stretch the rope out between yourself and another person.
2. Take one step toward the other person.
3. Have the other person hold the rope steady. Now move your end of the rope up and down once. Use a quick motion. Observe the wave the rope makes. Repeat this step.
4. Now move your hand up and down several times quickly. This should create a series of waves.
5. Have the other person repeat steps **1-4**. Hold your end of the rope steady.
6. Draw a diagram of the wave you observed. Label the crest and trough of the wave.

Questions

1. What is the top of the wave called? _____
2. What is the bottom of the wave called? _____
3. How did the wave move? _____

Review

I. Fill in each blank with the word that fits best. Choose from the words below.

transverse longitudinal crest trough moving height length cycle

Waves are repeated _____ patterns. Light moves in

_____ waves. The bottom of the wave is called the

_____ . The top of the wave is the _____ .

The _____ of a wave is called its amplitude.

II. Which statement seems more likely to be true?

A. _____ The greater the height of a wave, the greater its amplitude.

B. _____ The greater the wavelength of a wave, the greater its amplitude.

III. Fill in each blank with the word **amplitude, crest,** or **trough.**

A. Wavelength can be found by measuring the distance from one trough to

the next _____ .

B. The height of a wave determines its _____ .

C. The top of a wave is the _____ .

D. The bottom of a wave is the _____ .

IV. Answer in sentences.

Why are radio operators limited to certain frequencies?

_____

_____

_____

What Is Light?

Exploring Science

Laser, Laser Everywhere. The classic film *Star Wars* was released in 1977. Up to that time, the American public had little experience with lasers. The power of the light saber was startling.

Today, we experience lasers everywhere. The "smart bombs" of modern warfare depend on lasers. The eye surgeon repairing microscopic blood vessels depends on lasers. The astronaut measuring the distance to the earth depends on lasers. All of these uses depend upon the high accuracy and great energy of a laser beam. Lasers have many practical uses. Daily, the use of lasers for scanning has improved our lives. In the supermarket, the laser scans the bar codes on cans and boxes to speed up check-out. A package or piece of luggage is scanned to track its route as it travels. The surface of a compact disc is scanned to produce music. In fact, the development of the CD-ROM (Compact Disc Read-Only Memory) has changed broadcasting, libraries, and home entertainment.

Whether used to produce a special effect at a rock concert or to make a hologram (a 3-D image) on a credit card, lasers are essential to our high-tech world.

Star Wars introduced lasers to its audience in 1977.

● In your opinion, which use of laser technology will be most important in the future? Explain your answer.

A supermarket scanner uses a laser beam to "read" bar codes on products.

Light

A laser is a type of light. Objects that produce light are called **luminous** (LOOM-ih-nus) objects. Recall that the sun releases light energy because of fusion. A light bulb transforms electrical energy to light energy. Both the sun and the light bulb are luminous objects.

Objects on which a light shines are called **illuminated** (ihl-LOOM-ih-nay-ted) objects. The Earth is lit by the sun. A book can be read by lamplight. The Earth and the book are illuminated objects.

How does light travel from the sun to the Earth? Light travels in transverse waves. These waves travel in straight lines called **rays.** If a group of rays travels in the same direction, they form a **beam.**

Rays and beams can be directed. For example, if you were going to read a book, you could point a lamp toward the book. When you do this, you are directing the rays of the light.

Look at the diagram of the lamp. All of the waves in the diagram are coming from the same source—the lamp. But each wave differs from the other waves. How?

There are four things that can make light waves different from each other. Each of these things is discussed below.

● Velocity. **Velocity** (vuh-LAHS-uh-tee) refers to the speed and direction an object is moving. When light waves are moving through a vacuum, they travel at 300,000 kilometers per second. Air causes these waves to slow down. The waves travel even slower when passing through a transparent liquid. And waves are slowest when they move through a transparent solid, such as glass.

All of the waves in the diagram are moving at the same speed. How do we know this? Because all of the waves are travelling through the same matter—air. The matter that a wave travels through is called a **medium.**

● Amplitude. Recall that *amplitude* is the height of a wave. It is the distance that a wave rises or falls from its rest position. As the amplitude of a light wave increases, the light gets brighter. The brightness of a light is called its **intensity** (in-TEN-sih-tee).

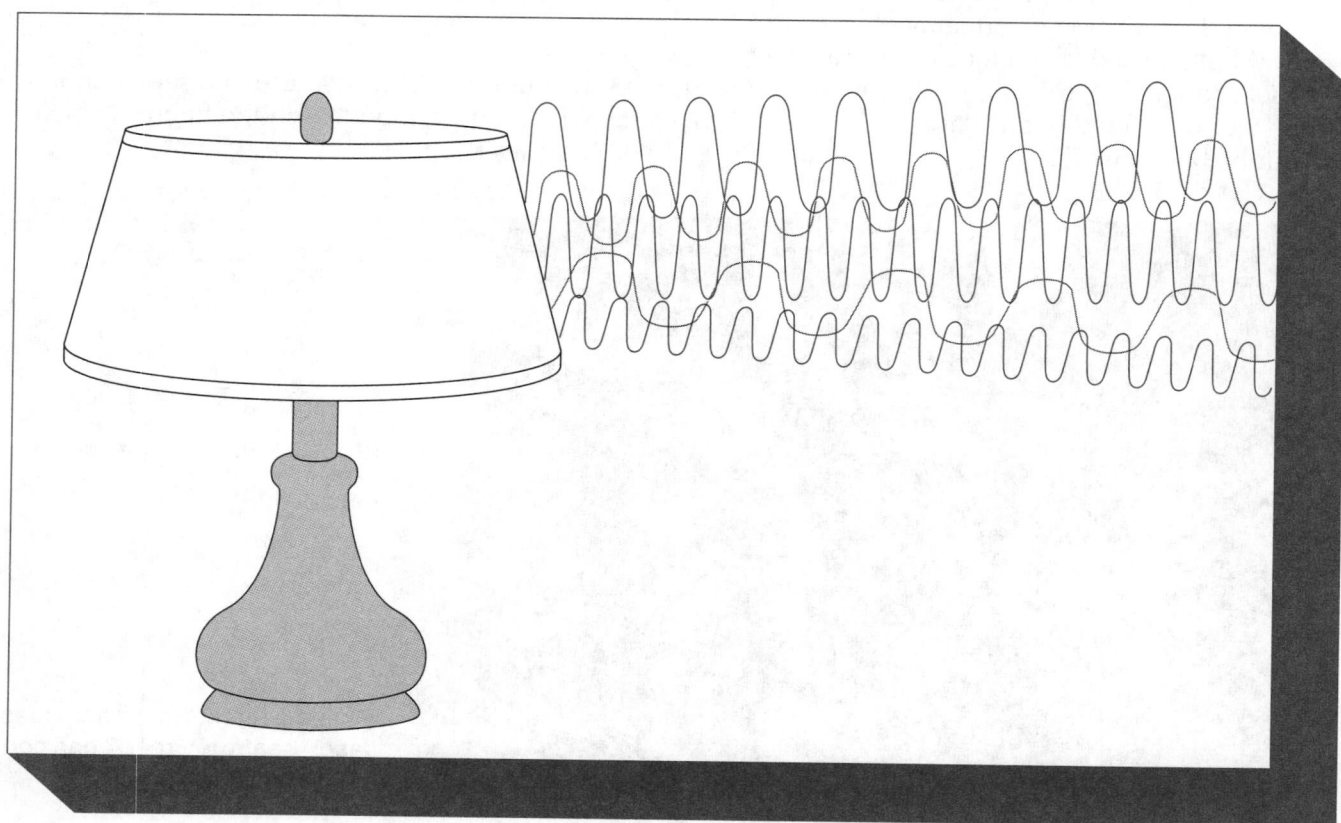

Look at the light waves from different flashlights in the diagram below. Wave **B** has twice the amplitude of wave **A**. As the amplitude of a light wave increases, its intensity, or brightness, increases. Thus, light from the lower flashlight is more intense than light from the upper flashlight. Wave **B** is brighter than wave **A**. The light from the lower flashlight is twice as bright.

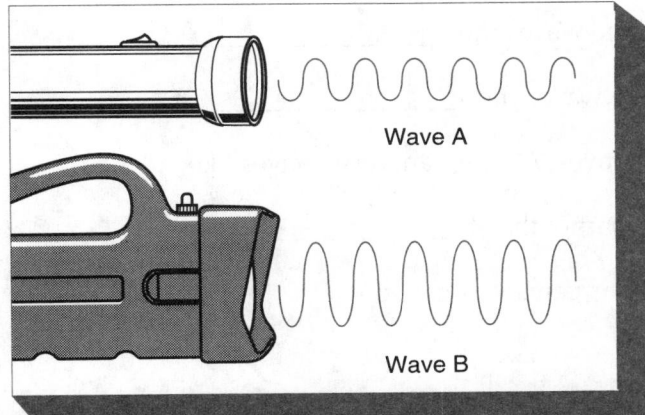

● **Wavelength**. Wavelength may be the most important characteristic of a light wave. The transverse waves of infrared light can be as long as 1 cm. Microwaves and radio waves are longer. An ultraviolet light wave may be as short as 0.000001 cm. An X ray or a gamma ray is shorter. Except for length, radio waves, microwaves, the light we can see (**visible light**), X rays, and gamma rays are all the same. The full range of these transverse waves is called the **electromagnetic spectrum** (i-lek-truh-mag-NET-ik SPEK-trum). Visible light is a tiny part of the electromagnetic spectrum between 0.0000004 cm and 0.0000007cm.

Look again at the diagram of the lamp on page 258. Why do the waves of visible light have different wavelengths? All the waves are part of white light from the lamp.

White light can be passed through a special triangular-shaped piece of glass. This piece of glass is called a **prism** (PRIZ-um). When white light is passed through a prism, it separates. Light of different color appears. These colors make up the **spectrum of visible light.**

The spectrum of visible light ranges from red through orange, yellow, green, blue, indigo, and violet. A prism reveals that all of these colors may be in white light.

Red light has the longest wavelength of visible light. Violet has the shortest wavelength of visible light. Another type of light, called **ultraviolet** (UHL-truh-VY-oh-let) **light** has waves too short to be seen. **Infrared** (IN-fruh-red) **light** has waves too long to be seen. Ultraviolet and infrared light are part of the electromagnetic spectrum.

● Plane of the Wave. If you hold a rope and move your arm up-and-down, you make a **vertical** (up-and-down) wave. If you move the rope from side to side, you make a **horizontal,** or back-and-forth, wave. These words refer to the plane of motion the wave is in. Light waves can move in any plane.

Light from an ordinary lamp has waves of different lengths. These waves are found in different planes. Lasers are a special kind of light. A **laser** is a light that has waves that are the same length. These waves also lie in the same plane.

These two characteristics give lasers great power. The power of such a beam is obvious when it cuts into a piece of steel.

The Spectrum of Visible Light

White Light

Prism

Red
Orange
Yellow
Green
Blue
Indigo
Violet

White light is made up of many colors. When white light is passed through a prism, the white light separates into its colors.

Review

I. Fill in each blank with the word that fits best. Choose from the words below.

**velocity   transverse   luminous   intensity   visible   infrared
ultraviolet   color   plane   electromagnetic**

Light travels in _____ waves. The type of matter that light travels

through affects the _____ of the wave. The _____

of light is determined by the frequency of the wave. The _____ of

light is determined by the amplitude of the waves. All the transverse waves like

light, including radio waves and X rays, are part of the _____

spectrum. The small part of that which humans can see is called

_____ light.

II. Which statement seems more likely to be true?

A. _____ All light beams are lasers.

B. _____ All light waves are transverse.

III. Match each term in column **A** with its description in column **B**.

A	B
_____ **1.** luminous object	**a.** the speed and direction of light
_____ **2.** velocity	**b.** the height of a wave
_____ **3.** beam	**c.** a flashlight
_____ **4.** intensity	**d.** a long wave
_____ **5.** microwave	**e.** brightness
_____ **6.** amplitude	**f.** a group of rays

IV. Answer in sentences.

How do we know that light can travel in a vacuum?

_____

_____

_____

How Is Light Absorbed?

Exploring Science

Curing Winter Blues. September brings changes for many people. Summer games come to an end. School begins again. And the hours of daylight get shorter and shorter.

Some people cannot adjust to changes in seasons. They become sleepy and depressed. They lose interest in almost everything. Many of these people suffer from a condition called SAD.

Doctors thought that exposure to light might help people with SAD. They knew that ultraviolet light could help people with rare forms of leukemia (loo-KEEM-ee-uh). Infrared light could relieve some muscle injuries. But would visible light help SAD cases?

SAD patients were exposed to bright light for several hours a day. The results were amazing. In two to four days, the symptoms reversed. When the light therapy was stopped, the symptoms returned.

Scientists have identified some chemical changes related to sunlight. They definitely recommend light therapy for SAD cases. They also warn us to keep light in our lives. Offices should have windows. Hospital patients and the elderly should have access to sunny areas.

● The prefix *photo-* means light. List three words that use this prefix and tell what they mean.

Absorption

An **opaque** (oh-PAKE) object does not allow light to pass through it. People are opaque. Mirrors are opaque. Black paper is also opaque.

Two things can happen when light strikes an opaque object. The light can be **reflected,** or bounced off, the object. Or, the light can be **absorbed** (ab-ZORBD), or taken in, by the object. The ability to take in and store substances is called **absorption** (ab-ZORP-shun).

The SAD research reminds us that we are always absorbing light. One result of people absorbing light is sunburn. Sunburn is caused by the absorption of ultraviolet light. Recall that ultraviolet light, or **UV,** is not a visible form of light.

Some opaque surfaces, such as black paper, absorb most of the light that strikes them. Other opaque substances, such as mirrors, reflect most of the light. In either case, the light waves are not **transmitted,** or allowed to pass through the object.

Look at the diagrams. These sketches are called **ray diagrams.** The **ray** is the path of the transverse light wave. A ray diagram of an opaque surface shows that no light passes through the surface. But, what happens to the light?

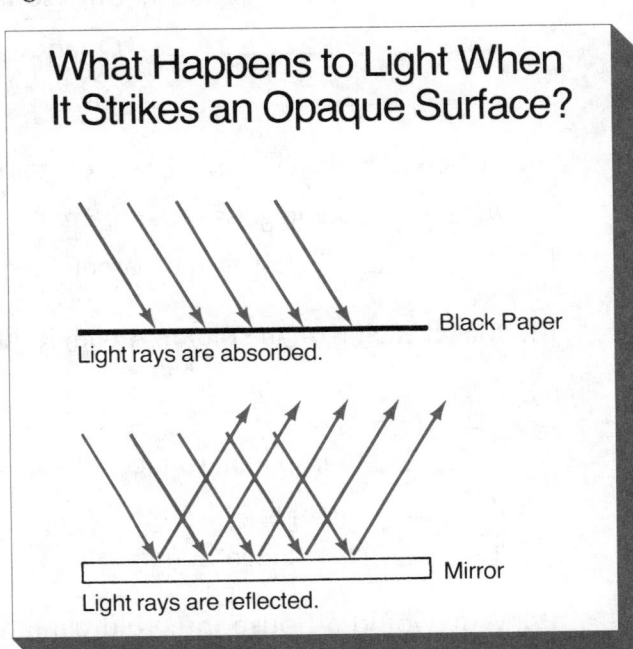

What Happens to Light When It Strikes an Opaque Surface?

Black Paper

Light rays are absorbed.

Mirror

Light rays are reflected.

If the surface were completely black, all of the waves would be absorbed. Recall the Law of Conservation of Energy. This law states that energy cannot be created or destroyed. However, energy can be transformed. Absorbed light is usually transformed into heat.

Many surfaces are not opaque. Some surfaces allow most light to pass through them. These surfaces are **transparent.** You can see through a transparent surface. One example of a transparent surface is glass.

Some surfaces allow only some light to pass through them. These surfaces are called **translucent** (tranz-LOO-sent). You cannot see through a translucent surface. Wax paper is an example of a translucent surface.

Look at the ray diagrams. Diagram **A** shows what happens to light when it strikes a transparent surface, such as glass. Diagram **B** shows light striking a translucent surface, such as wax paper. Can you think of five transparent objects and five translucent objects?

What Happens to Light When It Strikes a Surface That Is Not Opaque?

Transparent Surface

A   Glass

Light rays pass through surface.

Translucent Surface

B   Waxed Paper

Some light rays pass through surface.

Review

I. Fill in each blank with the word that fits best. Choose from the words below.

translucent   transparent   opaque   transmit   absorb

_____ objects do not transmit light. A piece of glass is a _____ surface. Transparent surfaces _____ light. A surface that allows only some light to pass through it is _____ . Opaque objects and translucent objects _____ light.

II. Which statement seems more likely to be true?

A. _____ Some glass is transparent.

B. _____ All glass is transparent.

III. Match each term in column **A** with its description in column **B**.

A	B
1. _____ transparent surface	**a.** wax paper
2. _____ opaque surface	**b.** black paper
3. _____ translucent surface	**c.** pane of glass

IV. Why would a house in a cold climate have a black roof?

Lesson Six

How Is Light Reflected?

Exploring Science

Telescope Technology. In 1672, Sir Isaac Newton was observing the sky. He realized that his data was limited because of his telescope. It was the best telescope available. But some of the starlight was separating into its colors. He needed a more accurate instrument.

Newton designed a new telescope. This telescope used a curved mirror instead of a lens. The mirror magnified the image. And it reflected all colors of light equally. This characteristic made the **image,** or picture, very clear.

Newton's design is still used in small telescopes. The mirror provides clear images of the moon and the closest planets. But astronomers need greater power. They must be able to see deep into other galaxies. So, the design of Newton's telescope has been changed. The mirrors have a new shape. And, the mirrors have been made much larger.

The technology of mirror-making has improved. Today, spinning ovens can make mirrors more than 8 meters wide. A computer-controlled system created 10-meter mirrors. These are made from 36 *hexagonal,* or six-sided pieces. Two 10-meter reflectors are used by the Keck Observatory in Hawaii. In Chile, a telescope called the Very Large Telescope will use four of the giant spin-cast mirrors working

Telescope spin-casting at the University of Arizona

together. Reflectors permit astronomers to see images that were not visible before.

● Why is Newton's telescope called a reflecting telescope?

Reflection

Telescopes used by astronomers must be perfect reflectors. The telescope must have perfect mirrors. But what makes a mirror perfect? A perfect mirror must not absorb light. It must have an exact shape. The shape must produce an **undistorted** (un-dih-STOHR-ted), or true, image.

Think about a mirror in a fun house. The mirror is not perfect. It produces a very distorted image. This is what makes the mirror fun. But it would not be fun for an astronomer to see a distorted image of a distant star.

Think about the mirrors in your home. These mirrors are flat, or **plane,** mirrors. What happens to light when it strikes a mirror? Let's look at the reflection from a plane mirror.

The light that strikes the mirror is called the **incident** (IN-sih-dent) **ray.** The light that is bounced off, or reflected from, the mirror is called the **reflected ray.**

Look at the illustration. The incident ray is labeled I. The reflected ray is labeled R. Light (I) strikes the mirror's surface. Then it is reflected (R) off the mirror. The light *is not* absorbed.

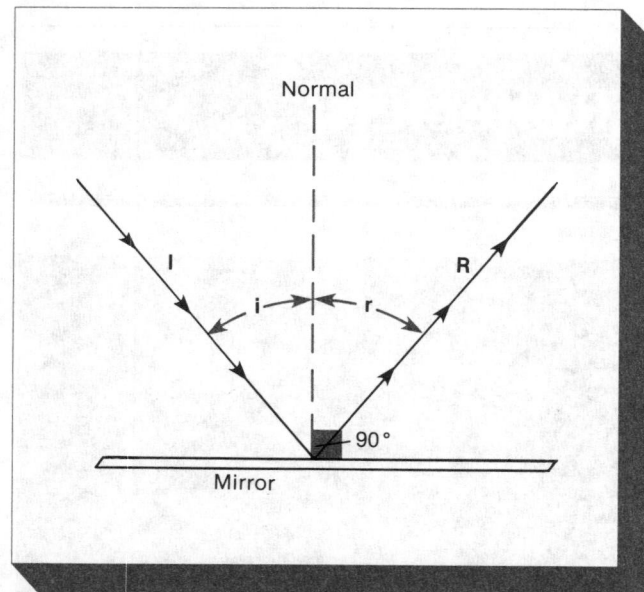

The broken line is called the normal. The normal is an imaginary line. It is drawn at a 90°, or right, angle to the plane of the mirror. The **normal** is used to measure the angles of the rays that strike the mirror.

The number of degrees between the incident ray and the normal is called the **angle of incidence.** This angle is labeled **i.** The number of degrees between the normal and the reflected ray is called the **angle of reflection.** This angle is labeled **r.**

Look at the diagram again. Do you notice any similarities between the angle of incidence and the angle of reflection? If you measured each of these angles, you would find that they are equal. This relationship is explained by the **Law of Reflection.** This law states that the angle of incidence must be equal to the angle of reflection. In other words, angles i and r must be equal.

Let's look at the reflection of three parallel (PAR-uh-lel) incident rays. **Parallel** rays travel in the same direction and are always the same distance apart. Parallel rays *never* meet.

Suppose three parallel incident rays struck a plane mirror. What do you think the reflected rays would look like? The reflected rays would be parallel also. This is true of all plane mirror reflection.

There are two other characteristics of plane mirrors that are important. The images reflected in a plane mirror are the same size as the object being reflected. The image is also reversed.

Not all mirrors are plane mirrors. The mirror that Newton used is called a concave (kon-KAYV) mirror. A **concave** mirror is rounded inward. Its shape is similar to that of a bowl or a spoon.

A concave mirror reflects light differently than a plane mirror. Let's look at what happens to light rays when they strike a concave mirror.

The Law of Reflection still holds. The angle of incidence is still equal to the angle of reflection. But our parallel incident rays do not become parallel reflected rays. Concave mirrors produce reflected rays that meet, or **converge.**

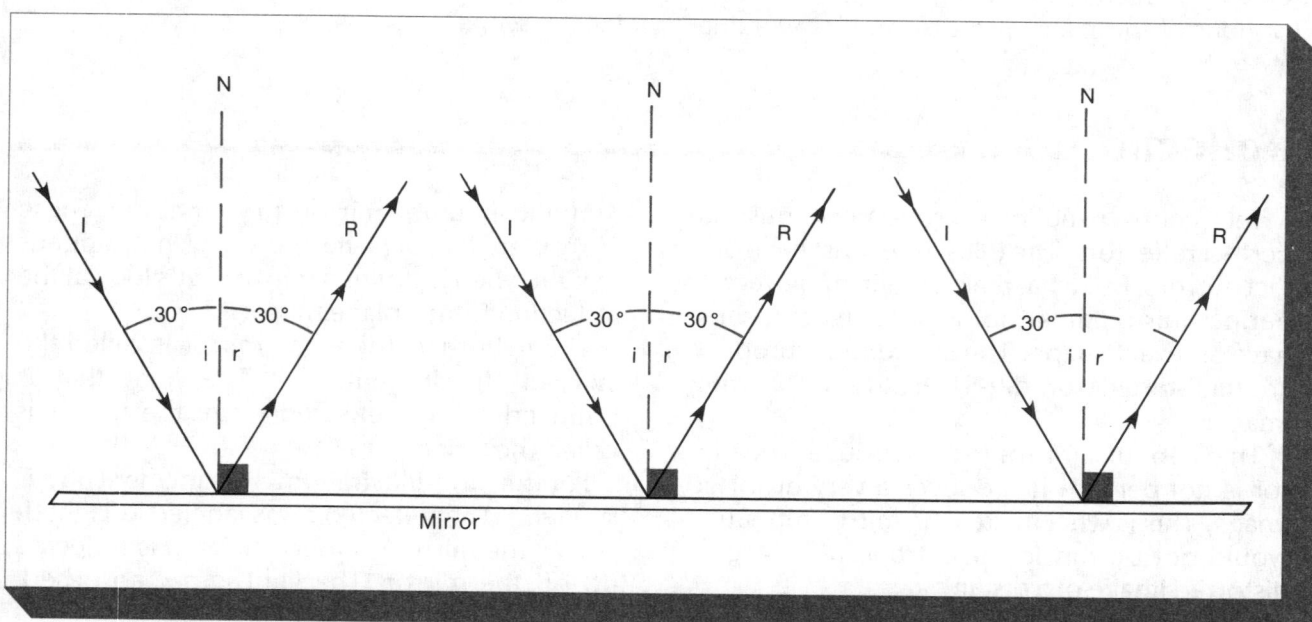

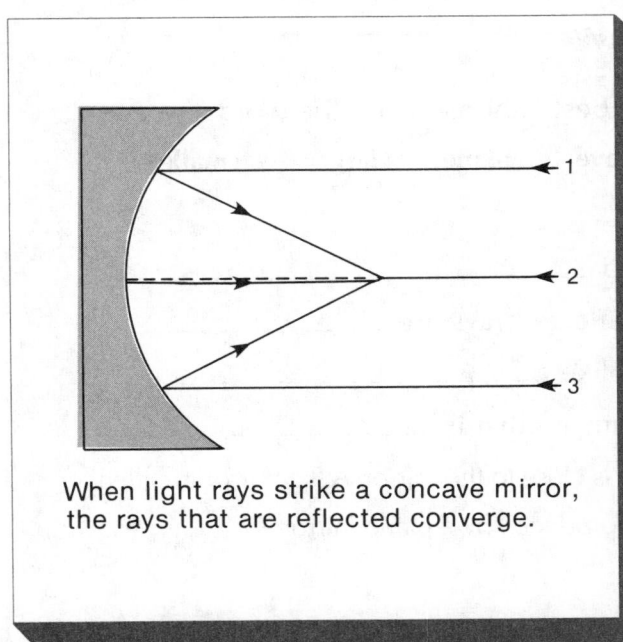

When light rays strike a concave mirror, the rays that are reflected converge.

the surface of this paper is very rough. What would happen if parallel incident rays struck this rough surface?

Look at the diagram below. The diagram shows the difference between light rays striking a rough surface and light rays striking a smooth, shiny surface. How are the rays reflected from the rough surface different from those striking the smooth surface?

The reflected rays show no definite pattern. They are not parallel. They do not meet, or converge. They do not spread apart, or **diverge.** The reflected rays are scattered. Because no pattern is formed by the rays, no image is produced. The reflection is diffused.

It is fortunate that some objects produce a diffused reflection. Imagine what life would be like if every object produced an image. It would be like living in a world made only of mirrors.

A concave mirror gathers light from stars and concentrates it. This is why astronomical telescopes use concave mirrors. But images of objects close to such mirrors are enlarged and turned upside down.

Mirrors are not the only objects that reflect light. In fact, all surfaces reflect some light. This is why we can see objects.

Look around you. The sunlight or overhead lighting is reflecting off objects. This light is reflected back to your eyes. When this reflected light strikes your eyes, an image forms. This is why you can see.

Look at your hands. Light is reflected from your hands to your eyes. This reflected light forms an image. So, you can see your hands. We will look more closely at vision in the next lesson.

Although all objects reflect light, not all objects produce an image. Look at one page of your book. Light is bouncing off the book to your eyes. You can see the page of the book. But you cannot see yourself in the book. The book does not reflect an image.

The reflection of light that does not produce an image is called **diffused** (dih-FYOOSD) **reflection.** Most rough objects produce a diffused reflection.

Suppose the surface of a piece of construction paper was magnified. What would the surface of the paper look like? You would see that

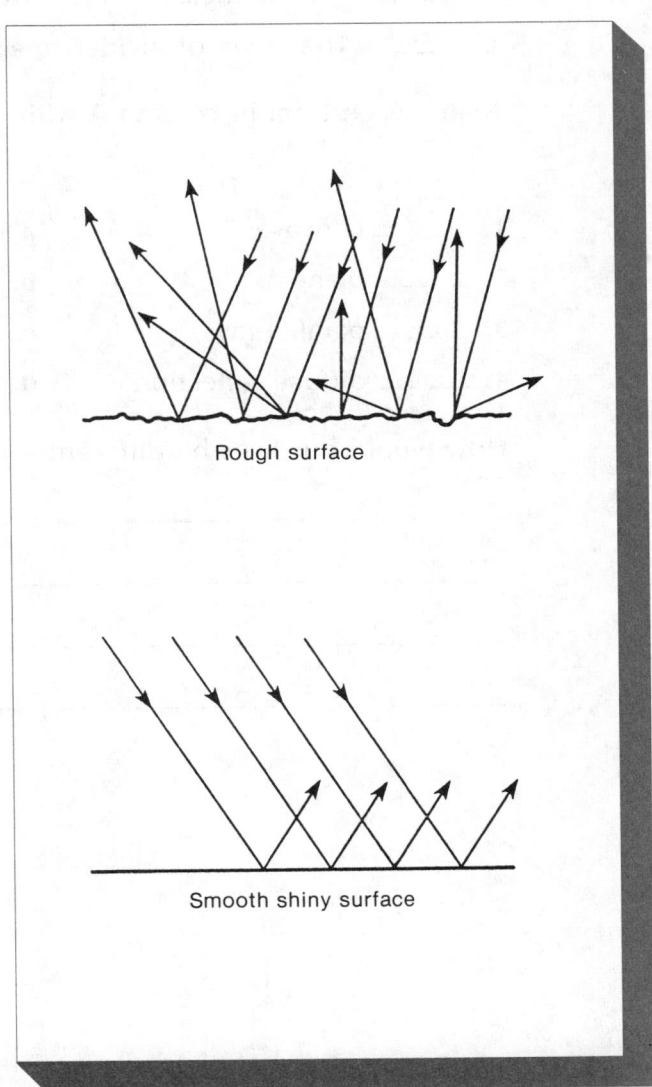

Rough surface

Smooth shiny surface

Review

I. Fill in each blank with the word that fits best. Choose from the words below.

**reflected   incident   diffuse   concave   plane   larger   smaller
reversed   upside down   convex**

The rays that strike an object are called _____ rays.

When rays strike a piece of wood, the reflected rays are _____ .

The inside of a polished silver bowl acts like a _____

mirror. This type of mirror produces an image that is _____

than the object is it reflecting when the object is close to the mirror. A flat mirror is called a

_____ mirror. The image produced by a plane mirror is

II. _____.

Which statement seems more likely to be true?

A. _____ Parallel incident rays always become parallel reflected rays.

B. _____ The angle of incidence always equals the angle of reflection.

III.

Match each term in column **A** with its description in column **B**.

A	**B**
1. _____ concave	a. flat
2. _____ plane	b. produces no image
3. _____ parallel rays	c. curves inward
4. _____ diffuse reflection	d. always the same distance apart

IV.

How would our lives be different if mirrors did not exist?

_____

_____

_____

_____

How Is Light Transmitted?

Exploring Science

Restored vision. A new question is found on many driver's license applications. "Do you want to take part in an organ donor program?" This is one way to get organs and tissues needed for transplants.

Donor tissue is beginning to meet transplant needs in one area—cornea replacement. The **cornea** (KOR-nee-uh) is found at the front of the eye. It is a transparent tissue. This tissue allows light to pass through it.

Disease or injury can make the cornea opaque. Light cannot pass through opaque tissue. This causes a loss of vision. But this tissue can be replaced. Surgery can be performed. A healthy cornea can be transplanted from a donor.

More than 20,000 cornea transplants are performed in the United States each year. Imagine how many people have been saved from blindness! But success still involves donors. The replacement tissue must be a healthy human cornea. No material has been developed to replace this living tissue.

● What other organs or tissues can be transplanted because of organ donations?

Transmission and Refraction

How does the human eye work? The eye forms images because of light. Look at the diagram. It shows the path of light rays as they travel through the eye.

Light rays enter the eye through the cornea. Recall that the cornea is transparent. It can transmit light. This light is transmitted through the pupil. The **pupil** is a small opening in the iris of the eye. After passing through the pupil, the rays reach the lens.

The **lens** of the eye has a convex shape. **Convex** things bulge outward. The back of a spoon has a convex shape. The lens transmits these rays to the fluid of the eyeball.

As the rays pass through this fluid, they change direction. The rays are **refracted** (ree-FRAK-ted), or bent. Finally the rays reach the retina. The **retina** absorbs the rays. An image forms on the retina. This image is communicated to the brain.

An image on the retina is **inverted,** or upside-down. The image is also smaller than the object being looked at. For example, the image of an object one meter high may only be a few millimeters high on the retina. The refraction of light by the lens causes this change.

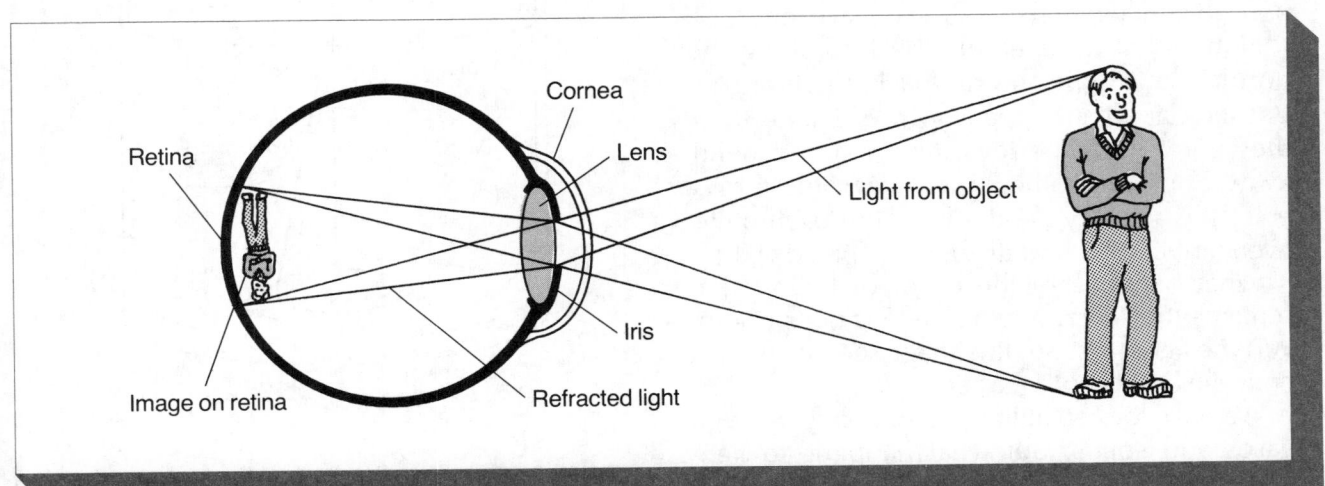

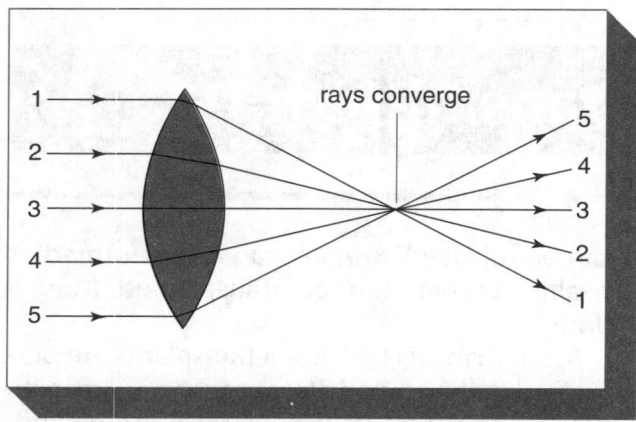

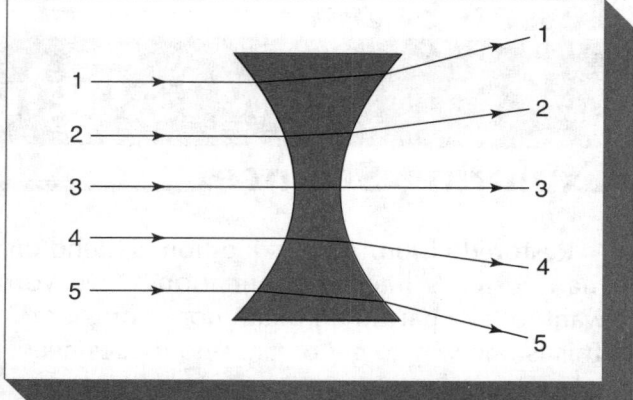

Recall that the lens has a convex shape. Let's look at what happens to parallel rays of light when they enter a convex glass lens.

The only ray that is not refracted is ray 2. It enters the lens along the normal. Light that enters along the normal is not refracted.

Ray 1 bends twice. As it moves from the air into the glass, it bends slightly down. When it enters the air again, it bends down for the second time. Ray 3 acts in the same way, but it bends up each time. The result is the **converging**, or meeting, of the three rays.

Look at the diagram again. What do you notice about the order of the rays after they meet? The order has reversed. Convex lenses invert light rays. This is why the retina receives an inverted image.

When rays pass through a perfectly shaped convex lens, they meet at one point. This point is called the **focal point.** Light energy is concentrated, or focused, on the focal point.

What happens when parallel light rays enter a concave lens? Recall that concave lenses curve inward. Let's use a diagram to show what happens to light when it enters a concave lens.

Notice that ray 2 enters the lens along the normal. Ray 2 is the only ray that is not refracted. But look at what happens to rays 1 and 3. As they enter the glass lens, rays 1 and 3 bend away from each other. Then when these rays enter the air, they bend away from each other even farther. The rays **diverge**, or spread apart.

What happens to the order of the rays? It remains the same. A concave lens does not invert the light rays. So, the image seen through a concave lens is not inverted.

We have looked at light rays passing through gases and solids. But what happens to light when it passes through a liquid? Suppose a ray of light strikes the surface of water at a 45° angle. The ray is passing from a less dense medium (air) to a more dense medium (the water). This change in density changes the velocity of the light ray. It also causes a change in direction. The ray is refracted toward the normal.

Now suppose that an object beneath the surface of the water reflects the ray of light back out of the water. The light travels from a more dense medium to a less dense medium. Again, the ray changes direction. But this time the ray is refracted away from the normal.

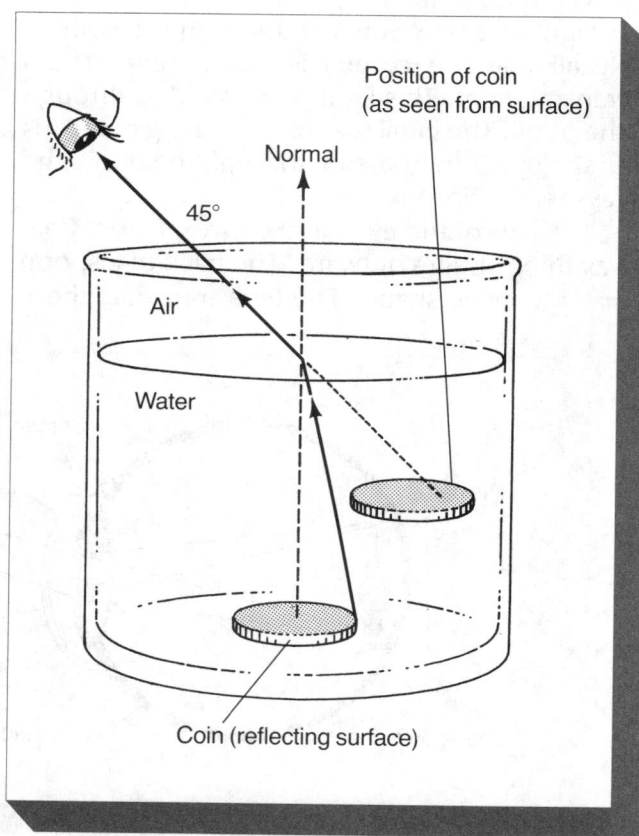

Refraction is affected by the speed of light. Recall that the velocity of light is slowed down as it passes through more dense media. As it enters a less dense medium it speeds up.

Imagine a car moving along a path. Suppose the path is like that of the light ray in the last diagram. The left wheel of the car would reach the new medium before the right wheel. For an instant, the left wheel would travel faster than the right wheel. That small difference in speed would cause the car to change direction.

If the car were entering the new medium at a 90° angle (along the normal), there would be no change in direction. Both wheels would reach the new medium at the same time. Light reacts in the same way.

From these examples, we can make some general statements about refraction.

● Refraction occurs when light moves from one medium to another.

● If light enters a new medium *along the normal*, it is *not refracted*.

● If light enters a *more dense* medium, it is refracted *toward the normal*.

● If light enters a *less dense* medium, it is refracted *away from the normal*.

Understanding refraction has helped scientists correct vision problems. We cannot correct a cornea problem without a transplant. But we can correct lens problems. Many different lens shapes are available. They are used to correct defects in the shape of the lens of the eye. Imagine how many people can see well because of eyeglasses and contact lenses.

To Do Yourself   How Do Convex and Concave Lenses Differ?

You will need:

Convex lens, concave lens, flashlight, white paper, pencil, tape

1. Print your name on the paper.
2. Hold a convex lens over your name. Slowly lift the lens toward your eye. Observe your name through the lens.
3. Repeat step 2 using a concave lens.
4. Tape the paper to the wall. Use this paper as a screen.
5. Hold the flashlight 10 cm from the paper. Place the convex lens over the lighted flashlight. Observe what happens.
6. Repeat step 5 using the concave lens.

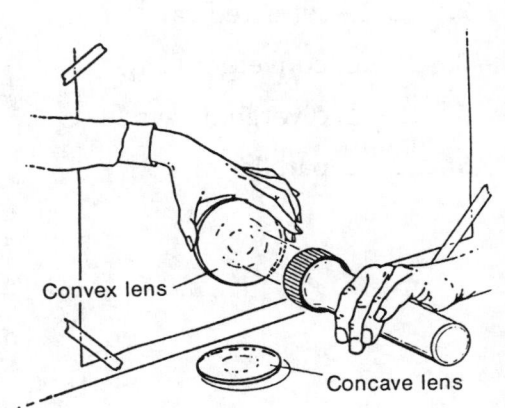

Convex lens

Concave lens

Questions

1. What did the convex lens do to your name? _____

2. What did the concave lens do to your name? _____

3. What did the convex lens do to the light beam? _____

_____

4. What did the concave lens do to the light beam? _____

_____

5. Which lens is used in magnifying lenses? _____

Review

I. Fill in each blank with the word that fits best. Choose from the words below.

velocity   smaller   inverted   converge   diverge   reflected   refracted

A convex lens produces a(n) _____ image. This image

may be _____ than the object being looked at. When

light rays pass through a convex lens they _____ at the

normal. Light entering a lens along the normal is not _____.

Light rays change direction when the _____ of the light
changes.

II. Which statement seems more likely to be true?

A. _____ All lenses refract light.

B. _____ All transmitted light must be refracted.

III. Match each term in column **A** with a diagram from column **B**.

A	**B**
1. _____ reflected ray	a. 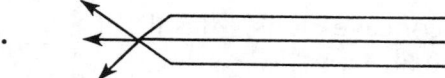
2. _____ refracted ray	
3. _____ convergent rays	b. 
4. _____ divergent rays	c. 
5. _____ parallel rays	d. 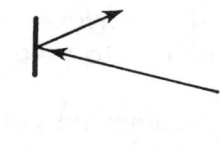
	e.

IV. Answer in sentences.

What would the path of a ray of light travelling at a 45° angle look like if it
passed from air through water and then through a piece of glass. Draw a ray
diagram of the path.

Lesson Eight

What Is Color?

Exploring Science

Mood and Color. "Bubble gum pink" is the name of a color of paint. It was not very popular before 1980. Then research found a link between color and behavior. "Bubble gum pink" became known as "passive pink." This color seemed to have a calming affect on people.

Many hospitals and institutions painted rooms this color. Disturbed patients were placed in these rooms. The patients became calm. Many fell asleep. Violent criminals calmed down after spending time in a pink holding cell.

One clever football coach had an idea. The visiting team's locker room was painted bubble gum pink. Some people thought this was unfair. If the pink shade calmed the visiting team, the home team would have an advantage.

● Why would the home team have an advantage because of the color of the visiting team's locker room?

Color

Many colors seem to have an affect on people. For example, red has been found to increase appetite. Some restaurants have painted their walls red. They hope it will make customers feel hungry and eat quickly.

But what gives an object color? Let's begin by answering this question by reviewing what we know about color so far.

● White light is made up of many colors.

● A prism can separate white light into its colors.

● The spectrum of visible light is made up of bands of colors. These colors are red, orange, yellow, green, blue, indigo, and violet.

● Color is determined by wavelength.

Think about a red object. The object looks red because only the wavelength for red light reaches our eyes. In other words, the object only reflects light of red wavelengths. All other wavelengths are absorbed.

Now, suppose a surface is white. A **white** surface reflects all wavelengths to our eyes. In fact,

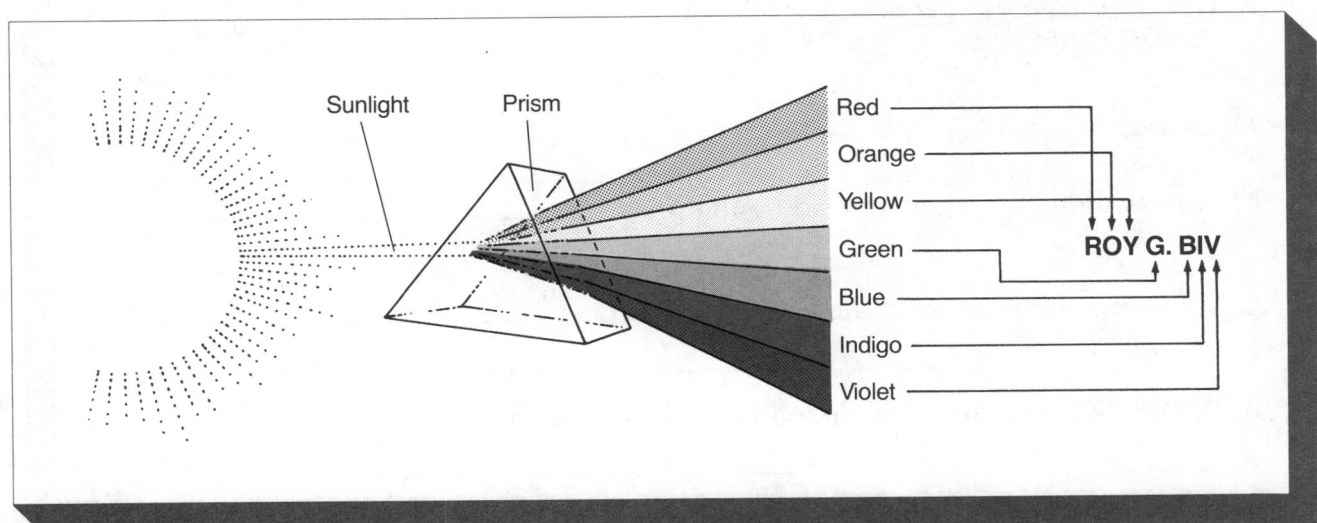

Sunlight Prism

Red
Orange
Yellow
Green
Blue
Indigo
Violet

ROY G. BIV

as long as the wavelengths for red, green, and blue are reflected, we see the surface as white. Red, green, and blue are called the **primary colors** of light.

There is a trick to help you remember the colors of the spectrum and the primary colors. Think of the name *Roy G. Biv*. Each letter in this name stands for a color. The initials RGB stand for the primary colors (red, green, and blue).

There is no black wavelength. So, what makes an object black? **Black** is the absence of color. An object appears black if it absorbs all wavelengths. But, nothing is perfectly black.

Some luminous objects emit single wavelengths of light instead of white light. Recall that a luminous object emits light. Recall that the sodium vapor lamp emits a golden-white light. It is a luminous object.

Suppose we wanted to make a blue light from a source of white light. How could we transmit only the blue wavelengths? We could use a filter. A **filter** separates the colors of light. Many filters transmit wavelengths of only one color.

Stage spotlights use filters. As the filters on a spotlight are changed, the color of the light is changed. What happens if we focus a blue spotlight on a red object? Recall that a red surface absorbs other wavelengths. In fact, it absorbs all wavelengths, except red. The object appears black because no light is reflected to the viewer.

To Do Yourself What Happens to Light Passing Through a Prism?

You will need:

Prism, flashlight, white paper, tape, colored pencils

1. Tape the paper to a wall. The paper will be used as a screen.
2. Stand the prism on one end.
3. Direct a beam of light toward the prism. Rotate the prism so that the light is directed toward the center of the paper.
4. Use the colored pencils to make a sketch of the image made on the paper.

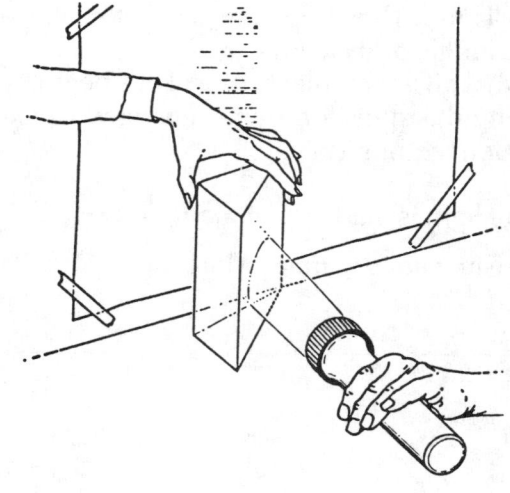

Questions

1. How is the light beam affected by the prism? _____

_____

2. What was the order of the colors on the paper? _____

_____

3. What would the light look like if the colors were mixed together? _____

_____

Review

I. Fill in each blank with the word that fits best. Choose from the words below.

**absorbs   reflects   wavelength   blue   white   red   black
colors**

White light is made up of many _____ . The color of a
ray of light depends on the _____ of the light.
A surface that looks blue reflects _____ light and
_____ light of other colors. A white surface usually
_____ all wavelengths. A _____ surface
absorbs all colors.

II. Which statement seems more likely to be true?

A. _____ A green filter transmits light of all wavelengths.

B. _____ A green filter transmits only green wavelengths.

III. Complete each statement with the word **reflects, transmits,** or **absorbs.**

A. A black surface _____ all wavelengths of light.

B. An opaque red object _____ red wavelengths.

C. A purple filter _____ only purple wavelengths.

D. A white surface _____ all wavelengths.

IV. Answer in sentences.

Is it better to wear black clothes or white clothes in winter? In summer?
Explain your answers.

_____

_____

_____

_____

What Is Sound?

Exploring Science

Dophin's Clicks. Researchers have studied the sounds of dolphins for many years. Dolphins do not have vocal cords like humans. They make noises that sound like clicks and whistles. Many hypotheses have been made about these sounds. Some researchers believe that the sounds are used for communication between dolphins. They believe the clicks are used as words or messages.

Kenneth Norris is a dolphin specialist. He believes that dolphins produce sound that stun smaller fish. This would explain how faster swimming fish can be caught by dolphins. There have been many reports of fish losing speed and direction when dolphins are nearby.

Other researchers do not accept Norris' idea. Most of the field reports did not include sound recordings of the dolphins. So they do not link sound with fish behavior.

But one set of sound recordings and films of dolphins does exist. Two of Norris' students recorded dolphins feeding in the Indian Ocean. The dolphins approached small, fast swimming fish. They made sounds like a rifle firing. The small fish jumped into the air and fell back to the surface. They were quickly captured and eaten by the dolphins.

Researchers have been studying the sounds of dolphins for years. Why do you think researchers are so interested in the sound these animals make?

It is very difficult to perform tests in nature. It may take years to collect data. Until this data is collected, Norris' hypothesis cannot be accepted or rejected.

● Why is it difficult to perform tests in nature?

Sound

Dolphins can make a wide range of sounds. They can also be taught to make new sounds. It is possible that one of these sounds could stun other fish. Let's look at how this might happen.

Recall that sound is a form of energy. Sound is made by vibrating objects. A **vibration** (vy-BRAY-shun) is a rapid back-and-forth motion.

Humans make sounds with their vocal cords. The vocal cords vibrate. Place your fingers on your throat. Hum. You should be able to feel these vibrations.

Different vibrations make different sounds. But what makes one sound different from an-other? Sounds can differ in **intensity,** or loudness. Sounds can also differ in pitch. **Pitch** is the tone of the sound. For example, some sounds are high. Others are low.

Few objects produce sounds of only one pitch. But a tuning fork does. If you strike a tuning fork hard, the sound is loud. If you strike it gently, the sound is soft. The amount of mechanical energy used to vibrate an object determines its intensity. The object itself determines the pitch.

We can yell by forcing a lot of air through our vocal cords quickly. If we move a small amount of air through our vocal cords, we can whisper. A dolphin has vibrating tissues. It can produce loud whistles. It can also produce soft clicks.

Once a sound is produced, it must be carried. Sound needs molecules to travel. It cannot travel through a vacuum. The molecules that sound travels through are called the **transmitting medium.** When we speak, the transmitting medium is air. When a dolphin clicks, water carries the energy.

Recall that sound travels in longitudinal waves. The molecules in a longitudinal wave move back-and-forth. The molecules move in the direction of the wave motion.

The cycle of a sound wave has two parts. One half of the wave has molecules that are spread apart. This area is called the **rarefaction** (rayr-uh-FAK-shun). The other half of the wave has molecules that are tightly packed. This area is called the **compression** (KOM-preh-shun). Look at the sound wave in the diagram (page 276). Can you identify the compression and the rarefaction?

This is a still picture. Remember that a sound wave is *always* in motion. The speed of that motion depends upon the transmitting medium. In air, sound travels at about 340 meters per second. In water, sound travels at about 1500 meters per second.

To Do Yourself   How Can Sound Energy Be Observed?

You will need:

Tuning fork, string, small pan or large plastic cup, paper, pencil, water

1. Strike the tuning fork against the heel of your shoe. Place your thumbnail against the tuning fork. Observe what happens.
2. Repeat step **1.** This time touch the tuning fork to the end of your nose. Observe.
3. Tie the pencil to a piece of string. Hold the pencil so that it is hanging from the string. Stike the tuning fork against your heel. Touch the tuning fork to the pencil. Observe what happens.
4. Place water in the small pan or cup. Strike the tuning fork against the heel of your shoe. Place the tuning fork at the surface of the water. Observe what happens.

Questions

1. Was any sound produced by the tuning fork? _____

   How do you know? _____

2. What happened when you touched the tuning fork to your nose? _____

3. What happened when you touched the tuning fork to the pencil? _____

4. What happened when you touched the surface of the water with the tuning fork? _____

5. What kind of energy was produced by the sound? _____

   How do you know? _____

Recall that transverse waves, such as light, slow down as they enter a denser material. Sound waves do the opposite. Sound waves usually travel faster in a denser medium. The compressions and rarefactions move faster as the number of molecules increases.

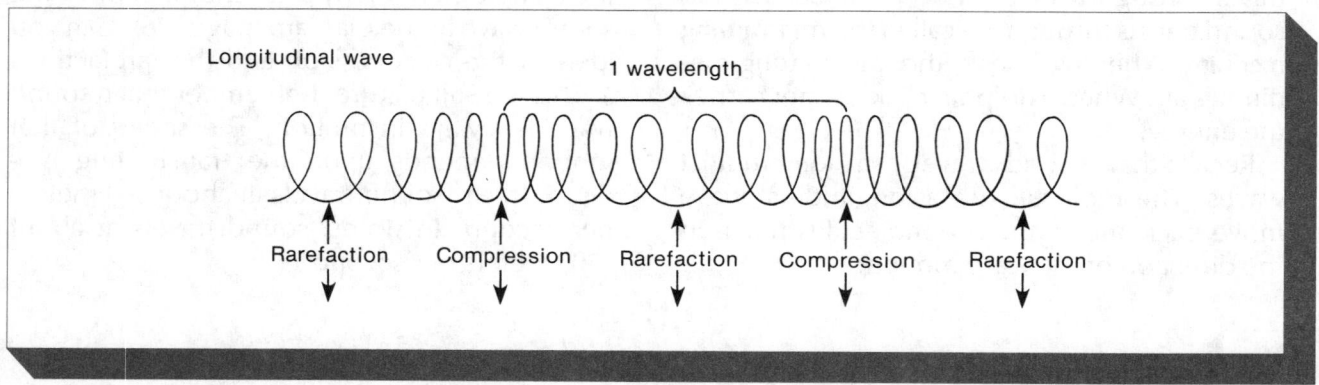

Review

I. Fill in each blank with the word that fits best. Choose from the words below.

longitudinal **transverse** **compression** **energy** **molecules**
rarefaction **intensity**

Sound is a form of _____ . It travels in

_____ waves. The area of closely packed

_____ is called the _____ . Mole-

cules are spread apart in the _____ .

II. Which statement seems more likely to be true?

A. _____ A tuning fork can produce sounds of many pitches.

B. _____ A tuning fork can produce sounds of many intensities.

III. Match each term in column **A** to its description in column **B**.

A	B
1. _____ intensity	**a.** tightly packed molecules
2. _____ rarefaction	**b.** loudness
3. _____ compression	**c.** molecules spread apart
4. _____ pitch	**d.** only one pitch
5. _____ tuning fork	**e.** high or low

IV. Answer in sentences.

Describe three ways that sound differs from light.

Lesson Ten

How Is Sound Measured?

Exploring Science

Voice Prints. Voices are similar to fingerprints. Every person has a unique voice. This knowledge has led to new developments. In fact, your unique voice may be a key in your future.

Your voice may actually replace many keys. It will allow you to unlock doors. It could give you access to a computer. Or, it may permit you to withdraw money from your bank account.

The name of this new technology is called **voice activation** (AK-tih-vay-shun). It is not science fiction. Researchers are working on this technology now.

Voice prints may someday be as important as fingerprints. Today, a technician may be able to identify your voice pattern from hundreds of others. But in the future, it might be chosen from over a million others.

The human voice has many variations. It can vary in loudness, pitch, and quality. Our speech also has patterns. Your voice may be very much like another person's. But your patterns of speech may be very different.

● What problems might arise in voice activated products?

Decibels and CPS

The intensity of sounds is measured in **decibels.** A whisper is less than 1 decibel. A jet engine is a million times louder than a whisper. It has over 120 decibels. All of the sounds that humans can hear fall into this range.

Pitch is related to frequency. **Frequency** is the number of cycles per second a vibrating body makes. Frequency is measured in **cycles per second,** or **cps.**

Suppose that tuning fork A vibrates 256 times in one second. Tuning fork B vibrates 512 times per second. The faster a vibration, the higher the pitch. Therefore, the pitch of sound B is higher than the pitch of sound A.

An **oscilloscope** (aw-SILL-uh-skohp) can be used to make graphs of sounds. It allows us to compare waves made by different sounds. Let's see how the graphs of two sounds made by the same tuning fork compare. Look at the illustrations on the next page.

An oscilloscope makes graphs of sound waves.

Graph A has a greater amplitude than graph B. Recall that amplitude is the height of a wave. The sound in graph A is louder. It is more intense. But both graphs have the same wavelength. The wavelength shows the pitch of a soundwave. The sounds in graphs A and B have the same pitch.

Let's look at another pair of graphs. The sounds pictured here are identical. Both waves have the same amplitude. Therefore, the intensity of the waves is the same. Both waves have the same wavelength. So the pitches of the sounds are also the same.

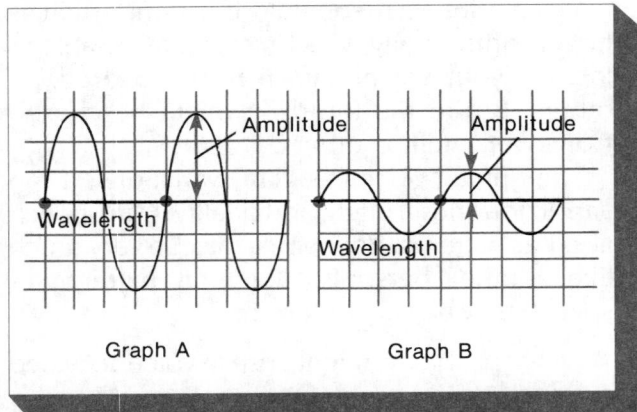

Graph A   Graph B

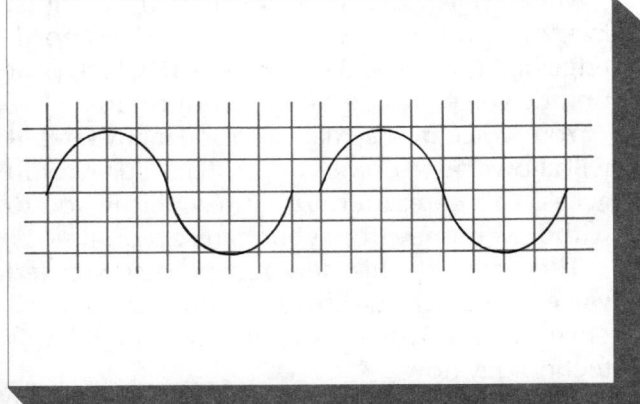

To Do Yourself   How Can the Pitch of Sound Be Changed?

You will need:

5 test tubes, test tube rack, metric ruler, water

1. Place different amounts of water in each test tube. Use the ruler to measure the amounts. You should use 2 cm, 4 cm, 6 cm, 8 cm, and 10 cm of water.
2. Place the test tubes in a test tube rack.
3. Blow air across the top of each test tube. Put the test tubes in order from the lowest pitch to the highest pitch. Observe the order of the test tubes.
4. Close your eyes. Have a classmate switch the order of the test tubes.
5. Keeping your eyes closed, try to put the test tubes back in order by the sound of their pitches. Observe the order of the test tubes.

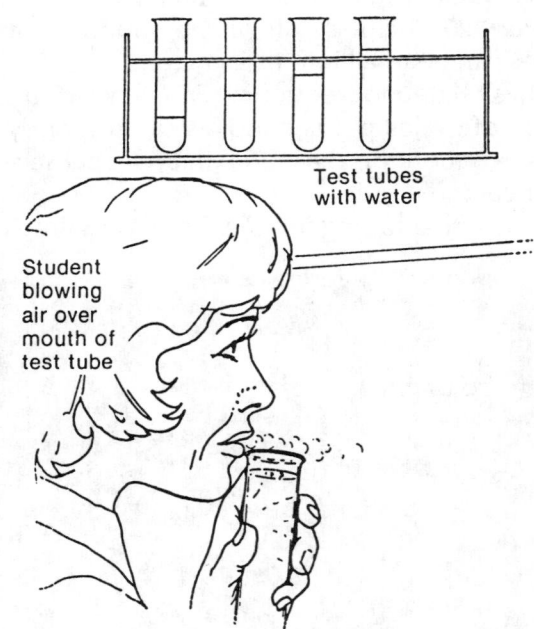

Test tubes with water

Student blowing air over mouth of test tube

Questions

1. Which test tube had the lowest pitch? _____

2. Which test tube had the highest pitch? _____
3. What affects the sound that each test tube makes the amount of water in the tube

or the amount of air? Explain your answer. _____

Now, let's go back to tuning fork A and tuning fork B. Tuning fork A had 256 cycles per second, or cps. Tuning fork B had 512 cps. Look at the diagrams of the two waves. The wavelength of B is shorter than the wavelength of A. In fact, wavelength B is one-half the length of wavelength A.

The shorter a wavelength, the greater the frequency. The greater the frequency, the higher the pitch. Look back at the diagrams. Which wave has the greater frequency and the higher pitch? Wavelength B.

The graph an oscilloscope makes of a human voice is much more complex. We cannot produce the single unchanging note of a tuning fork. In fact, many of the sounds made by humans are not even heard. Some sounds are too quiet to hear. Others are too high or too low in pitch. Humans can only hear sounds between 20 and 20,000 cps.

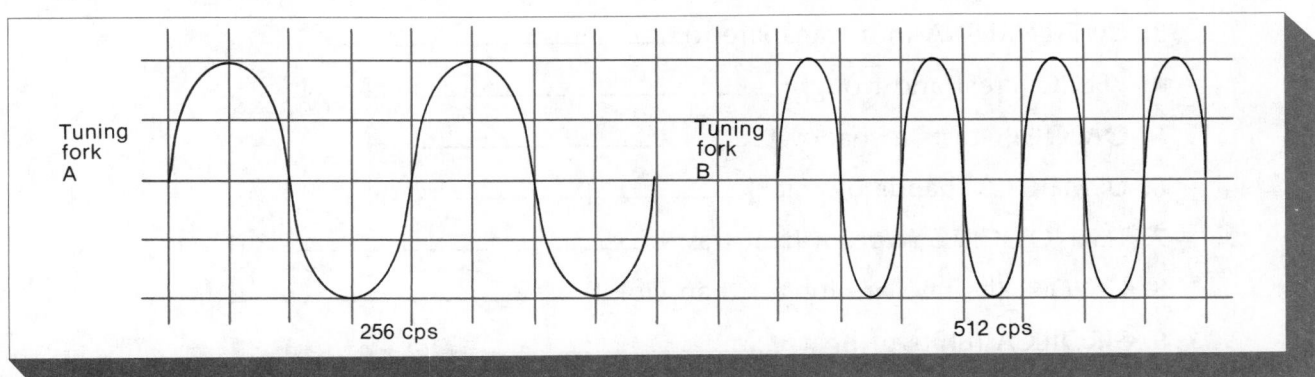

Review

I. Fill in each blank with the word that fits best. Choose from the words below.

**decibels   pitch   cps   intensity   oscilloscope   tuning fork
calories**

The loudness of a sound is called its _____ . Loudness is

measured in _____ . The highs and lows of sounds are called

_____ . Pitch is measured in _____ .

Sound can be graphed on an instrument called a(n) _____ .

II. Which statement seems more likely to be true?

A. _____ A sound of 50,000 cps would be very loud.

B. _____ A sound of 50,000 cps would not be heard by humans.

III. Match each term in column **A** with its description in column **B**.

A	B
1. _____ decibels	a. used to measure pitch
2. _____ oscilloscope	b. used to measure loudness
3. _____ intensity	c. loudness
4. _____ cps	d. instrument that makes a graph of a sound

UNIT 10

Review What You Know

A. Unscramble the groups of letters to form science words. Write the words in the blanks.

1. GOLDNNIUTLAI (a type of wave) _____

2. UCVMAU (absence of matter) _____

3. CUTTENRLSNA (not transparent) _____

4. YELCC (crest and trough) _____

5. QAEOUP (casts a shadow) _____

6. UPMERTCS (bands of color) _____

7. TENNOOCIVC (type of heat travel) _____

8. GOIDNI (its rays are longer than violet) _____

9. SIROILCA (units of heat) _____

10. HMERETREOMT (measures temperature) _____

B. Write the words (or words) that best completes each statement.

1. _____ A measure of the average kinetic energy of all the molecules in a substance is **a.** heat **b.** temperature **c.** calorie

2. _____ A measure of the total kinetic energy of all the molecules in a substance is **a.** heat **b.** temperature **c.** calorie

3. _____ Temperature is measured in **a.** calories **b.** kilocalories **c.** degrees Celsius

4. _____ The amount of heat needed to raise the temperature of 1000 grams of water from 2°C to 3°C is **a.** 1 calorie **b.** 1 kilocalorie **c.** 3 calories

5. _____ A thermometer is used to find **a.** heat **b.** temperature **c.** calories

6. _____ A substance that allows heat to pass through it easily is **a.** a conductor **b.** an insulator **c.** a current

7. _____ Heat travels through liquids by **a.** conduction **b.** convection **c.** radiation

8. _____ The only way for heat to travel through a vacuum is by **a.** conduction **b.** convection **c.** radiation

9. _____ A repeated pattern of motion that moves energy from one place to another is a **a.** decibel **b.** wave **c.** trough

10. _____ The top part of a wave is the **a.** crest **b.** trough **c.** cycle

280

11. _____ The distance from the trough on one cycle to the next trough is the **a.** wavelength **b.** frequency **c.** amplitude

12. _____ The number of cycles a wave travels in one second is its **a.** wavelength **b.** amplitude **c.** frequency

13. _____ A group of light rays is called a **a.** wave **b.** focal point **c.** beam

14. _____ An example of a luminous object is a **a.** star **b.** shadow **c.** mirror

15. _____ Sound travels most slowly through **a.** air **b.** water **c.** glass

16. _____ Light travels most slowly through **a.** a vacuum **b.** water **c.** glass

17. _____ White light can be separated into its colors with a **a.** convex lens **b.** concave lens **c.** prism

18. _____ Light rays converge if passed through a **a.** convex lens **b.** concave lens **c.** vacuum

19. _____ The reflection of sound causes **a.** absorption **b.** an echo **c.** a prism

20. _____ Sound travels in **a.** transverse waves **b.** longitudinal waves **c.** transverse and longitudinal waves

21. _____ Amplitude refers to a wave's **a.** length **b.** width **c.** height

22. _____ If 3 parallel light rays strike a plane mirror at an angle of 30°, the angle of reflection will be **a.** 10° **b.** 300° **c.** 30°

23. _____ An object that is red reflects light of **a.** all wavelengths **b.** only red wavelengths **c.** green and red wavelengths

24. _____ Sound is measured in **a.** decibels **b.** amperes **c.** meters

25. _____ The two parts of a sound wave are the **a.** crest and trough **b.** trough and compression **b.** compression and rarefaction

C. Apply What You Know

1. Study the diagrams. Then answer the questions on page 282.

A   B   C   D   E

a. A longitudinal wave is shown in drawing _____ .

b. The transverse wave with the longest amplitude is shown in drawing _____ .

c. A compression of a longitudinal wave is shown in drawing _____ .

d. The transverse wave with the lowest frequency is shown in drawing _____ .

2. Study the ray diagrams. Then answer the questions.

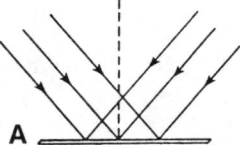 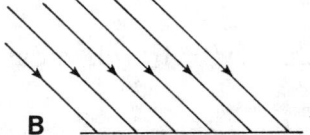 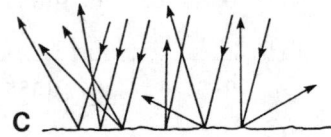

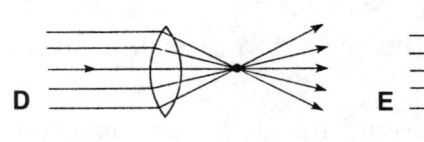 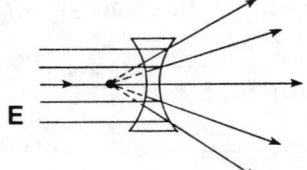

a. Rays are converging in diagram _____ .

b. Rays are reflected off a plane mirror in diagram _____ .

c. A ray is refracted toward the normal in diagram _____ .

d. A convex lens is shown in drawing _____ .

e. A concave lens is shown in drawing _____ .

f. Rays are absorbed in diagram _____ .

g. The Law of Reflection is best illustrated by diagram _____ .

h. A black object is shown in drawing _____ .

i. A diffused reflection is shown in drawing _____ .

D. Find Out More

1. Visit a local radio station. Find out how radio transmission is accomplished. Prepare a report of your findings. Present your findings to the class.
2. Find out what FCC regulations are. How do these regulations apply to radio stations? How do they apply to "ham" radio operators? Present your findings to the class.
3. Make a small hole in a piece of cardboard. Tape the piece of cardboard over the lens of a flashlight. Set up several mirrors. Place the mirrors at different angles. Observe how the light is reflected off the mirrors. Draw a sketch of your results.

Glossary

absolute zero: according to scientists' predictions, the temperature at which molecules will stop moving (-273°C), 80

absorb: to take in, 261

absorption (ab-ZORP-shun): ability to take in and store substances, 261

acid: substance that gives off hydrogen ions when placed in water; a substance with a pH of less than seven, 72

acidic: description of any substance with a pH of less than seven, 73

alloy: mixture of metals, 88

alpha (AL-fuh) **particle:** combination of two protons and two neutrons, 138

ammeter: device used to measure the strength of electrical current, 225

amp: an ampere, 225

ampere: unit for measuring electrical current, 225

amplitude: height of a wave, 254

angle of incidence: number of degrees between the incident ray and the normal, 264

angle of reflection: number of degrees between the normal and a reflected ray, 264

atomic mass: sum of the masses of the protons and neutrons of an atom, 32

atomic mass unit (amu): unit for measuring the total mass of the protons and neutrons of an atom, 29

atomic number: number of protons found in the nucleus of an atom, 33

base: substance that gives off hydroxide (OH⁻) ions when placed in water; a substance with a pH higher than seven, 73

basic: description of any substance with a pH higher than seven, 73

beam: group of light rays that travel in the same direction, 258

beta particle: an electron omitted from the nucleus of an unstable element, 138

black: apparent color caused by the absorption of all wavelengths of color by an object, 272

blue litmus paper: pH indicator that consists of paper that turns red when it contacts an acidic substance, 72

bond: force that holds the atoms in a molecule together, 57

byte: in computer language, a code used to present a letter or number, 229

calorie: unit for measuring heat; the amount of heat needed to raise the temperature of one gram of water one degree Celsius, 246

centi-: prefix meaning 1/100, 3

chain reaction: reaction that continues by itself as long as there are more atoms of an element present, 149

chemical change: change in which a new substance is formed, 113

chemical energy: energy stored in chemical bonds, 162

chemical symbol: shorthand way of writing the name of an element; the abbreviation for the name of an element, 24

chemist: scientist who studies matter and its properties, 4

circuit (SUR-cut): path taken by a current, 229

closed circuit: uninterrupted path taken by a current, 229

coefficient (koh-uh-FISH-unt): large number written before a chemical formula which indicates the number of molecules present in a substance, 119

compass: magnet that responds to the magnetic pull of the Earth, 216

compound: substance made up of two or more elements that have combined chemically, 53

compression (KOM-preh-shun): part of a sound wave in which molecules are close together, or tightly packed, 275

concave (kon-KAYV) **mirror:** mirror that is rounded inward, 264

conclusion: solution to a scientific problem, 6

condensation (kon-den-SAY-shun): change in the phase of matter from a gas to a liquid, 80

conduct: ability of a substance to carry electricity, 15

conduction (kon-DUK-shun): movement of heat caused by atoms or molecules bumping into each other; the way that heat is transferred in solids, 250

conductor (kon-DUK-tohr): material that electricity can pass through easily, 223

controlled variable: aspect of an experiment that a scientist works to keep from changing, 5

convection (kon-VEK-shun): way heat travels in liquids and gases, 250

convection current: circular pattern of heat as it is transferred through liquids or gases, 250

converge: to meet, 268

convex: shape that bulges outward, 267

cornea (KOR-nee-uh): transparent tissue found at the front of the eye, 267

covalent (koh-VAY-lent) **bond:** chemical tie formed by the sharing of a pair of electrons, 65

crest: top of a wave, 254

cryogenic (kreye-uh-JEN-ik) **liquid:** liquid that forms at extremely low temperatures, 11

cubic centimeter: basic unit of measurement of volume for small, solid objects, 2

cycle: complete pattern of one crest and one trough of a wave, 254

data: collection of facts and observations made during a scientific experiment, 5

decibel: unit for measuring the intensity, or loudness, of a sound, 277

decomposition (dee-kom-puh-ZISH-un) **reaction:** chemical reaction in which the bonds in a compound are broken resulting in a compound that is broken down into its elements, 121

degrees Celsius: unit scientists use for measuring temperature, 246

density: measurement of the amount of mass contained in one unit of volume of a substance; density has the formula D=m/v, 8

diatomic element: molecule formed when two atoms of the same element are bonded together, 56

diffused (dih-FYOOSED) **reflection:** reflection of light from a rough surface that does not produce an image, 265

direct union reaction: chemical reaction that results in the formation of a compound from two or more elements, 118

discharge: release of electrical charges from an object, 223

displace: to take the place of, 5

distillation (dis-tuh-LAY-shun): process of separating a mixture into its parts using evaporation and condensation, 91

diverge: to spread apart, or move away from, 265

domain (doh-MAYN): unit consisting of one positive charge and one negative charge, 216

double bond: chemical tie formed when two pairs of electrons are shared by two atoms, 69

double replacement reaction: chemical reaction in which the elements in two compounds changing places with each other, 126

ductile (DUK-teyel): ability of a substance to be drawn out into a thin wire, 15

effort: applied force, 192

effort arm: arm of lever that is doing the lifting, 196

electrical current: flow of electrons from one point to another, 224

electrical (ee-LEK-trih-kul) **energy:** energy created by electrons moving through a conductor, 162

electrodes (ee-LEK-trohds): positive and negative poles through which electrical current flows, 224

electromagnet (ee-LEK-troh-mag-net): device that uses electricity to make magnetism, 237

electromagnetic spectrum (i-lek-truh-mag-NET-ik SPEK-trum): full range of transverse light waves, 259

electromotive (ee-LEK-troh-moh-tiv) **force (EMF):** push that moves electrons, 227

Electromotive Series of Metals: list of metals that are placed in order of their chemical activity, 124

electron (ee-LEK-tron): subatomic particle with a negative charge that orbits the nucleus of an atom, 29

electroscope (EE-lek-troh-skohp): device that shows where the electrical charges in an object are located, 222

element: simplest kind of matter that cannot be broken down into any other kind of matter by ordinary chemical means, 23

emit: to give off, 37

emulsion (ee-MULL-shun): mixture that can stay uniformly mixed for long periods of time, 106

energy: ability to do work, 162

equilibrium (ee-kwuh-LIB-ree-um): balanced, 196

evaporation (ee-vap-or-AY-shun): process by which a liquid becomes a gas if it is uncontained in the presence of heat, 79

experiment: test of a hypothesis, 5

family of elements: group of elements having similar properties that are arranged vertically (in columns) on the Periodic Table of the Elements, 43

filter: device that can separate the colors of light and transmit wavelengths of one only one color, 272

fission (FIZ-shun): splitting of a large nucleus into two smaller nuclei, 148

fissionable materials: heavy atoms that can be split into smaller nuclei, 153

fixed pulley: pulley that is attached to a surface, 206

focal point: point at which rays meet, 268

force: push or a pull, 172

formula: shorthand notation of the kind and number of atoms of each element in a compound, 56

freezing: change in the phase of matter from a liquid to a solid, 80

frequency (FREE-kwun-see): measure of the number of cycles of a wave that pass through a point in one second, 254; measure of the number of cycles per second a vibrating body makes, 277

friction (FRIK-shun): opposing force created by the rubbing of one body against another body, 189

fulcrum (FUL-krum): point where a lever pivots, 195

fusion (FYOO-shun): joining together of two light atoms to create one heavier atom, 150

gamma ray: form of energy that is released during a nuclear change, 139

gas: matter with no definite volume and no definite shape, 11

Geiger (GEYE-gur) **counter:** device that detects radioactivity, 140

gram: basic unit of mass measurement, 2

gravity (GRAV-uh-tee): force of attraction that exists between two objects, 176

half-life: measure of the amount of time it takes for one-half of the atoms of a radioactive element to decay, 143

heat: measure of the total kinetic energy of all the molecules in a substance, 246

heat conductivity (kon-DUK-tiv-ih-tee): measure of how fast heat travels in a substance, 15

heat energy: energy that is created by molecular motion, 162

heat insulators: substances that do not conduct heat well, 250

horizontal: across, or back-and-forth, 259

hypothermia (heye-poh-THUR-mee-uh): condition of reduced body temperature, 249

hypothesis (heye-POTH-ih-sis): "an educated guess" as to a solution to a scientific problem, 4

ideal mechanical advantage (IMA): mechanical advantage that ignores the effect of friction, 201

illuminated (ih-LOOM-ih-nayt-ed): an object on which a light shines, 258

incident (IN-sih-dent) **ray:** the ray of light that strikes a surface, 263

inclined (in-CLEYEND) **plane:** simple machine made by a sloping surface, 199

induced current: current that is made by the forced flow of electrons, 238

inert (IN-uhrt): inactive, 62

infrared (IN-fruh-red) **light:** form of light that has waves that are too long to be seen, 259

insoluble: material that does not dissolve well in water or other substances, 99

insulator (IN-suh-lay-tohr): material that does not allow heat or electricity to pass through it easily, 223

intensity (in-TEN-sih-tee): brightness of a light or loudness of a sound, 258

inverted: upside-down, 267

ion (EYE-on): an atom that has an electrical charge because it has gained or lost electrons, 46

ion propulsion (pruh-PUL-shun): force created by a large number of charged atoms moving in the same direction, 45

ionic bond: chemical tie created by the force of attraction between the positive and negative electrical charges of atoms, 58

ionizing radiation: type of radiation in which radioactive particles change the atoms they hit into ions, 139

isotopes (EYE-suh-topes): atoms of the same element having the same atomic number, but different atomic masses, 35

joule: metric unit for work; equal to one Newton-meter, 186

kilo-: prefix meaning 1000, 2

kilocalorie: 1000 calories; the unit used of measuring food calories, 247

kinetic energy: energy of motion, 170

kinetic-molecular theory: theory that states that molecules are always in motion and constantly bumping into each other, 80

laser: special form of light that has waves of equal lengths that lie in the same plane, 259

Law of Conservation of Matter: law that states that matter can neither be created nor destroyed, only changed, 119

Law of Magnetic Poles: law that states that unlike poles attract, and like poles repel, 216

Law of Reflection: law that states that the angle of incidence must be equal to the angle of reflection, 264

lens: convex-shaped part of the eye that transmits light from the iris through the fluid of the eyeball, 267

lever: simple machine made of a bar or board that is supported at one place, 195

liquid: matter that has definite volume, but no definite shape, 11

liter: basic unit of measuring liquid volume, 3

longitudinal (LAWN-jih-tood-ih-nul) **wave:** wave in which the particles move back and forth in the direction of the wave motion, 253

LOX: abbreviation for liquid oxygen, 11

luminous (LOOM-ih-nus): object that produces light, 258

magnet: any object that produces a magnetic force; an iron bar that produces a magnetic force, 216

magnetic field: area around a magnet where the magnetic force is found, 220

magnetic pole: part of a magnet where the force is strongest, 216

magnetism: force created by the attraction of unlike electrical charges between two substances, 215

malleable (MAL-ee-uh-bul): property of a substance that allows it to be bent, or changed in shape, without breaking, 15

mass: measure of how much matter an object contains, 1

mass spectrometer (spek-TRAHM-uh-ter): machine that counts and measure the number and kind of different atoms in a substance, 32

matter: anything that has mass and volume, 2

mechanical advantage (MA): amount that any machine multiplies an applied force, 193

mechanical energy: energy of moving things, 162

medium: matter that a wave travels through, 258

metallized (MET-uh-lized) **plastics:** plastics coated with metal, 14

metalloids (MET-ul-oyds): group of elements having 4 or 5 electrons in their valence shells and properties of both metals and nonmetals, 41-43

metals: group of elements having 1, 2, or 3 electrons in their valence shells, 41-43

milli-: prefix meaning 1/1000, 3

mixture: matter containing two or more substances that are not chemically combined, 88

molecular motion: constant random movement of molecules, 78

molecules (MAHL-uh-kyoolz): smallest natural units of compounds, 56

moveable pulley: pulley that is not attached to a surface, 206

neutral (NOO-trul): having no electrical charge, 29

neutralization (new-trul-ih-ZAY-shun): process in which the pH of a substance is changed to 7 through the proper balancing of acids and bases, 73

neutralization reaction: double replacement reaction involving an acid and a base, 127

neutron (NOO-tron): neutral subatomic particle, with a mass of 1 amu, found in the nucleus of an atom, 29

Newtons (N): metric unit of force, 173

noble gases: group of elements having 8 electrons in their valence shells, that are characterized as being chemically inactive; sometimes called inert gases, 41-43

nonmetals: elements having 6 or 7 electrons in their valence shells, 41-43

normal: imaginary line at a right angle to the plane of a reflecting surface, 264

North Magnetic Pole: point where the magnetic pull of the Earth is strongest, 216

nuclear energy: energy stored in the nucleus of an atom, 162

nuclear (noo-CLEE-ar) **force:** force that holds the particles of radioactive elements together, 138

nuclear reactor: plant in which a controlled (nuclear) fission chain reaction takes place, 152

nucleus (noo-KLEE-us) [pl. **nuclei**]: dense center of at atom where the protons and neutrons are located, 29

ohm: unit for resistance, 233

Ohm's Law: law of electrical energy which states that electrical current is equal to electromotive force (EMF), or voltage, divided by resistance (), 233

omega (OH-may-guh): symbol () for ohm, 233

opaque (oh-PAKE): an object that does not allow light to pass through it, 261

open circuit: break in the path of a current, 229

orbit: path of an electron around a nucleus, 29

organic compounds: compounds that contain the element carbon, 68

oscilloscope (aw-SILL-uh-skohp): device that makes graphs of sound waves, 277

parallel (PAR-uh-lel) **circuit:** an electrical circuit containing more than one path through which a current can flow, 230

parallel rays: rays that travel in the same direction and always remain the same distance apart, 264

Periodic Table of the Elements: chart listing all of the known elements in order of their atomic number (horizontally) and according to their similar properties (vertically), 41–43

phase: form, or state, or matter, 11

physical change: change in which no new substances are formed, 115

physical scientist: scientist who studies problems involving matter and energy, 4

physicist (FIZ-ih-sist): scientist who studies the forces of nature and the forms of energy, 4

pivot: to move at one place, 195

plane mirror: flat mirror, 263

plasma: phase of matter in which the particles that make up the matter are electrically charged, 11

polymers (POL-ee-merz): organic substances made up of large molecular units that are bonded together in repeated patterns that can continue indefinitely, 69

potential (poh-TEN-chul): likelihood of atoms bonding together under the right conditions, 63

potential energy: energy that is stored, 170

power: rate at which work is done, 186

precipitate (prih-SIP-ih-tayt): an insoluble solid that forms in a solution, 126

primary colors: red, green, and blue; the three colors of light necessary for an object to be seen as white, 272

prism (PRIZ-um): triangular-shaped piece of glass that separates white light into its colors, 259

products: end materials after a chemical reaction has taken place, 118

properties: characteristics of matter, 1

proton: positively charged subatomic particle, with a mass of 1 amu, found in the nucleus of an atom, 29

pulley: simple machine consisting of a wheel-and-axle turned by a rope, 206

pulley system: combination of pulleys connected by supporting ropes or cables, 207

pupil: small opening in the iris of the eye, 267

radiant (RAY-dee-unt) **energy:** way that heat travels from the Sun to the Earth, 251

radiation (RAY-dee-ay-shun): transfer of heat without the involvement of matter, 250

radical: group of atoms that can act as a single charged unit, 70

radioactive decay: spontaneous change of an atom of one element into an atom of another element, 138

radioactive isotope: an unstable form of an element with the same atomic number but a different atomic mass, 140

radioactive series of elements: series of 14 elements beginning with the decay of U-238 until it reaches a form of stable lead Pb-206, 138

radioactivity (ray-dee-oh-ak-TIV-ih-tee): spontaneous release of particles and energy, 138

random motion: constant unpredictable movement of molecules, 79

rarefaction (rayr-uh-FAK-shun): part of a sound wave in which the molecules are spread apart, 275

ray: group of light waves that travel in straight lines, 258

ray diagram: sketch of the path of a transverse light wave, 261

reactants (ree-AK-tunts): raw materials or beginning materials in a chemical reaction, 118

red litmus paper: pH indicator that consists of paper that turns blue when it contacts a basic substance, 73

reflect: to bounce off, 261

reflected ray: ray of light that is bounced off, or reflected from, a surface, 263

refract: to bend, 267

repel: to push apart or away from, 28

resistance: anything that slows down, or prevents, motion, 233

resistance arm: arm the resistance is resting on, 196

retina: part of the eye where an image is formed as a result of the absorption of light rays, 267

salt: compound (precipitate) formed by the positive ion of a base and the negative ion of an acid, 76

saturated (sa-choor-AY-tud): solution that can hold no more solute, 97

schematic (skee-MAT-ik): diagram of an electrical system, 231

series circuit: an electric circuit with only one path for a current to follow, 230

simple machine: mechanical device that changes an applied force, 192

single replacement reaction: chemical reaction that involves one element taking the place of another element, 124

solid: type of matter with definite volume and definite shape, 11

soluble: able to dissolve, 99

solute (sahl-YOUT): in a solution, the substance that is dissolved, 94

solution (suh-LOO-shun): uniform mixture, 94

solvent (sahl-VENT): substance that is used to dissolve another substance, 94

sound energy: energy produced by vibrating objects, 163

spectrum of visible light: full range of light, from red through orange, yellow, green, blue, indigo, and violet, that can be seen by the human eye, 259

static electricity: build-up of electrical charges, 222

subatomic particles: different particles that make up an atom, 29

superconductor: material that does not oppose the flow of electrical current, 233

supersaturated: solution in which the amount of solute that can be held has increased, 97

suspension (suhs-PEN-shun): liquid mixture in which the particles in the mixture are large enough to block light, 106

switch: simple device that opens or closes an electrical circuit, 230

symbolic equation: an equation that contains the chemical symbols and formulas of the reactants and products in a chemical reaction, 118

synthesis (SIN-thih-sis) **reaction:** chemical reaction in which two or more elements unite to form a compound, 118

technology (TEK-nawl-uh-gee): applied science; basic scientific knowledge applied to everyday needs, 15

temperature: measure of the average kinetic energy of all the molecules in a substance, 246

thermometer: an instrument used for measuring temperature, 245

thermonuclear (thur-moh-NOO-klee-ur): nuclear reaction that requires heat to take place, 150

toxic: poisonous, 121

translucent (tranz-LOO-sent): surface that cannot be seen through because it only transmits some light waves, 262

transmit: allow to pass through, 261

transmitting medium: molecules that a wave travels through, 275

transparent: surface that can be seen through because it transmits most light waves, 262

transverse wave: way in which the particles in the wave move at right angles to the direction of the wave motion, 253

triple bond: chemical tie formed when three pairs of electrons are shared by two atoms, 69

trough (TROF): bottom of a wave, 254

ultraviolet (uhl-truh-VEYE-oh-lit) **light:** form of light with waves that are too short to be seen, 259

unbalanced equation: an equation where the number of atoms on the left side of the equation (the reactants) is not equal to the number of atoms on the right side of the equation (the products), 119

undistorted (un-dih-STOHR-ted): true, or accurate, image, 263

unsaturated: solution that can hold more solute, 97

vacuum (VAK-yew-um): absence of matter, 251

valence number: number of electrons in an atom that are available for bonding, 63

valence shell: outermost energy level of an atom, 41,43

vapor: matter in the gas phase, 14

variable (VAR-ee-uh-bul): an aspect of an experiment that could change, 5

vector (VEK-tur): direction in which a force is applied, 173

velocity (vuh-LAHS-uh-tee): speed and direction an object is moving, 258

vertical: up-and-down, 259

vibration (veye-BRAY-shun): rapid back-and-forth motion, 274

voice activation: type of technology in which a machine responds to the voice of a specific person, 277

volt: unit for measuring voltage, 227

voltage: electromotive force, 227

voltmeter: device used for measuring voltage, 227

volume: measure of the amount of space an object occupies, 1

watt: unit for power; equal to 1 joule of work per second, 186

wave: repeated pattern of motion that carries energy from one place to another, 253

wavelength: distance from a point on one wave to the same point on the next wave, 254

weight: measure of the Earth's gravitational pull on the mass of an object, 176

wheel and axle: simple machine that is made from two different-sized wheels that turn together, 203

white: appearance of color of any surface that reflects all wavelengths of light to our eyes, 271

word equation: way of writing a chemical reaction, 77

work: force that move an object over a distance, 186

Index

copper sulfate, 102
cornea, 267
covalent bonds, 65–67
CPS, 277–79
crest, 254
criminology, science and, 113
cryogenic, 11, 78, 233
   temperatures, 78
crystals, 58, 96–97
   space vs. earth, 96
Curie, Marie, 23
current
   convection, 250
   electrical, 224–225, 227, 233–235
   increasing and decreasing, 233, 235
   induced, 237–238
   magnetism and, 237–238
current electricity, 224–25
cycle, 254
cycles per second (CPS), 277–79

D

Dalton, John, 54
Dalton's rule, 55, 57
data, 5
dead zones, 129
decibels, 277–79
decomposition reaction, 121–22
decreasing current, 233–35
degrees Celsius, 246
density, 8–9
desalinization, 90–91
deuterium, 34, 35
diameter, 204
diamond, 68
diet, fad, 121
diffuse reflection, 265
direction, sense of, 219–21
direct union, 118
discharge, 223
Discovery, 165
displace, 5
distillation, 91
diverge, 265, 268
dolphin, 274, 275, 276
domains, 216
double bond, 69
double replacement reaction, 126–28
ductile, 15

E

earth crystals, 96
EDTA (calcium disodium edetate), 117
effort, 192
effort arm, 196
Einstein, Albert, 23, 150
einsteinium, 23, 24
electrical current, 224
electrical energy, 162
electric circuits, 229–31
electricity, 122
   current, 224–25
   magnetism and, 237–238
   static, 222–23
electrodes, 224–25
electromagnet, 237–38
electromagnetic spectrum, 259
electromotive force (EMF), 227
electromotive series of metals, 124–25
electron, 29
   configuration, 38
   shells, 37–39
electrophoresis, 224
electroscope, 222–23
elements, 23–24, 110, 41
   periodic table of, 41–43
   radioactive series, 138
elevator, 206
EMF, 227
emit, 37, 138
The Empire Strikes Back, 257
emulsion, 106
energy
   and chemical change, 114
   forms, 162–63
   kinetic, 170
   law of conservation of, 166, 262
   levels, 37
   potential, 170
   radiant, 251
equilibrium, 196
evaporation, 79
   and physical change, 11–12, 115
experiment, designing, 5

F

family of elements, 43
Fermi, Enrico, 148
filter, 272

filtration, 91
fission, 148–50
fissionable, 153
fixed pulley, 206
flight, solar-powered, 165
floating train, 236
flourine (F), 46, 60, 63
focal point, 268
force, 172–74
formula, 56–57
   structural, 65
freezing, 80
frequency, 254, 277
friction, 189–91
fuels, 162
fulcrum, 195
fusion, 150

G

gallium, (Ga), 42
gamma radiation, 138–39
   ray, 139
gas(es), 11
   noble, 43, 46, 62
Geiger counter, 140
Gerney, Brigette, 169
gold, 23, 88–89, 124, 223
Gossamer Albatross, 14
Gossamer Penguin, 165
gram, 2–3
gravity, 176
Great Pyramid, 199
green, 272

H

H-2, 35
half-life, 142–44
heat, 246
   conductivity, 15
   energy, 162, 246
   insulators, 250
   measuring, 245–47
heavy water, 34, 35
helium, (He), 24, 46, 62, 150, 233
Hevesy, George Karl, 140
homogenized, 106
horizontal, 43, 259
hydrochloric acid, (HCl), 72, 113–14, 125, 127
hydrogen, (H), 23, 24, 53, 54, 65–66, 67, 114, 119, 122, 125, 150

hydrogen chloride, 57, 119
hydrogen sulfide, 57
hypothermia, 249
hypothesis, 4–5

I

ideal mechanical advantage (IMA), 201
illuminated, 258
IMA, 201
immiscible, 95
incident ray, 263
inclined plane, 199–201
increasing current, 233–35
indicators, 72
indium, (In), 42
induced current, 238
inert, 62
information, gathering, 5
infrared light, 259, 261
insoluble, 99
insulator, 15, 223
    electrical, 223
    heat, 15, 250
intensity, 258, 274
inverse square law, 145–46, 176
iodine, (I), 99
ion, 46–47
ionic
    bonds, 58–60
    compounds, 58–60
ionizing radiation, 139
ion propulsion, 45
iron, (Fe), 23–24, 215
    pumping, 185
isotopes, 35, 140–41

J

Joliot-Curie, Jean, 34
joule, 186

K

Keck Observatory, Hawaii, 263
Kekulé, Friedrich August, 65, 67
ketones, 121
kilo-, 2–3
kilocalorie, 247
kinetic, 80

kinetic energy, 170, 246
kinetic-molecular theory, 80, 246
Knorr, research ship, 99
krypton, (Kr), 62

L

laser, 162–63, 257, 259
law
    of conservation of energy, 166, 262
    of conservation of matter, 119
    inverse square, 145–46, 176
    of magnetic poles, 216
    Ohm's, 233
    of reflection, 264
lead, (Pb), 118
    hazard, 117
lead carbonate, 99
lead poisoning, 117
lead sulfide, 118
lens, 267
leukemia, 261
lever, 195–98
levitation, 236
Light, Tony, 137
light, 258–59
    absorption, 261–62
    reflection, 263–65
    transmission and refraction, 267–69
light energy, 162–63
lightning, 222
limestone, 93–95
liquid, 11
liquid oxygen (LOX), 11, 26
liter, 2–3
lithium, 87
longitudinal wave, 253
Los Alamos, Mexico, 87, 89
luminous, 258

M

MA, 193
machine, simple, 192–93
MacReady, Paul, 14, 165
Macy's Thanksgiving Day parade, 1
Maglev, 236
magnesium, (Mg), 46, 54, 63, 118

magnesium chloride, 118
magnesium hydroxide, 71
magnesium oxide, 54, 55
magnet, 216
magnetic
    domains, 216
    field, 220–21
    levitation, 236
    pole, 216
magnetism, 215–16
magnetometer, 215
malleable, 15
March of Dimes, 137
mass, 1
mass deficit, 150
mass spectrometer, 32
materials engineer, 51
materials technician, 51
matter, 1–2
    law of conservation of, 119
    measuring, 2–3
    phases, 11
    properties, 15
measurements, 5
mechanical, 91
    advantage (MA), 193
    energy, 162
    ideal mechanical advantage (IMA), 201
medium, 258
Meitner, Lise, 148
melting, and physical change, 115
Mendeleev, Dimitri, 41–42
mercuric oxide, 121
mercury, (Hg), 37, 121
metallized plastics, 14
metalloids, 43
metallurgist, 51
metals, 43, 46
    electromotive series of, 124–25
meteorite, 32
meter, 2–3
methane, 65–66
microscope, 26
migrations, 219
milli-, 3
mirror, 263
mirror-making, 263
miscible, 94–95
mixtures, 87–89
molecular motion, 78–80
molecules, 56–57
molybdenum, (Mo), 87, 89
mood, and color, 271

X

xenon, 62, 63

Y

yell, 274

Z

zinc, (Zn), 63, 113–14, 124, 125
zinc chloride, 114
zinc fluoride, 63
zinc oxide, 99, 124
zirconia, 26
zirconium, (Zn), 26

Photo Credits